Electronic Communications Engineering

Electronic Communications Engineering

Edited by **Bob Tucker**

WILLFORD **P**RESS

New York

Published by Willford Press,
118-35 Queens Blvd., Suite 400,
Forest Hills, NY 11375, USA
www.willfordpress.com

Electronic Communications Engineering
Edited by Bob Tucker

International Standard Book Number: 978-1-68285-091-6 (Hardback)

Contents

Preface

This book was inspired by the evolution of our times; to answer the curiosity of inquisitive minds. Many developments have occurred across the globe in the recent past which has transformed the progress in the field.

Electronic communications engineering is an emerging field of study that focuses upon understanding the technology of electronic communications and networks. It plays a key role in simplifying business management processes in the current scenario. This book attempts to decipher diverse aspects of electronic communications and networks including their architecture, frameworks, modeling, controlling, etc. It compiles researches and case-studies by eminent academicians and experts from various parts of the globe that provide an in-depth analysis of the different networks and communication systems, their construction and control mechanism, evaluation and assessment of different communication networks and systems, etc. Students, researchers and engineers will find this book helpful as it will serve as a comprehensive source of information.

This book was developed from a mere concept to drafts to chapters and finally compiled together as a complete text to benefit the readers across all nations. To ensure the quality of the content we instilled two significant steps in our procedure. The first was to appoint an editorial team that would verify the data and statistics provided in the book and also select the most appropriate and valuable contributions from the plentiful contributions we received from authors worldwide. The next step was to appoint an expert of the topic as the Editor-in-Chief, who would head the project and finally make the necessary amendments and modifications to make the text reader-friendly. I was then commissioned to examine all the material to present the topics in the most comprehensible and productive format.

I would like to take this opportunity to thank all the contributing authors who were supportive enough to contribute their time and knowledge to this project. I also wish to convey my regards to my family who have been extremely supportive during the entire project.

Editor

Parameter estimation for propagation along random rough surface by using line of sight data

Kazunori Uchida[a,*], Masafumi Takematsu[b], Jun-Hyuck Lee[b] and Junichi Honda[c]

[a]*Department of Information and Communication Engineering, Fukuoka Institute of Technology, Higashi-Ku, Fukuoka, Japan*
[b]*Graduate School of Engineering, Fukuoka Institute of Technology, Higashi-Ku, Fukuoka, Japan*
[c]*Surveillance and Communications Department, Electronic Navigation Research Institute, Tokyo, Japan*

Abstract. This paper is concerned with a numerical procedure to estimate two parameters, amplitude modification factor α and propagation order of distance β, for electromagnetic (EM) propagation in complicated natural environments such as random rough surface (RRS). These two parameters are key parameters when we simulate field distributions along various types of RRS based on 1-ray model. We assume that the former parameter α can be evaluated by the visual planar angles of illuminated lines in case of 1D RRS and by the visual solid angles of illuminated planes in case of 2D RRS. We also assume that the latter parameter β can be estimated not only by base station (BS) antenna height, similar to Okumura-Hata model simulating EM propagation in urban and suburban areas, but also by mobile station (MS) antenna height. In order to demonstrate validity of the proposed parameter estimation, we compare the numerical field distributions obtained by the 1-ray model using estimated two parameters with those computed by discrete ray tracing method (DRTM) which is an effective EM field solver. It is shown that both numerical results are in good agreement.

Keywords: Propagation, random rough surface, 1-ray model, field estimation, discrete ray tracing method

1. Introduction

Recently, demands for wireless communications, such as cellular phones, wireless local area networks (LAN), ad hoc networks as well as sensor networks, have been rapidly increasing. Base stations (BS) and/or mobile stations (MS) of these systems are often located in complicated natural or artificial propagation environments, such as desert, hilly mountain, sea surface, urban or suburban area, and so forth. Consequently, it is important for wireless communication engineers to estimate propagation characteristics accurately in order to construct reliable wireless networks in these complicated propagation environments [1].

So far, we have introduced 1-ray and 2-ray models from which we can easily estimate electric field distributions in many complicated propagation environments by using two parameters, amplitude modification factor α and propagation order of distance β [2,3]. The ray models have been devised so that

---

*Corresponding author: Kazunori Uchida, Department of Information and Communication Engineering, Fukuoka Institute of Technology (FIT), 3-30-1 Wajiro-Higashi, Higashi-Ku, Fukuoka 811-0295, Japan. E-mail: k-uchida@fit.ac.jp.

we can obtain electric field intensities corresponding to the path loss provided by Okumura-Hata model for propagation in urban, suburban and open areas as listed in Appendix A. The Okumura-Hata model indicates that the propagation order of distance β is dependent only on BS antenna height while the amplitude modification factor α is dependent on communication areas, BS and MS antenna heights, and operating frequency [5–7].

The ray models have also been successively applied to EM wave propagation along various types of random rough surfaces (RRS) [8–10]. The most interesting feature of the 1-ray model is that we can evaluated EM field distributions by choosing α and β appropriately, and thus we can also evaluate communication distance easily [1,9]. The parameters α and β seem to be closely related to the statistical quantities of RRS, that is, deviation of height dv and correlation length cl, because many spectral functions for RRS are well described by dv and cl [11]. At the present stage, however, we have no explicit mathematical expressions which could adequately describe the relationship between (α, β) and (dv, cl) [5,6]. The propagation order of distance β might be dependent on BS and MS antenna heights. The amplitude modification factor α, on the other hand, might be strongly associated with the height deviation dv and correlation length cl of RRS. In this context, it is significant to propose a procedure to estimate α by introducing equivalent planar or solid angle of illuminated regions of RRS and β by using the field matching factor γ numerically [5,6].

The main purpose of this paper is to check the accuracy of the proposed method by comparing present numerical results with those computed by discrete ray tracing method (DRTM) [1]. First, we review the 1-ray and 2-ray models which play an important role in computing electric field distributions in complicated propagation environments. Second, we discuss how to solve the propagation order of distance β and we show that β can be estimated in terms of field matching factor γ having correlation with BS and MS antenna heights. Third, we propose an estimation procedure for α by using the equivalent angle corresponding to illuminated planar and solid angles for 1D and 2D RRSs, respectively. Finally, we propose tentative analytical expressions for estimating α and β, and we also show some numerical examples for electric field distributions to demonstrate validity of the proposed method.

2. Field expressions

Now we review the 1-ray and 2-ray models characterized by introducing amplitude modification factor α and propagation order of distance β. It should be noted that the models are simple but we can estimate field distributions in many complicated propagation environments, such as urban, suburban and rural areas as well as RRSs. In other words, we can always evaluate electric field distributions in any propagation environments with the appropriately estimated parameters α and β. It should also be emphasized that the 1-ray model enables us to estimate communication distance in such complicated propagation environments.

2.1. 1-ray and 2-ray models

The results are summarized in far zone ($r \gg \lambda$) as follows [2,3]:

$$\boldsymbol{E_1} = 10^{\alpha/20} 10^{(\beta-1)\gamma/20} r^{(1-\beta)} \boldsymbol{E_i}, \quad \boldsymbol{E_2} = 10^{\alpha/20} 10^{(\beta-2)\gamma/20} r^{(2-\beta)} \boldsymbol{E_t}. \tag{1}$$

The first and second equations in Eq. (1) correspond to the 1-ray and 2-ray models, respectively.

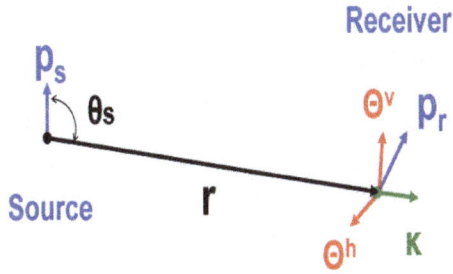

Fig. 1. Incident ray.

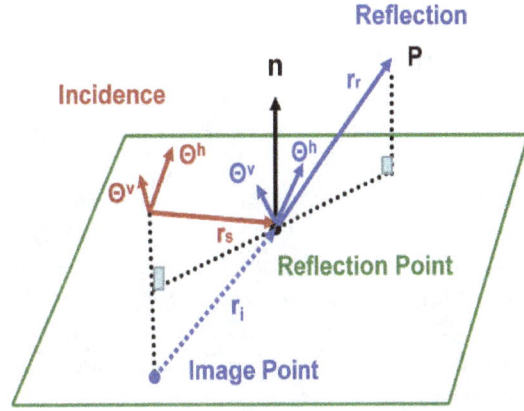

Fig. 2. Reflection ray.

The incident electric field E_i in free space and the total electric field E_t above a ground plane are employed for the first and second models, respectively. These two electric fields are expressed as follows [12,13]:

$$E_i = \sqrt{30 G_s P_s} \sin\theta \frac{e^{-j\kappa r}}{r} \Theta^v(r, p_s), \quad E_t = E_i + \sqrt{30 G_s P_s} \sin\theta_0 \frac{e^{-j\kappa r_0}}{r_0} e_r \quad (2)$$

where the time dependence $e^{j\omega t}$ is assumed. Moreover, $\omega = 2\pi f$ is an angular frequency with operating frequency f, $\kappa = 2\pi/\lambda$ is a wave number with wave length λ in free space, and r is a position vector from a source to a receiver as shown in Fig. 1. The electric field vector for a reflected wave from the ground plane is given by

$$e_r = R^v(\theta_i)[\Theta^v(r_1, p_s) \cdot \Theta^v(r_1, n)]\Theta^v(r_2, n) + R^h(\theta_i)[\Theta^h(r_1, p_s) \cdot \Theta^h(r_1, n)]\Theta^h(r_2, n)] \quad (3)$$

where $r_0 = r_s + r_r$ and r_s is the distance from source to reflection point and r_r is the distance from reflection point to receive point P as shown in Fig. 2. The vertical and horizontal unit vectors used in the above relation are defined by

$$\Theta^v(r, p_s) = \frac{((r \times p_s) \times r)}{|(r \times p_s) \times r|}, \quad \Theta^h(r, p_s) = \frac{(r \times p_s)}{|r \times p_s|}. \quad (4)$$

Moreover, the field matching factor $\gamma\ [dB]$ used in Eq. (1) and the field matching length $\Gamma\ [m]$ is defined by use of Eq. (2) as follows:

$$\gamma = 20\log_{10}(\Gamma)\ [dB], \quad \Gamma = (r|E_t|/|E_i|)\ [m]. \quad (5)$$

When the distance d between a base station (BS) and a mobile station (MB) is much larger than BS antenna height h_b and MS antenna height h_m, the field matching factor can be well approximately in a simple expression as follows [7]:

$$\gamma \simeq 20\log_{10}(2\kappa h_b h_m)\ [dB], \Gamma \simeq 2\kappa h_b h_m\ [m], (d >> h_b, h_m). \quad (6)$$

The Fresnel reflection coefficients for the vertical and horizontal electric field components are given by [12,13]

$$R^h(\theta_i) = \frac{\cos\theta_i - \sqrt{\epsilon_c - \sin^2\theta_i}}{\cos\theta_i + \sqrt{\epsilon_c - \sin^2\theta_i}} \ , \ \ R^v(\theta_i) = \frac{\epsilon_c\cos\theta_i - \sqrt{\epsilon_c - \sin^2\theta_i}}{\epsilon_c\cos\theta_i + \sqrt{\epsilon_c - \sin^2\theta_i}} \tag{7}$$

where $\epsilon_c = \epsilon_r - j\sigma/\omega\epsilon_0$ is the complex permittivity of the ground plane with dielectric constant ϵ_r and conductivity σ together with permittivity ϵ_0 in free space. In the subsequent numerical examples we have employed the parameters such that $\epsilon_r = 5.0$ and $\sigma = 0.0023 \ [S/m]$ corresponding to a dry soil constituting the ground plane. Moreover, we have assumed that gains of source and receive antennas are $G_s = G_r = 1.5$, directivity of source antenna is $\sin\theta$, directivity of receive antenna is $\sin\theta_0$ in Eq. (2), respectively, and θ_i in Eq. (7) is an incident angle of the incident ray at a reflection point. These angles are given by

$$\sin\theta = \frac{|\boldsymbol{p_s} \times \boldsymbol{r}|}{r}, \sin\theta_0 = \frac{|\boldsymbol{p_s} \times \boldsymbol{r_s}|}{r_s}, \sin\theta_i = \frac{|\boldsymbol{n} \times \boldsymbol{r_s}|}{r_s} \tag{8}$$

where the distance $r = |\boldsymbol{r}|$ between source and receive antennas should be much larger than λ.

2.2. Communication distance

Since the direction vectors of the source and receive antennas are $\boldsymbol{p_s}$ and $\boldsymbol{p_r}$ as shown in Fig. 1, received power of a small dipole antenna can be expressed as follows [12,13]:

$$P_r = \frac{\lambda^2 G_r}{4\pi} \cdot \frac{|\boldsymbol{E} \cdot \boldsymbol{p_r}|^2}{Z_0} \ \ [\text{W}] \tag{9}$$

where the intrinsic impedance in free space is given by $Z_0 \simeq 120\pi \ [\Omega]$ [12]. From Eqs (1) and (2), on the other hand, the 1-ray model yields the following relation

$$|\boldsymbol{E_1}| = 10^{\frac{\alpha}{20}} \cdot 10^{\frac{(\beta-1)\gamma}{20}} \cdot \frac{\sqrt{30 G_s P_s}}{r^\beta} \tag{10}$$

where antenna orientation is assumed to be arranged so that the maximum received power may be obtained.

Let $E_{\min}$ be the minimum detectable electric field intensity and D_c be the maximum communication distance. Then, Eq. (10) yields the following relation

$$E_{\min} = 10^{\frac{\alpha}{20}} \cdot 10^{\frac{(\beta-1)\gamma}{20}} \cdot \frac{\sqrt{30 G_s P_s}}{D_c^\beta} \ . \tag{11}$$

Consequently, Eq. (11) leads to the maximum communication distance D_c given by

$$D_c = 10^{\frac{\alpha}{20\beta}} \times 10^{\frac{(\beta-1)\gamma}{20\beta}} \times (B_{\min})^{\frac{1}{2\beta}} \tag{12}$$

where the minimum parameter $B_{\min}$ is defined by using the minimum detectable electric field intensity $E_{\min}$ or the minimum detectable receiving power $P_{\min}$ as follows:

$$B_{\min} = \frac{30 G_s P_s}{(E_{\min})^2} = \frac{\lambda^2 G_r}{4\pi} \cdot \frac{30 G_s P_s}{Z_0 P_{\min}} . \tag{13}$$

It is evident from Eq. (9) that the minimum detectable power is related to the minimum detectable electric field intensity as follows:

$$P_{\min} = \frac{\lambda^2 G_r}{4\pi} \cdot \frac{(E_{\min})^2}{Z_0} . \tag{14}$$

3. Path loss for propagation in urban areas

Assuming ideal isotropic source and receive antennas with unit gain $G_s = G_r = 1$ in accordance with the Okumura-Hata model [4], the received power in Eq. (9) can be rewritten as follows:

$$P_r = \frac{\lambda^2}{4\pi} \frac{|\boldsymbol{E}|^2}{Z_0} . \tag{15}$$

This equation can be expressed in [dB] as follows:

$$P_r \, [\text{dBW}] = |\boldsymbol{E}| \, [\text{dBV/m}] - 10 \log_{10} Z_0 + 10 \log_{10}(\lambda^2/4\pi). \tag{16}$$

Since the path loss L_p expressed in [dB] is given by the difference between the transmitted power and the received power [4], the path loss can be described in [dB] as follows:

$$L_p \, [\text{dB}] = P_s \, [\text{dBW}] - P_r \, [\text{dBW}] = P_s \, [\text{dBW}] - |\boldsymbol{E}| \, [\text{dBV/m}] + 10 \log_{10} Z_0 - 10 \log_{10}(\lambda^2/4\pi). \tag{17}$$

Equation (17) indicates that the received electric field can be obtained by the path loss L_p as long as the input power P_s is specified.

Assuming unit input power $P_s = 1$ [W] for the 1-ray model in Eq. (1), the path loss in Eq. (17) is rewritten as follows:

$$L_p \, [\text{dB}] = -\alpha - (\beta - 1)\gamma + 20\beta \log_{10} r - 10 \log_{10} 30 + 10 \log_{10} Z_0 - 10 \log_{10}(\lambda^2/4\pi). \tag{18}$$

Thus the 1-ray model ensures that a relation between α and β can be explicitly described by Eq. (18), as far as L_p is known experimentally or theoretically and Eq. (6) is satisfied. The path loss in urban, suburban or open area can be derived from the Okumura-Hata model as listed in Appendix A. One of the important features is that β depends only on the source antenna height h_b [m] as follows:

$$\beta = \beta(h_b) = (44.9 - 6.55 \log_{10} h_b)/20 . \tag{19}$$

As a result, based on the Okumura-Hata model, it is evident that propagation order of distance β is a monotonically decreasing function with respect to BS antenna height h_b alone. Combining Eqs (18) and (19) leads to the explicit expression for α as follows:

$$\alpha = -L_p \, [\text{dB}] - (\beta - 1)\Gamma + 20\beta \log_{10} R + 20 \log_{10} f_c + 60\beta + 20 \log_{10}(4\pi/300) \tag{20}$$

where some conversions in unit have been made in accordance with the Okumura-Hata model; that is, f [Hz] $\rightarrow f_c$ [MHz] and r [m] $\rightarrow R$ [Km]. It should be noted that the above relation is restricted only to the urban, suburban and open areas. In other regions, however, we must interpolate α approximately [7].

4. Estimation of α and β for propagation along RRS

The Okumura-Hata model model shows that the propagation order of distance β can be evaluated by BS antenna height h_b alone and the modification factor α can be obtained by the path loss in the urban, suburban and open areas. In this section we propose a procedure to estimate α and β numerically in case of EM propagation along RRSs [5,6].

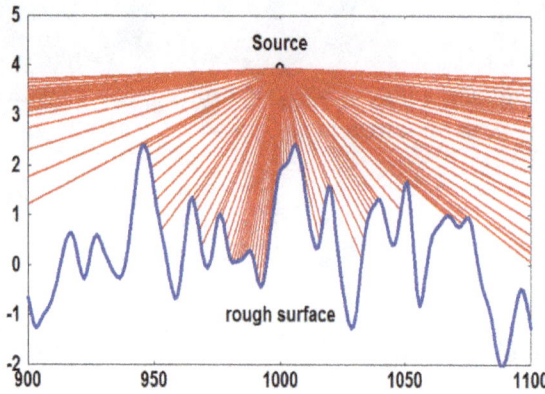

Fig. 3. Incident rays on 1D RRS.

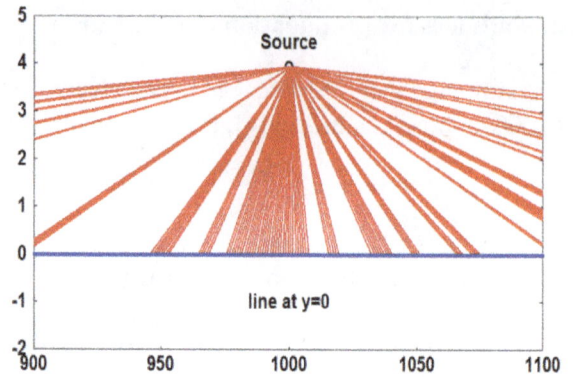

Fig. 4. Illuminated regions for 1D RRS.

Considering the EM propagation along a uniform ground plane with parameters $\alpha = 0$ and $\beta = \beta_0 = 1$ in Eq. (1), Eq. (12) yields the following communication distance

$$D_c = 10^{\frac{(\beta_0 - 1)\gamma}{20\beta_0}} \times \left(B_{\min}\right)^{\frac{1}{2\beta_0}} . \tag{21}$$

The above communication distance is increased for a large BS antenna height, since the field matching factor γ is increased for large h_b as shown in Eq. (6). When the BS antenna height is increased from h_{b0} to h_b, the propagation order of distance is changed from β_0 to β and the field matching factor is varied from γ_0 to γ. Then, assuming that the analytical property of Eq. (21) with respect to β is unchanged together with γ, we have the following relation

$$10^{\frac{(\beta_0 - 1)\gamma}{20\beta_0}} \times \left(B_{\min}\right)^{\frac{1}{2\beta_0}} \simeq 10^{\frac{(\beta - 1)\gamma_0}{20\beta}} \times \left(B_{\min}\right)^{\frac{1}{2\beta}} \tag{22}$$

which can be solved for β as follows:

$$\beta = \beta(\gamma) \simeq \frac{\beta_0(\gamma_0 - 10\log_{10} B_{\min})}{\beta_0\gamma_0 - (\beta_0 - 1)\gamma - 10\log_{10} B_{\min}} . \tag{23}$$

It has been demonstrated that β computed by Eq. (23) is fairly in good agreement with the Okumura-Hata model [5]. Anyway the most interesting feature is that β is given by a function of γ or $\beta(\gamma)$ which has a monotonically decreasing property, in contrast to the Okumura-Hata model where β is given by a function of h_b or $\beta(h_b)$ as shown in Eq. (19). As is evident from Eq. (6), it is concluded that the propagation oder of distance is dependent not only BS antenna height but also MS antenna height as well as operating frequency.

We propose a procedure to estimate α by considering the effective planar angles for illuminated segments. First, we approximate propagation characteristic by taking account of the incident rays as shown in Fig. 3. Then we project the illuminated segments on the straight line of at $y = 0$ as shown in Fig. 4, and we sum up the planar angles corresponding to the illuminated segments. We denote the total angle as equivalent angle θ_e of incidence. Next, we assume that the equivalent angle concentrates near the nadir of the source antenna as shown in Fig. 5 where R is a maximum communication distance and R_e is an equivalent communication distance [5]. The equivalent communication distance is defined by

$$R_e = 2h_b \tan(\theta_e/2). \tag{24}$$

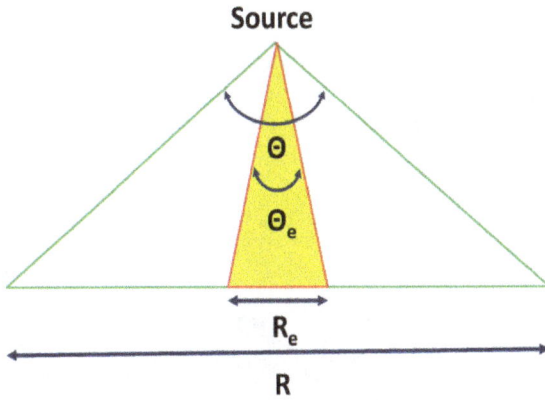

Fig. 5. Equivalent angle for 1D RRS.

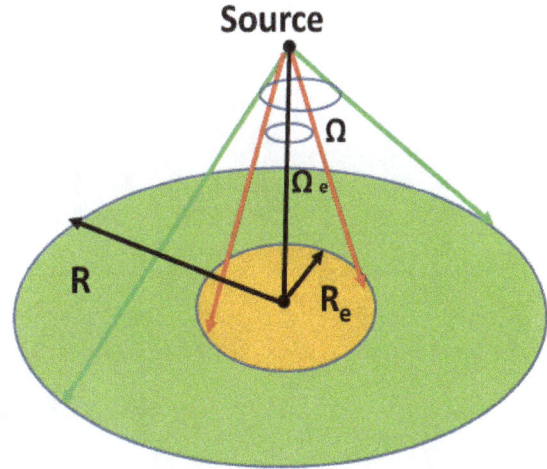

Fig. 6. Equivalent angle for 2D RRS.

Finally we assume that α might be given by a function of the equivalent length as follows:

$$\alpha = K \log_{10}(R_e/R) \tag{25}$$

where K is a constant which might be in proportion to wave number κ.

In case of 2D RRS, extension from the planar angle to the solid angle [14] is enough to obtain the equivalent angle θ_e by summing up each solid angle corresponding to every illuminated triangular plate projected on the average plane at $z = 0$. The analytical relation between a planar triangle and corresponding solid angle is shown in Appendix B.

The total solid angle Ω and the planar angle θ are related by

$$\Omega = 2\pi(1 - \cos\theta) \tag{26}$$

and the illuminated equivalent angle for 2D RRS is given by

$$\theta_e = \cos^{-1}(1 - \Omega_e/2\pi) . \tag{27}$$

Thus, the amplitude modification factor α in case of 2D RRS can be estimated by use of Eqs (25) and (27). The physics of this theory is that electric field distributions along RRSs are mainly influenced by the optical situation whether receiving point is in line-of-sight (LOS) or in non-line-of-sight (NLOS).

5. DRTM

We calculate EM field distributions along RRS by using DRTM in order to check the accuracy of the 1-ray model which includes two parameters α and β estimated by the numerical procedure proposed in this paper. RRS generation, on the other hand, is performed by the convolution method which is prescribed by two parameters, height deviation dv and correlation length cl, together with type of rough surface spectrum [11]. In this paper we deal with only the Gaussian type of spectrum.

The first step of DRTM is to discretize RRS profile in terms of straight lines for 1D case and rectangular plates for 2D case, respectively. Thus, the RRS profile can be approximated by piece wise linear

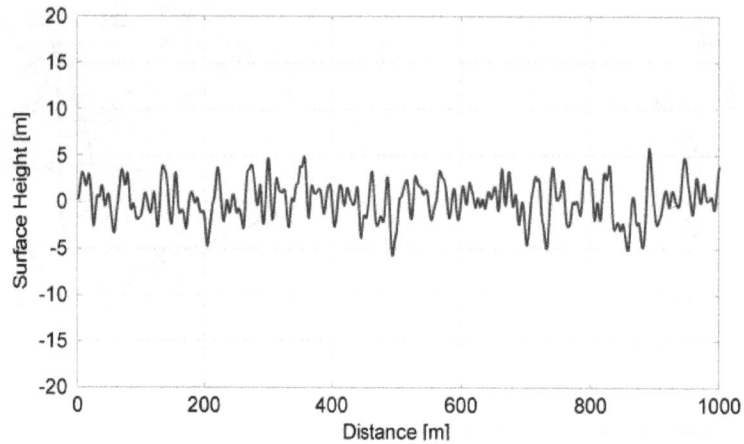

Fig. 7. An example of 1D RRS.

lines or piece wise planar plates, and this discretization should be performed so that computer memory together with numerical errors could be as small as possible. The second step is to discretize the procedure for searching rays traveling along RRS. To do this, we assume that arbitrary two lines (1D) or plates (2D) composing a RRS profile are in LOS if a representative point of one line or plate is in LOS with that of another line or plate, and otherwise, they are in NLOS. This assumption enables us to simplify ray searching drastically, resulting in saving much computation time [1]. It is worth noting that this algorithm can be modified to achieve more accurate rays.

Once necessary rays are searched for BS and MS antennas arbitrarily located along a generated RRS, we can evaluate EM fields in terms of searched ray data together with RRS data; the ray data are reflection and diffraction points composed of representative points of discretized lines or plates, and the RRS data are position and normal vectors of lines or plates of a discretized RRS profile. Detailed discussions are omitted here, but the electric field E at a receive point is formally expressed in the following dyadic and vector form [1]:

$$E = \sum_{n=1}^{N} \left[\prod_{m=1}^{m=M_n^r} (R_{nm}) \cdot \prod_{k=1}^{k=M_n^d} (D_{nk}) \cdot E_0 \right] \frac{e^{-j\kappa r_n}}{r_n} \tag{28}$$

where R_{nm} is the dyadic for reflection at the m-th reflection point of the n-ray, D_{nk} is the dyadic dor diffraction at the k-th diffraction point of the n-ray, and E_0 is the electric field of the n-th ray at the first reflection or diffraction point. Moreover, r_n is the distance of the n-th ray from BS antenna to MS antenna, N is the total number of rays considered, M_n^d is the number of times of source diffractions of the n-th ray, and M_n^r is the number of times of image diffractions of the n-th ray.

6. Numerical examples

Figure 7 shows a RRS example where parameters are chosen as $dv = 2\ [m]$ and $cl = 5\ [m]$. The BS antenna is located at $x = 0$ with height h_b above RRS, and the MS antenna height is 1 [m] above RRS. In the following numerical examples, operating frequency is selected as $f = 300$ [MHz] and material constants of RRS are chosen as $\epsilon_r = 5$ and $\sigma = 0.0023$ [S/m].

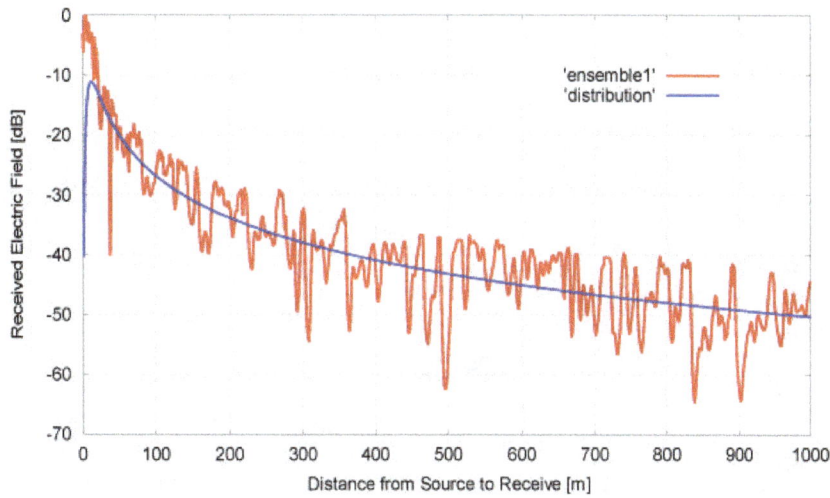

Fig. 8. Estimated field compared with DRTM.

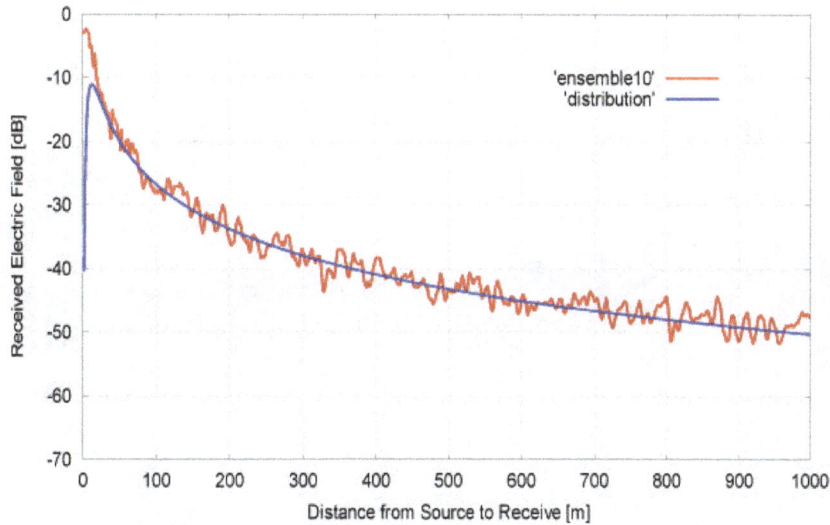

Fig. 9. Estimated field compared with DRTM.

Figure 8 shows electric field distribution computed by DRTM for one generated RRS with $dv = 2$ [m], $cl = 5$ [m] and $h_b = 10$ [m] in comparison with the 1-ray model in Eq. (1) where α and β are tentatively estimated by the following relations:

$$\alpha = K \log_{10}(R_e/R), \quad \beta = (2w^2 + \gamma^2)/(w^2 + \gamma^2) \tag{29}$$

where $K = 5$ and $w = 20$ have been assumed. This parameter setting is employed in all the numerical examples listed hereafter. It should be noted that the second relation in Eq. (29) is arranged so that $\beta \to 1$ for large antenna heights, that is, for small γ and $\beta \to 2$ for small antenna heights, that is, for small γ. Fig. 9 shows ensemble averaged electric field distribution computed by DRTM for ten generated RRSs with $dv = 2$ [m], $cl = 5$ [m] and $h_b = 10$ [m] in comparison with the 1-ray model using α and β

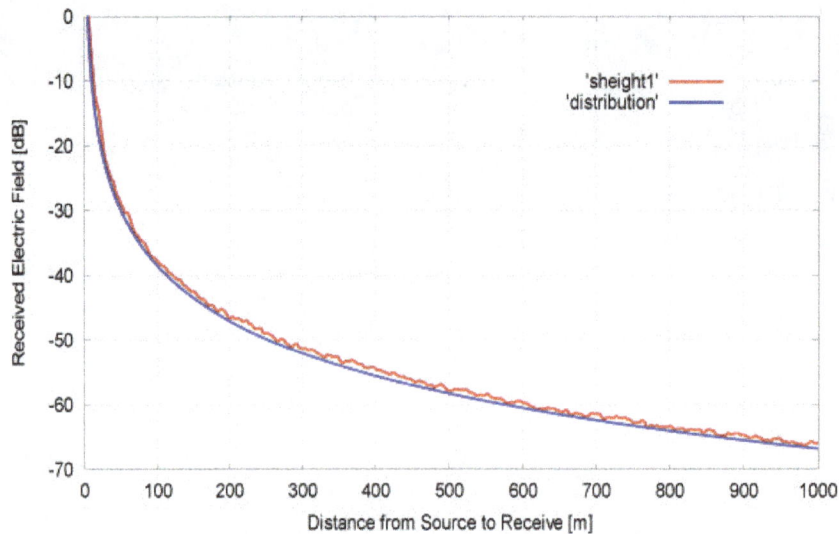

Fig. 10. Estimated field compared with DRTM.

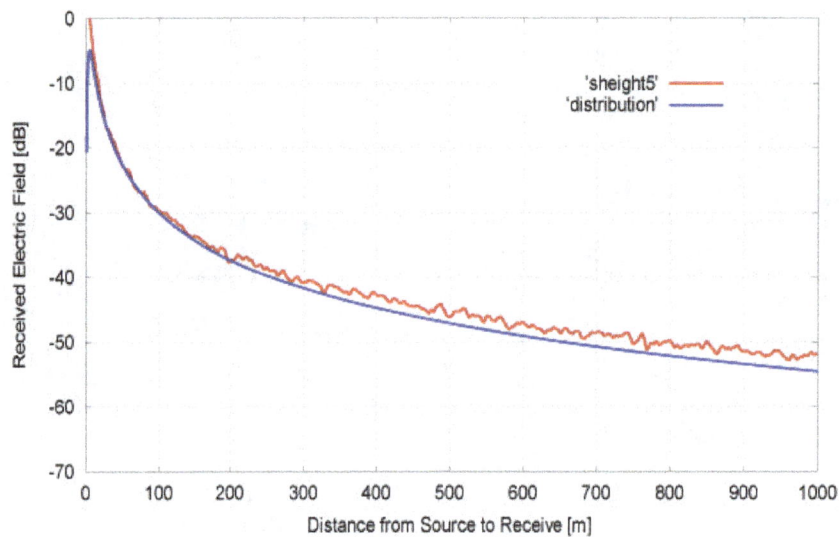

Fig. 11. Estimated field compared with DRTM.

estimated by Eq. (29). It is well demonstrated that the estimated field distributions based on the 1-ray model employing the proposed parameters are in good agreement with the ensemble average of field distributions computed by DRTM even if the number of RRS samples is only ten.

Figures 10, 11 and 12 show electric field distributions of 100 ensemble average of DRTM results in comparison with the estimated fields based on the proposed 1-ray model using parameters estimated by Eq. (29) for $h_b = 1$ [m], 5 [m] and 10 [m], respectively. Other parameters are chosen as the same as Figs 8 and 9. It is shown that the numerical results of the proposed field estimation procedure provide a good accuracy in the NLOS-dominated case ($h_b = 1$ [m]) as well as in the LOS-dominated case ($h_b = 10$ [m]), and also a fairly good accuracy in the grazing-angle case ($h_b = 5$ [m]).

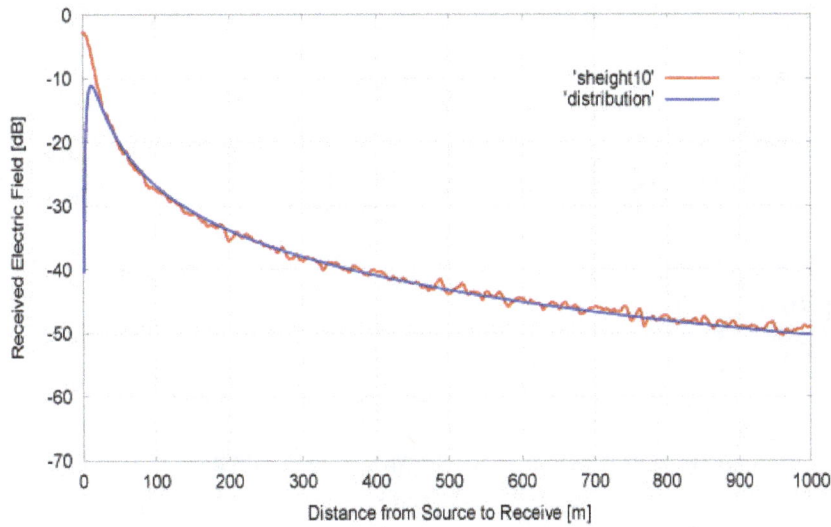

Fig. 12. Estimated field compared with DRTM.

7. Conclusion

First, we have reviewed the 1-ray and 2-ray models in order to estimate approximately propagation characteristics in urban, suburban and open areas as well as along RRS. Second, we have discussed two propagation parameters, amplitude modification factor α and propagation order of distance β, which play an important role for estimating communication distance in complicated propagation environments. Third, we have proposed an algorithm to estimate α by computing equivalent angle or length related to the illuminated portion of RRS, and we have also introduced a procedure to estimate β in terms of field matching factor γ which is used by in the 1-ray model. Finally, we compared the field distributions obtained by the proposed method with the ensemble averaged DRTM solutions.

It has been demonstrated that the proposed methods are useful for propagation estimation in complicated natural or artificial environments, because the proposed method can be applied effectively not only in LOS-region but also NLOS-region. It is needed to consider other situations with different frequency and other RRS parameters. This deserves as a future investigation.

Acknowledgments

The work was supported in part by a Grand-in Aid for Scientific Research (C) (24560487) from Japan Society for the Promotion of Science.

Okumura-Hata model

Empirical equations for the path loss in urban, suburban and open areas are summarized as follows [4]:

– Urban Area:

$$L_p\,[dB] = 69.55 + 26.16\log_{10} f_c - 13.82\log_{10} h_b - a(h_m) + (44.9 - 6.55\log_{10} h_b)\log_{10} R \quad (30)$$

where modifying factor for the height of the mobile antenna h_m is given by

$$a(h_m) = (1.1 \log_{10} f_c - 0.7)h_m - (1.56 \log_{10} f_c - 0.8) \quad \text{(in medium} - \text{small city)}$$

$$a(h_m) = 8.29(\log_{10} 1.54 h_m)^2 - 1.1; \quad f_c \leqslant 200 \text{ [MHz]} \quad \text{(in large city)}$$

$$= 3.2(\log_{10} 11.75 h_m)^2 - 4.97; \quad f_c \geqslant 400 \text{ [MHz]} \quad \text{(in large city)} \tag{31}$$

– Suburban Area:

$$L_{ps} \text{ [dB]} = L_p\{\text{Urban area}\} - 2\{\log_{10}(f_c/28)\}^2 - 5.4 \tag{32}$$

– Open Area:

$$L_{po} \text{ [dB]} = L_p\{\text{Urban area}\} - 4.78(\log_{10} f_c)^2 + 18.33 \log_{10} f_c - 40.94 \tag{33}$$

The carrier frequency f_c [MHz] is confined from 150 [MHz] to 1500 [MHz], the antenna height h_b [m] of BSs is limited from 30 [m] to 200 [m], the antenna height h_m [m] of MSs is ranged from 1 [m] to 10 [m], and the communication distance R [Km] is confined from 1 [Km] to 20 [Km].

B. Solid angle

Based on the spherical trigonometry [14], we can numerically compute the solid angle constituted by arbitrary three vectors $\boldsymbol{r_1}$, $\boldsymbol{r_2}$ and $\boldsymbol{r_3}$ in the following fashion [15]. First we compute three cosines in terms of inner products of the three vectors as follows:

$$\cos \theta_i = \frac{(\boldsymbol{r_j} - w_{ij}\boldsymbol{r_i}) \cdot (\boldsymbol{r_k} - w_{ik}\boldsymbol{r_i})}{|\boldsymbol{r_j} - w_{ij}\boldsymbol{r_i}| \cdot |\boldsymbol{r_k} - w_{ik}\boldsymbol{r_i}|}$$

$$i, j, k = \{1, 2, 3\}, \quad i \neq j, k, \quad j \neq k \tag{34}$$

where the weights are computed by the following inner products defined by

$$w_{ij} = \frac{(\boldsymbol{r_i} \cdot \boldsymbol{r_j})}{(\boldsymbol{r_i} \cdot \boldsymbol{r_i})}, \quad w_{ik} = \frac{(\boldsymbol{r_i} \cdot \boldsymbol{r_k})}{(\boldsymbol{r_i} \cdot \boldsymbol{r_i})}$$

$$i, j, k = \{1, 2, 3\}, \quad i \neq j, k, \quad j \neq k. \tag{35}$$

Then the required solid angle Ω in steradian is computed by taking the inverse cosines in Eq.(34) as follows:

$$\Omega = \theta_1 + \theta_2 + \theta_3 - \pi. \tag{36}$$

Thus we can numerically estimate the solid angle spanned by an arbitrary 3D surface by discretizing it in terms of triangular cells.

References

[1] K. Uchida and J. Honda, Estimation of Propagation Characteristics along Random Rough Surface for Sensor Networks, *Wireless Sensor Networks: Application-Centric Design, Geoff V Merret and Yen Kheng Tan (Ed.), InTech*, Chapter 13, (2010-12) 231–248.

[2] K. Uchida, J. Honda, T. Tamaki and M. Takematsu, Two-Rays Model and Propagation Characteristics in View of Hata's Empirical Equations, *IEICE Technical Report*, AP2011-14 (2011) 49–54.

[3] K. Uchida, J. Honda and Jun-Hyuck Lee, A Study of Propagation Characteristics and Allocation of Mobile Stations, *IEICE Technical Report*, IN2011-98, MoMuC2011-32 (2011) pp. 31–36.

[4] M. Hata, Empirical Formula for Propagation Loss in Land Mobile Radio Services, *IEEE Trans Veh Technol* **VT-29** (3) (1980) 317-325.

[5] K. Uchida, M. Takematsu and J. Honda, An Algorithm to Estimate Propagation Parameters Based on 2-Ray Model, *Proceedings of NBiS-2012*, The 15th International Conference on Network-Based Information Systems, Melbourne, Australia, (2012), 556–561.

[6] K. Uchida, M. Takematsu, J.H. Lee and J. Honda, Field Distributions of 1-Ray Model Using Estimated Propagation Parameters in Comparison with DRTM, *Proceedings of BWCCA-2012*, 2012 Seventh International Conference on Broadband, Wireless Computing, Communication and Applications, Victoria, Canada, (2012), 488–493.

[7] K. Uchida, K. Shigetomi, M. Takematsu and J. Honda, An Estimation Method for Amplitude Modification Factor Using Floor Area Ratio in Urban Areas, *Information Technology Convergence: Security, Robotics, Automations and Communication – Lecture Notes in Electrical Engineering 253*, Springer, ISBN 978-94-007-6995-3, (2013), 101–109.

[8] K. Uchida, J. Honda and K.Y. Yoon, Distance Characteristics of Propagation in Relation to Inhomogeneity of Random Rough Surface, *Proceedings of ISMOT 2009*, The 12th International Symposium on Microwave and Optical Technology, New Delhi, India, (2009), 1133–1136.

[9] J. Honda, K. Uchida and K.Y. Yoon, Estimation of Radio Communication Distance along Random Rough Surface, *IEICE Trans. ELECTRON* **E93-C**(1) (2010), 39–45.

[10] K. Uchida and J. Honda, An Algorithm for Allocation of Base Stations in Inhomogeneous Cellular Environment, *Proceedings of NBiS-2011*, The 14th International Conference on Network-Based Information Systems, Tirana, Albania, (2011), 507–512.

[11] K. Uchida, J. Honda and K.Y. Yoon, An Algorithm for Rough Surface Generation with Inhomogeneous Parameters, *Journal of Algorithms and Computational Technology* **5** (2) (2011), 259–271.

[12] Y. Mushiake, Antennas and Radio Propagation, *Corona Publishing Co., LTD. Tokyo*, 1985.

[13] R.E. Collin, Antennas and Radiowave Propagation, *McGraw-Hill Book Company, New York*, 1985.

[14] G.A. Korn and T.M. Korn, Mathematical Handbook for Scientists and Engineers, *McGraw-Hill Book Company* (1968), 888–890.

[15] A. Van Oosterom and J. Strackee, The Solid Angle of a Plane Triangle, *IEEE Transactions on Biomedical Engineering*, **BME-30** (2) (1983), 125–126.

Modelling spatio-temporal relevancy in urban context-aware pervasive systems using voronoi continuous range query and multi-interval algebra

Najmeh Neysani Samany[a], Mahmoud Reza Delavar[b,*], Nicholas Chrisman[c] and
Mohammad Reza Malek[d]

[a]*Department of Surveying and Geomatics Engineering, College of Engineering, University of Tehran,
Tehran, Iran*
[b]*Center of Exellence in Geomatic Engineering in Disaster Management, Department of Serveying and
Geomatic Engineering, College of Engineering, University of Tehran, Tehran, Iran*
[c]*Department of Geomatic Science, Laval University, Québec, QC, Canada*
[d]*Department of GIS, Faculty of Geodesy and Geomatic Engineering, K.N. Toosi University of
Technology, Tehran, Iran*

Abstract. Space and time are two dominant factors in context-aware pervasive systems which determine whether an entity is related to the moving user or not. This paper specifically addresses the use of spatio-temporal relations for detecting spatio-temporally relevant contexts to the user. The main contribution of this work is that the proposed model is sensitive to the velocity and direction of the user and applies customized Multi Interval Algebra (MIA) with Voronoi Continuous Range Query (VCRQ) to introduce spatio-temporally relevant contexts according to their arrangement in space. In this implementation the Spatio-Temporal Relevancy Model for Context-Aware Systems (STRMCAS) helps the tourist to find his/her preferred areas that are spatio-temporally relevant. The experimental results in a scenario of tourist navigation are evaluated with respect to the accuracy of the model, performance time and satisfaction of users in 30 iterations of the algorithm. The evaluation process demonstrated the efficiency of the model in real-world applications.

Keywords: Context-awareness, spatio-temporal relevancy, customization, multi interval algebra, range query, tourist

1. Introduction

Context appears as a fundamental key to enable systems to filter relevant information from what is available [5,9,40], to choose relevant actions from a list of possibilities [1,40], or to determine the optimal method of information delivery [8,41]. The major challenge of context-aware systems is the separation of the relevant from the irrelevant information [12,40]. This process requires finding an acceptable degree of information reduction, i.e., presenting as much information as needed and as little as required [40].

---

*Corresponding author: Mahmoud Reza Delavar, Center of Exellence in Geomatic Engineering in Disaster Management, Department of Serveying and Geomatic Engineering, College of Engineering, University of Tehran, Tehran, Iran.
E-mail: mdelavar@ut.ac.ir.

However, there are few reports concerning appropriate services to manage spatio-temporal relevancy parameters that determine whether a context is spatially related to the user or not.

Most of the current models for the spatio-temporal relevancy parameters in context-aware pervasive systems are based on the spatio-temporal relationships between the user and related objects. The main concentration of these models is on spatial relationships while in time dimension they usually applied temporal intervals [12,29,40]. Some studies have used the proximity relations between the user and the related contexts to model the spatial relevancy and utilised K-N neighbourhood or range queries [10, 29]. Such relations cover the inclusion of related contexts in a distinct area or range and the distance to other entities [10]. Stiller et al. [14] presented spatial user-item relations into recommender systems with distance functions and using weighting approaches. Holzmann and Ferscha [12] defined a Zone-Of-Influence (ZOI) for any entity with a specified distance and direction and used the RCC5 [3]. The position, direction and extension of both ZOI are also included in their model. The most important drawback of these systems is that they do not mention the characteristic of the user's movement in an urban network which typically follows a linear route with a specific direction and velocity [30,32,33]. However position, direction and velocity of a moving user have an important role in adaptation process. Moreover these approaches do not apply all of the topological relationships such as order relationships (e.g., behind or in- front- of), which could be useful in providing spatially relevant context-aware services. On the other hand in an urban network, the real distance between two objects is not the Euclidean distance but the actual network distance and applying K-N neighbourhood or range query without any modification will decrease the efficiency of such systems [15,18,26,27,31,35].

To model the spatio-temporal relevancy parameters in an urban context-aware system, which could cover the characteristics of the user's movement and utilise all of the spatial relationships (metric, directional and topologic) regarding to the time dimension, the approach proposed in this paper is organised into two main steps as follows:

(1) The quantitative representation of the moving user and his/her related contexts in space and time dimension with spatial and temporal intervals. The Spatial Interval of the user is Directed (DSI) and will be dynamically updated based on the position, velocity and direction of his/her movement. The Spatial Intervals (SIs) of the related contexts are assumed to be static and unchanging. The temporal intervals of the user (TIU) and related contexts (TIC) is not static and will be updated.

(2) Comprehensive representation of spatio-temporal relationships between the DSI and TI of the user and the SIs and TIs of the related urban contexts that can be modelled with the Voronoi Continuous Range Query (VCRQ) [25] and Multi Interval Algebra (MIA) [34].

The main contribution of this paper is the use of MIA for modeling spatio-temporal relevancy in an urban context-aware system. The selection of MIA in this research is considered for the following reasons: (1) when the user is moving in an urban network, he/she has a directional linear route. Therefore, the position of the moving user can be effectively modelled through a directed spatial interval [33,34] that is adaptive to the user's behaviour (it can be extended or shortened according to the user's velocity). (2) In an urban area, one usually encounters solid objects; therefore, most of the related contexts of the user are considered with their external views which is sufficient to abstract the contexts and show them with spatial intervals. In the proposed approach "related contexts" refers to the entities that are preferred by the user. For example, the related contexts of a tourist are the attractive areas such as monuments, gardens and museums. (3) The relationships between the spatial and temporal intervals of the moving user and related contexts can manage spatio-temporal relationships in such a way that in space dimension all of the including topological, metric and directional relationships are covered. Figure 1 shows the importance of each spatial relationship. Most of the related contexts in an urban network have a valid

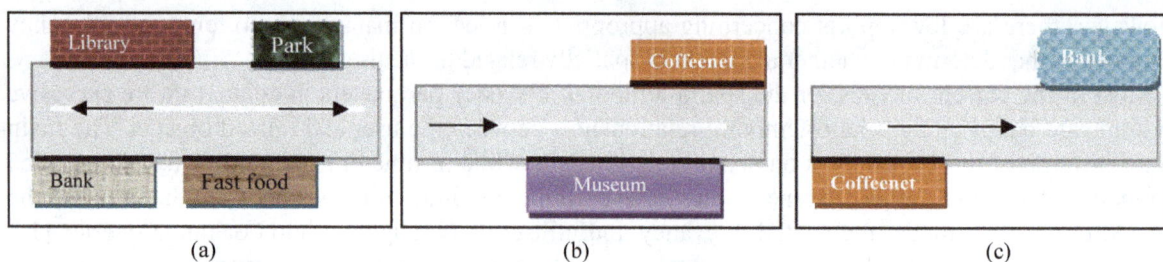

Fig. 1. The applications of MIA for specification of spatial relevant entities: a) the effect of the direction of the moving user, b) the effect of the distance between the moving user and related entities, c) the effect of topological relationships of the moving user and his/her related entities.

time to visit or to provide services, such as banks, shopping centers, libraries and etc. Therefore, it is necessary to specify temporal relationships between the moving user and his/her related contexts which lead to adapted services. (4) The algebra between spatial intervals and temporal intervals could introduce appropriate instructions (based on related contexts) to the user according to their arrangement in space. It should be noted that the algorithm is acted partially and followed up with the execution of a VCRQ [25]. Indeed, MIA assesses the spatio-temporal relationships between the user and the related contexts which are selected by the VCRQ.

Our approach is implemented in tow districts of Tehran, the capital of Iran, and we have focussed on an outdoor guided tour as an example. In this scenario, the user is a tourist who intends to visit some selected points of interest with a specified origin and destination. It is assumed that the tourist is equipped with a PDA or a laptop computer, and a GPS for positioning, and the route is constrained by a directed network. The origin and the destination of the user are recognised by the system or identified by the user. The proposed algorithm guides the user from the origin to destination. Guiding the user is adapted in modelling and representing procedures during the navigation task. The adaptation process is based on the time, position, direction and velocity of the user. The output of the system is the location and time information of hisher points of interest along the pre-defined route.

The evaluation process is based on three factors: the accuracy of the results, the time performance of the algorithm and the satisfaction of the users with the navigation process. The accuracy assessment of the model is based on the comparison of the expected and the actual spatio-temporally relevant objects detected by the model. The Chi-squared test is used to evaluate the goodness of fit. The experimental results show that the proposed approach can effectively model and accurately detect the spatio-temporally relevant contexts within a reasonable time frame. This approach also provides context-aware instructions for the user with a high percentage for user satisfaction.

The rest of this paper is structured as follows: Section 2 presents fundamental aspects concerning the spatio-temporal relevancy parameters and the principals of MIA theory with the approaches of customization. Then it describes the VCRQ and presents the research methodology and architectural design of the system. Section 3 presents a case study. The evaluation of the algorithm and the results obtained from the case study are explained in Section 4. Section 5 discusses the theoretical and practical issues of the proposed approach and attempts and gives a comprehensive comparison with related work. Finally, conclusions and directions of potential future research are considered in Section 6.

2. Materials and methods

In this section, the concept of spatio-temporal relevancy is described, followed by a description of the MIA theory and the approaches of customization. Then the principles of VCRQ are explained. Finally a description of the research methodology is introduced.

2.1. Spatio-temporal relevancy in context-aware systems

Saraceviec [41] offers a general definition of relevance derived from its general qualities: "Relevance involves an interactive, dynamic establishment of a relation by inference, with intentions towards a context. Relevance may be defined as a criterion reflecting the effectiveness of exchange of information between people (or between people and objects) in communication relation, all within a context". Three main relevancies in context-aware systems are identical relevancy, spatial relevancy and temporal relevancy [10,42]. Among these relevancies, the current position – 'the here' – is usually the centre of action, perception and attention. The identical information may be fully relevant at one position but irrelevant at another position [6,8,17,37,42]. Thus, the spatial relevancy is a parameter which has a fundamental role to provide context-aware services [4,6]. Similar to the 'here' in location humans are always at a certain point in time – the 'now'. This point of time is where one acts and also where one perceives the environment, analogous to the perception in space. The perception of contexts that are around is restricted to the point in time [7,20]. Therefore modeling of "spatio-temporal relevancy" is proved as an essential task in context-aware systems implementation.

2.2. Multi intervals algebra

Being similar to the well-known Interval Algebra developed for temporal intervals [9,21] it seems useful to develop spatial interval algebra for modelling spatial relationships especially in an urban traffic network. When both temporal and spatial intervals for modelling spatial and temporal relationships between objects are applied, we encounter with Multi-Interval Algebra. The properties of spatial and temporal intervals and the interval algebra in time and space dimensions are described in Sub-sections 2.2.1 and 2.2.2.

2.2.1. Temporal interval
The Interval Algebra (IA) describes the possible relationships between convex intervals on a directed line. The default application of the Interval Algebra is temporal, so the considered line is usually regarded to be the timeline. In reality every events and objects start their life (are born) at T_S, exists for a period of time (life span or duration) T_D, and ceases to exist (die or finish) at T_F. There are 13 temporal relations between two temporal intervals including before <, after >, meets m, met-by mi, overlaps o, overlapped-by oi, equals $\equiv$, during d, include di, starts s, started-by si, finishes f, and finished by fi [9].

2.2.2. Spatial interval
There are several differences between spatial and temporal intervals which have to be considered when extending the Interval Algebra towards dealing with spatial applications [38]. The most important characteristic of spatial interval is its direction. A spatial interval can have the same or the opposite direction [23]. This leads to the definition of Directed Interval Algebra (DIA) with 26 base relations given in Renz [23], which result from refining each relation into two sub-relations specifying either the same or opposite direction of the involved intervals, and of all possible unions of the base relations.

2.2.3. Spatio-temporal relationship in MIA
Multi interval algebra defines spatio-temporal relationships between objects which could be represented with specified intervals in space and time dimensions. As we have 13 temporal relationships between temporal intervals and 26 spatial relationships between spatial intervals, there are 26×13 spatio-temporal relationships in MIA which can be represented by MIA_{338}.

2.3. Customizing MIA_{338} to MIA_{72}

Regarding to the characteristics of the moving user and related contexts in urban traffic networks, MIA could model spatio-temporal relevancy in an effective way [34]. The problem is that using all of the 13 relations in temporal domain and 26 relations in spatial domain or reasoning based on MIA_{338} will reduce the speed of performance and it may decrease the efficiency of the context-aware system. Particularly when the user is moving with a specified velocity and he/she intends to make a decision due to the receiving messages of the system, the time of delivering appropriate instructions should be shortened as much as possible. Therefore it is necessary to reduce the existence spatio-temporal relationships in order to decrease computational complexity of the algorithm which leads to increase the time performance. Therefore the MIA_{338} should be customized. There are two ways for customizing the Multi Interval Algebra [20,24].

(1) To use macro relations [28], i.e., unions of base relations. Indeed combining IA base relations and use these macro relations as base relations is the approach of customizing. For example, It is possible to combine {m, o}, {mi, oi}, {s, d, f}, {si, di, fi} and use them together with the relations {<}, {>}, and {=} as new base relations in space and time dimensions. This corresponds to the algebra A_7 defined in [28]. Regarding to the spatial and temporal intervals and their difference in MIA there are 7 relations in temporal dimension and 14 relations in spatial dimension. Therefore we have 14×7 or 98 spatio-temporal relationships with this type of customization or we have MIA_{98}.

(2) To use only the relations which is needed, namely, the interval relations $<, >$, d, di, o, oi, $=$ and do not use m, mi, s, si, f, fi which correspond to intervals with common endpoints. This is similar to the algebra A_6 defined in [28] with the exception of the identity relation. In this approach there are 6 relations in temporal dimension and 12 relations in spatial dimension. Therefore we have 72 spatio-temporal relationships with this type of customization or we have MIA_{72}. This method of customization is applied in this paper.

2.4. Voronio continuous range query

This paper utilizes from VCRQ as *S–D continuous range search query* which is defined as: "Retrieving all objects of interest on any point during the moving of the query point from the start point (S) to the destination (D) in the networks" [25]. In the continuous environment, when the query point is moving, it will cause a series of changes on the pattern of expected searching range in respect to the moving distance of the query point during the movement. Some objects could be moved out, others could be moved in. Therefor the time of updating should be specified [25]. The voronio continuous range query is carried out based on the proposed algorithm of Voronoi-based Range Search (VRS) which is defined in [25].

2.5. The proposed spatio-temporal relevancy model

This section presents our proposed approach for modeling spatio-temporal relevancy which is based on VCRQ and MIA.

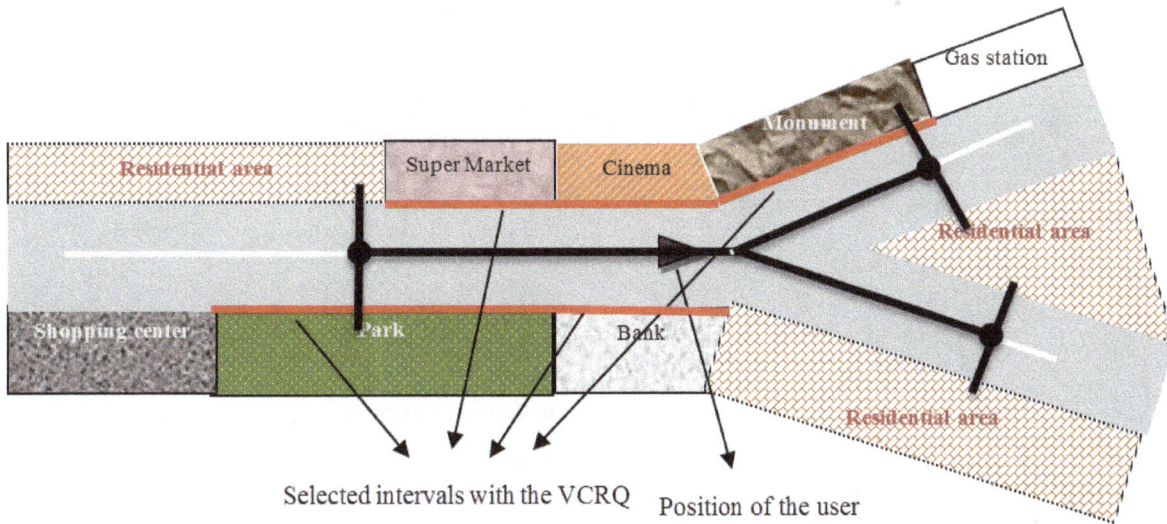

Fig. 2. The VCRQ for finding entities those are nearest to the user.

2.5.1. The proposed algorithm

The main steps of the proposed spatio-temporal relevancy model are summarised as follows:

1) The first step of the proposed algorithm is performing a voronio continuous range query [25], where the query point is the user's position. The results of this step are the related contexts which are near to the user based on the introduced searching range as illustrated in Algorithm 1. The searching range of VCRQ is changed based on the velocity of the user. It is computed as $e_{VCRQ} = V \subseteq t$, where V is the velocity of the user (m/s) at the moment of updating and t (s) is the duration time of updating which is considered to be 6 seconds in this research (6 seconds are the minimum time required by the user to make each decision during the navigation task) (Fig. 2).

Algorithm 1 modified VCR

Input: start point: S, destination point: D, (pre-defined path: SD), searching range: $e = V \subseteq 6$, (x_u, y_u): the position of user
Output: a set of SIs
1: VRS (S, e)
2: $Total_{dis} \leftarrow dis_{net}(S, D)$
3: **repeat every 6 s**
4:     VRS $([x_u, y_u], e)$
5:     Update $Total_{dis} \leftarrow disnet([x_u, y_u], D)$
6: **until** $Total_{dis} = 0$

In Algorithm 1, VRS (Voronoi Range Search) [25] extracts some related contexts which are determined based on dynamic searching range (e) in every updating. Also is the network distance between two points [25].

2) The next step is the specification of the characteristics of the temporal interval and directed spatial interval of the user including its extension and direction which should be updated when the user moves. The positive and negative directions of the directed intervals are specified as shown in Fig. 3, in which the direction of the interval is determined based on its bearing of the interval (the computing approach of the bearing is presented in Appendix). If the positive and negative bearing of the intervals is between 0° and 180° (0° =< bearing < 180°), then the direction of the interval is

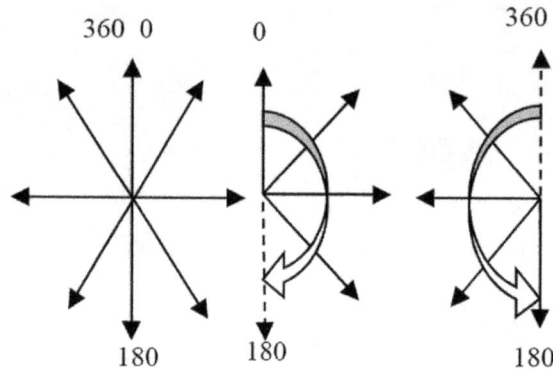

Fig. 3. Orientation relationships of the spatial directions.

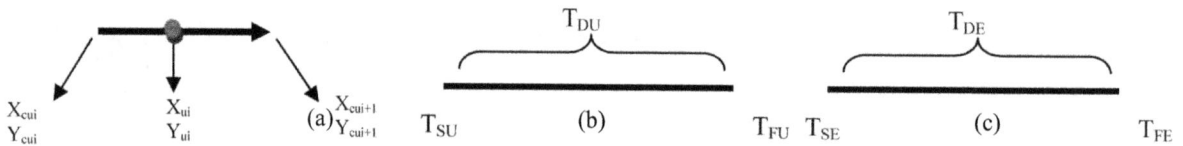

Fig. 4. A schematic view of (a) DSI (b) TIU (c) TIE.

positive ($\text{dir}_I > 0°$) and if the bearing of the intervals in between 180° and 360° (180° = < bearing < 360°), the direction of the interval is negative ($\text{dir}_I < 0°$). The extent of the spatial interval of the user is calculated using Eq. (1):

$$I_{ui} = x_{cui} = x_u - (V \times 6)\sin B_{i.i+1}$$
$$y_{cui} = y_u - (V \times 6)\cos B_{i.i+1}$$
$$x_{cui+1} = x_u + (V \times 6)\sin B_{i.i+1}$$
$$y_{cui+1} = y_u + (V \times 6)\cos B_{i.i+1}$$

(1)

Where I_{ui} is the moving interval of the mobile user, x_u and y_u are the coordinates of the user's position, $A_{i,i+1}$ is the bearing (B) of the direction$_{i,i+1}$, (x_{cui}, y_{cui}) and (x_{cui+1}, y_{cui+1}) are the coordinates of the start and end points of the directed interval respectively. As the velocity of the moving user in an urban traffic network varies, we consider V as the velocity of the user at the moment of an update and assume 6 seconds as the minimum time required by the user to make each decision during the navigation task. $V \times t$, which is equal to distance travelled during decision making process, is used the coefficient of the bearing. Figure 4(a) shows a schematic view of the directed intervals.

The extent of temporal interval of the user is specified as shown in Fig. 4(b) which is updated every 6 seconds. Where TSU is the start time of TIU which is equal to the current time of the moving user, TFU is finish time of TIU which is equal to the finish time of navigation which is introduced by the user at first and TDU is distance time between the TSU and TFU.

The appropriate spatio-temporal relationships between the DSI, SIs, TIU and TICs (which are determined by step1) are specified based on MIA$_{72}$. The important aspect of the achieved instructions is that they will be sorted based on the distance to the user which is ranked after the execution of the VCRQ on step 1. It is fundamental that the user meets the nearer contexts before meeting the objects that are farther away.

The procedure is repeated after every message update that is provided by the movement of the user. In this research we periodically update the position of the moving user every 6 seconds [16]. However the update of the position is done only when the user is on the move; when the user stops no update is performed. The direction of spatial interval of the user is updated based on the position update of the user.

Whenever the position of the user is updated, the spatial and temporal intervals of the user are constructed and the spatio-temporal relationships between the user and the nearest contexts (which are determined by the VCRQ) are evaluated. Based on the spatio-temporal relationships between the user and the detected contexts, the appropriate instruction is sent to the user. For example, if the spatial relationship is 'met from behind' and the temporal relation is 'contains' then the instruction is 'now you will arrive at place, you can visit it till 2:15 hours later", where "2:15 hours later" is the time duration that the related context will be closed to visit.

Temporal intervals of the related contexts are updated based on their collected information including opening times and closing times in every days. If we consider temporal interval of the contexts as Fig. 4(c), then T_{SC} is the opening time of the related contexts from 0:0 to 24, T_{FC} is the first closing time of the related objects after T_{SC} from 0:0 to 24 and T_{DC} is the distance time between the T_{SC} and T_{FC}. It should be noted that this duration should be continuous. To achieve this goal, after any closing with opening of the related contexts, the temporal interval is updated. The pseudo-code for the proposed spatio-temporal relevancy is given in Algorithm 2.

Algorithm 2 Finding spatio-temporal relevant contexts

Inputs: start point: S, destination point: D, finish time of navigation (t_{end}) and the user's preferences.
Outputs: Guiding instructions based on the spatio-temporal relevant entities on the route of the user.
1: Determination of the related entities in all of the procedures based on the user's preferences (SIs).
2: For each 6 seconds in current time
3: While ($V_t > 0$)
4: Selection of the near SIs through a VCRQ with the centre at [x_u, y_u] and updating the DSI and TI of the user (according to section 2.5.1).
5: Comparison the spatio-temporal relations between DSI, SIs, TIU and TIE based on 72 relationships in Customized MIA theory.
6: Specification of the spatio-temporal relations and sending appropriate instructions to the user.
7; End while
8: Next

2.5.2. Architectural design of the system

The proposed spatio-temporal relevancy model is independent of the software and the programming language. In this study the user is a tourist with a specified scenario and the software is designed accordingly.

2.5.2.1. System scenario

In this research, the user is a tourist who is supposed to be guided from a hotel, which is his/her origin; if the location of the origin is not known, the user should introduce his/her current location to the system, also he/she should specify the finish time of the navigation process according to his/her schedule. After inputting the information about the origin, the user selects his/her point of interest (destination) based on his/her preferences (the characteristics of the places are introduced textually).

The system determines the optimum route between the origin and the destination. While moving, the tourist can be provided with location and time information of the other points of interest that are along

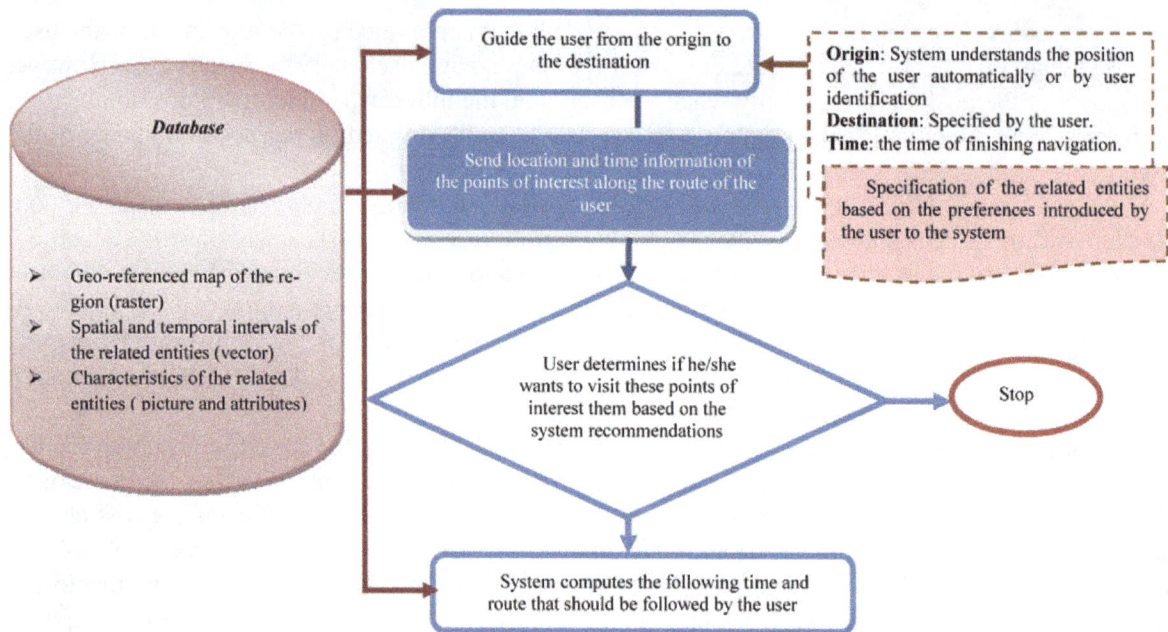

Fig. 5. A schematic view of the tourist scenario.

the route of the user (including the characteristics of the places shown on the user screen). Keeping track of the locations of services along the route, the tourist can obtain get an overview of the place where the points of interest are located. Moreover the user-adaptive system can direct the tourist when he/she is near such a spot. Figure 5 depicts a schematic view of the tourist scenario.

2.5.2.2. Hardware architecture of the system

The hardware architecture of the system consists of three main units for the correct display of context-aware services (Fig. 6).

3. Implementation and case study

The proposed spatio-temporal relevancy model for an urban context-aware system is implemented in the windows application environment with the Vb.Net. programming language, using a four-stage configuration wizard (Fig. 7(a),(b), (c) and (d)), and is delivered as the STRMCA software in the form of an exe. (or set-up). The set-up file has a feature for downloading the data of the region of interest. The applied spatial data are in vector format; however, we also considered a raster map as a background. The required data can be classified into static and dynamic data which are described as follows:

3.1. Spatial data

Static data are the information whose spatial characteristics are fixed during the navigation process. We have a raster map of the region as the background and a generalised georeferenced vector map with the following layers (all data used are at the scale of 1:2000):

– *Graph of road network:* It consists of the centreline of urban roads directed based on the urban rules (such as one-side and two-side).

Modelling spatio-temporal relevancy in urban context-aware pervasive systems using voronoi continuous...

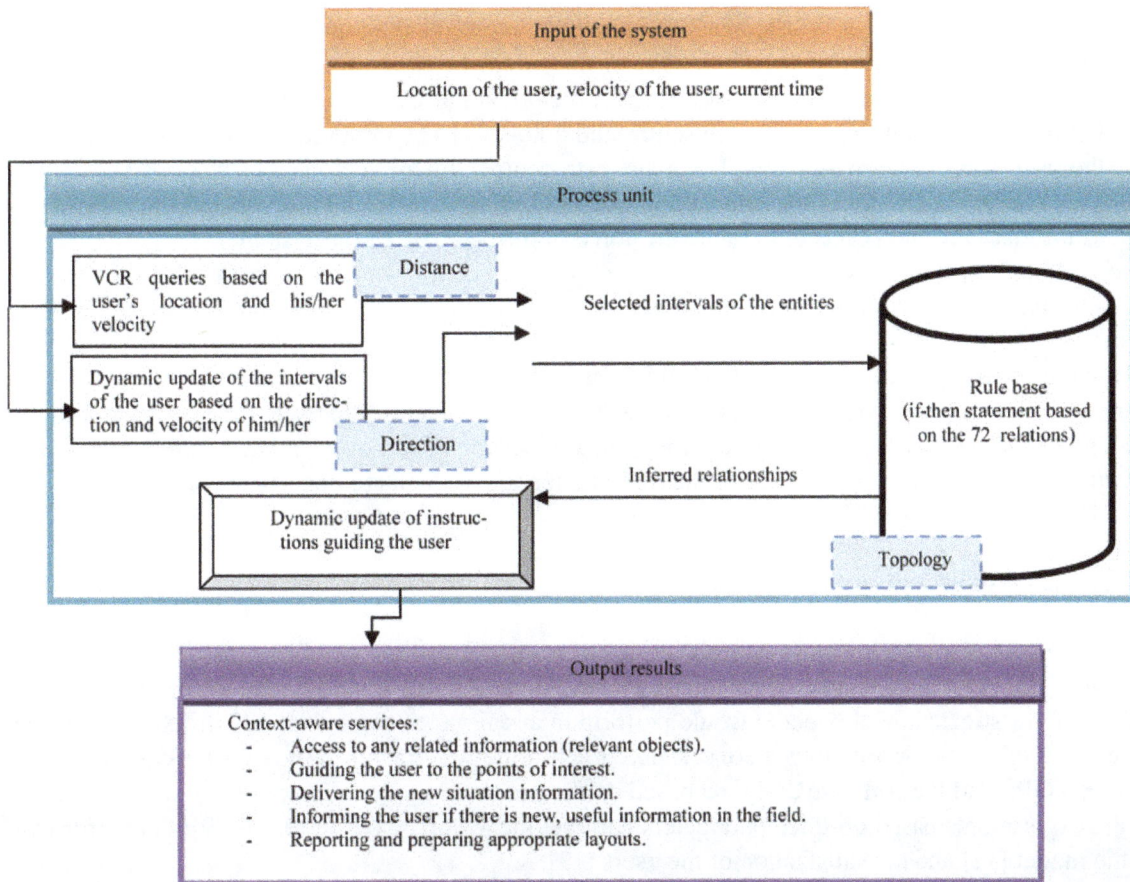

Fig. 6. Overview of the system architecture.

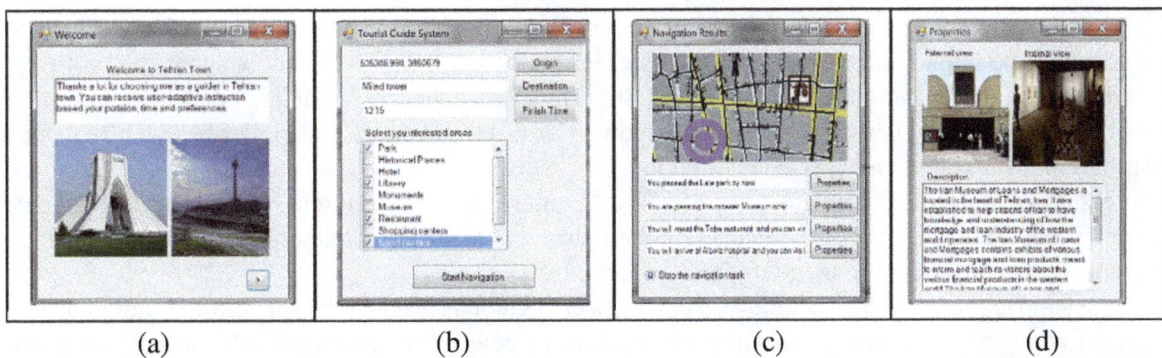

(a)	(b)	(c)	(d)

Fig. 7. The configuration wizard of the implemented system: a) welcome page, b) introduction of the origin and the destination to the user, specification of the preferences and the start of the navigation task, c) representation of context-aware instructions to the user and highlights of the spatially relevant contexts, d) illustration of pictures and characteristics of the selected area.

– *Spatial intervals of the tourist points of interest and urban facilities*: Because the spatial relationships are calculated with respect to the spatial intervals of the related contexts, they should be stored in the system to have a context-aware instruction. The SIs are represented by straight lines that cover the external boundaries of artefacts. The related contexts of the tourist are categorised as: (1) the tourist's points of interest, which consist of restaurants, coffee shops, parks, green spaces, shopping centers, downtown centers, monuments, historical places, cultural heritage sites, universities, libraries, exhibitions, sport sites, museums and hotels. (2) The urban facilities that may be used by the user, such as petrol stations, hospitals, care centres, metro stations, bus stations and airports. The external boundary of the related contexts is digitised as a spatial interval, and the coordinates of the start and end points of the line are stored in the database of the system.

3.2. Dynamic data (real-time data)

Dynamic data are information that should be updated periodically with the user's movement. The computation process is carried out based on variations in the data. In this study the position, time and velocity of the user are the dynamic (real-time) data. Since almost all of the related contexts in an urban network have a valid time to visit. So the temporal intervals of them are updated accordingly.

4. Experimental results and evaluation

The spatio-temporal relevancy model is implemented in urban network of two districts of Tehran, the capital of Iran. Districts 3 and 6 cover some attractive areas for a tourist and are considered for our case study. To investigate how the model would perform in a real-world application, 30 different routes with different origins and destinations were selected. Each route was traversed by a visitor equipped with a laptop, a GPS and the software designed based on the proposed model. This paper evaluates the results of the experiments based on three parameters namely; the accuracy of the results [9], the performance of the model [12] and the satisfaction of the users [13].

4.1. Accuracy of the results

The metric employed for accuracy assessment of the proposed model is based on a comparison of the count of the number of contexts detected by the model and the related contexts that should be selected in the environment (control contexts) [11,36]. The Chi-squared ($\chi2$) statistic is selected for testing the proposed approach. Chi-squared test of goodness of fit establishes whether an observed frequency distribution differs from an expected distribution.

To test this parameter, 30 tourists traversed 30 different routes with different origins and destinations. In each route the related contexts selected by the tourist via the user's preference options are specified as control points regarding to their temporal constraints.

The system is run while the user moves, and the user is guided based on the spatio-temporally relevant contexts with ordered instructions. Then, the number of detected contexts in each route is compared with the control contexts. Figure 8 graphically depicts the difference between the two diagrams of the detected contexts and the control contexts.

Chi-squared goodness- of- fit test was used to compare the expected spatio-temporally relevant contexts; the results demonstrate the efficiency of the model based on the accuracy of the detected relevant contexts at 95% and 99% confidence levels. Table 1 shows the values of Chi-squared tests at 95% and 99% confidence levels. A comparison between the value of Chi-squared statistics shown in columns 3 and 4 of the Table 1 specifies the accuracy of the Chi-squared proposed algorithm.

Table 1
Results of Chi-squared goodness of fit

Number of iterations	DF	$\chi 2$	95% Confidence level	99% Confidence level
30	29	2.423	17.708	14.256

Table 2
Time in (s) for selecting related entities based on the distance parameter with the VCRQ

The number of related entities	1–5	5–10	10–15
Time (second)	0.03–0.30	0.3–0.48	0.48–0.59

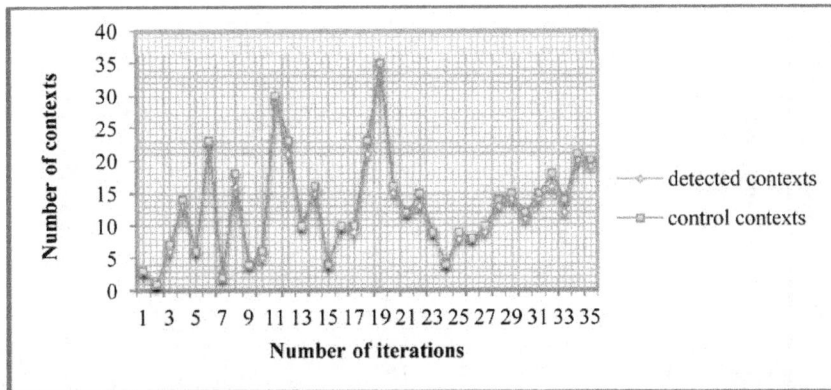

Fig. 8. Comparison between the control entities and detected the entities.

The results of the comparison between the values shown in Table 1 indicate that in 30 iterations of the algorithm in 30 different routes, the p-value of the Chi-squared test is significant. In other words, the value of $\chi 2$ which is equal to 2.423, is smaller than of errors at a 95% confidence level (17.708) and 99% confidence level (14.256) with the Degree of Freedom (DF) equal to 29. Thus, the statistics demonstrate that the proposed approach can accurately model spatio-temporal relevancy in a context-aware system.

It should be noted that the introduced model is utilizing from VCRQ despite of Dynamic Range Neighbour Query (DRNQ) which was applied in our previous work [33]. The achieved results based on the comparison between the current approach and previous research demonstrated that using VCRQ increases the efficiency of the system by decreasing the number of false or undetected contexts (the details of this comparison are not mentioned in this paper).

4.2. Performance of the model

In this section, the results of tests that have been performed to show the run-time efficiency of the algorithm are presented. Three performance tests were conducted, for which a Windows 7 Ultimate system (Intel® Atom (TM) CPU N270 and 2GB RAM) was used. The first set of results shows how much time is needed for updating a VCRQ based on the user's position (Table 2). The second evaluation reveals the measured time that is required for providing context-aware instructions, which consists of updating the directed spatial interval and temporal intervals based on the time, position, direction and velocity of the moving user (Table 3). The final performance evaluation parameter is the required time for implementing the proposed model in any updating procedure including (1) running a VCRQ, (2) updating the DSI and TI of the user, specifying of the type of spatio-temporal relationships between the

Table 3

Time in (s) for updating DSI and TI of the user and introducing spatio-temporal relevant entities to the user in any update

The number of spatio-temporal relevant entities	1–5	5–10	10–15
Time (second)	0.03–0.18	0.18–0.27	0.27–0.34

Table 4

Time in (s) for introducing spatio-temporal relevant entities to the moving user in every update

Total time (second)	1–5 instruction/s	5–10 instructions	10–15 instructions
Time	0.06–0.58	0.58–1.03	1.03–1.25

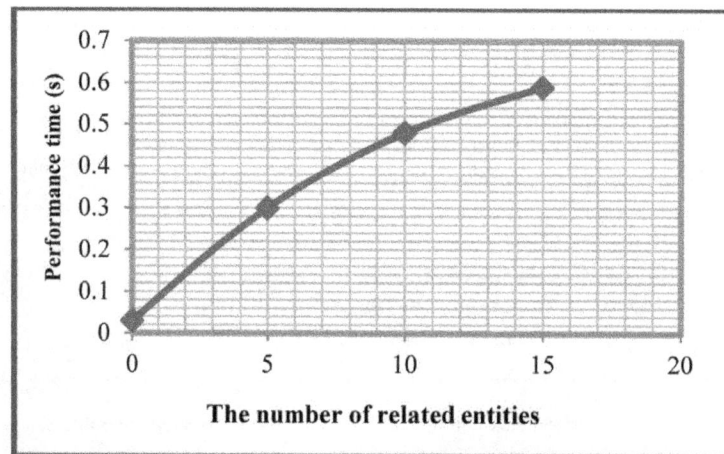

Fig. 9. Time taken for the VCRQ.

DSI, SIs, TIU and TIC and giving appropriate instructions to the user (Table 4). It should be noted that the updating of TI of the related contexts is performed by the system software whenever it is needed.

The archived results reveal that the processing time depends on the number of related contexts (SIs) around the user; however, similar times have been measured for a smaller number of SIs with a correspondingly large number of related contexts.

Up to 15 SIs in the study area around the user area have been considered (based on the constraints in an urban network, 15 is the maximum number of SIs that are detected in the updating algorithm). The measured time for building a VCRQ, updating the DSI and TI of the user and providing context-aware instructions and the total time for the whole procedure are shown in Figs 9, 10 and 11. The aim of the performance evaluation was to investigate how much time is needed for a particular test system for performing the respective tasks with variable numbers of SIs. The results achieved demonstrate that the total computation time of STRMCA performance in the study area is less than 1 s and users have 5 s for decision making.

4.3. Satisfaction of the users

The final parameter for the assessment of the system model is the satisfaction of the user with the system assistance procedure. The user's satisfaction with the services provided is a key issue in modeling context-aware systems [11,13]. It is obviously crucial for adoption and acceptance of such technologies [12,39]. This study considered seven elements for estimating user satisfaction from the model and

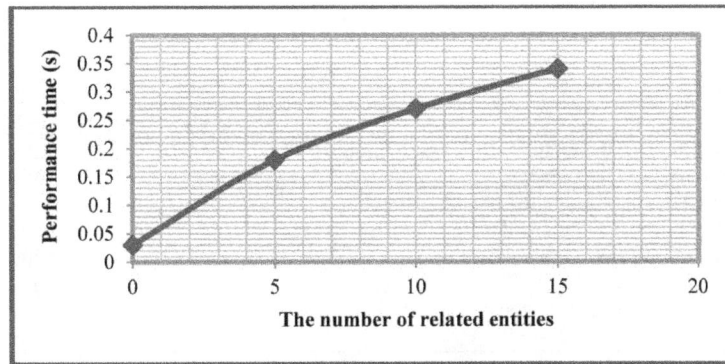

Fig. 10. Time taken for updating, the DSI and TIU and providing user-adaptive instructions according to the number of SIs.

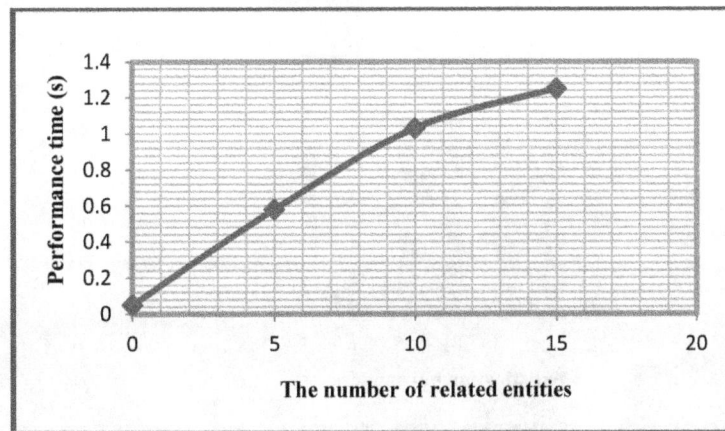

Fig. 11. Time performance of the proposed algorithm.

context-aware services, namely: (1) the usefulness of the prediction in the user's decision making process for choosing his/her related contexts [2,13], (2) the usefulness of the presentation of contexts that have spatial relationships behind the user and in the opposite direction, (3) the effect of the contexts introduction based on the velocity (this factor is derived from the "adapt to varying situations of the user" parameter pointed by [12], (4) the order of the contexts presentation and (5) the responding time of the algorithm to the user [3,12], (6) number of navigation steps [13], and (7) task success [13].

To achieve this goal, a questionnaire was designed and given to each visitor. All 30 tourists completed the questionnaire at the end of the process and gave a yes or no score for each parameter. For every parameter, the number of visitors who agreed or disagreed with the satisfaction factor was obtained. Table 5 shows the percentages of the visitors who agreed and those who disagreed for each parameter.

The statistical analysis of the results obtained from questionnaires completed by the visitors demonstrated that on average more than 90% of the visitors agreed with the method implemented in STRMCAS for presenting approach of navigation instructions (Fig. 12). The results shown in Table 6 indicate a high level of user satisfaction with the proposed model.

Table 5

The percentage of the satisfaction of the users with each parameter

No.	Satisfaction parameter	Score	
		Yes	No
1	Usefulness of the related entities prediction	92%	8%
2	Usefulness of the presentation of entities that have spatial relationships behind the user and in the opposite direction	84%	16%
3	Effect of the entities introduction based on the velocity	88%	12%
4	The order of the entities presentation	92%	8%
5	Response time	85%	15%
6	Number of navigation steps	88%	12%
7	Task success	95%	5%

Table 6

Classification of the level of user satisfaction level of the users

Percentage of satisfaction	0–25	25–50	50–75	75–100
Level of satisfaction	Low	Moderate	Good	High

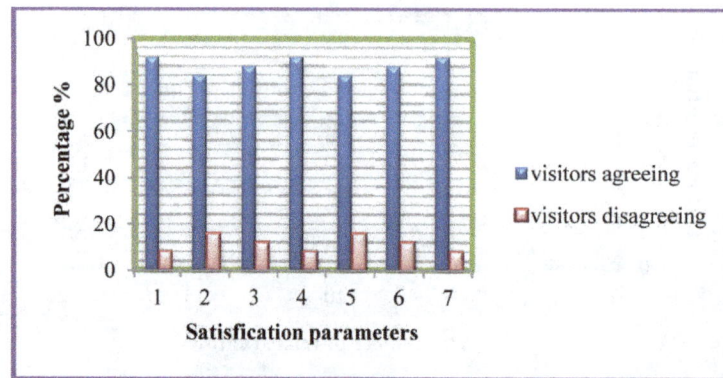

Fig. 12. The comparison of the rate of satisfaction.

5. Discussion

This study proposes and verifies a new approach to model the spatio-temporal relevancy parameter in context-aware systems constrained by directed urban traffic network. The proposed model has some specific characteristics that make this model distinctly different from current models. The first key feature of this model is the consideration of the influence of the moving user with a directed spatial interval despite of considering it as a point or a region. A directed spatial interval can model the movement characteristics of the user in an urban network effectively because it includes the direction of the user, which is needed for deciding about continuing/returning/stopping the route. Such a model can also characterise the velocity of the user's movement by decreasing and increasing the size of the user's directed spatial interval; thus, whenever the velocity increases, the DSI is extended, and when the velocity decreases, the DSI is shortened. Therefore, in the former case, the more SIs are considered to have spatial relationships with the DSI and the user have sufficient time to make a decision about whether to visit the related contexts. In the latter case, because of the low velocity of the user, the fewer contexts are found, and the user can decide to visit a place for a longer time (Fig. 13).

To evaluate the model in real-world applications, three metrics were considered (Sections 4.1, 4.2 and

Table 7
The comparison between a few related projects

Related work	SAPM (2008)	Relate Project (2009)	ZOI (2010)	Proposed model
Parameters				
Distance relations	√	√	√	√
Topological relations	9-Interaction Model	left-of, right-of, approaching, moving away	RCC5	RCC8 and ordered through MIA
Directional relations	√	√	√	√
Velocity of the user	×	×	×	√
Consideration of the influence domain	×	√	√	√
Arrangement of the contexts in spatial dimensions	×	√	√	√
Temporal relations	Time is considered as an attribute	Time is considered as an attribute	Interval Algebra	Interval Algebra

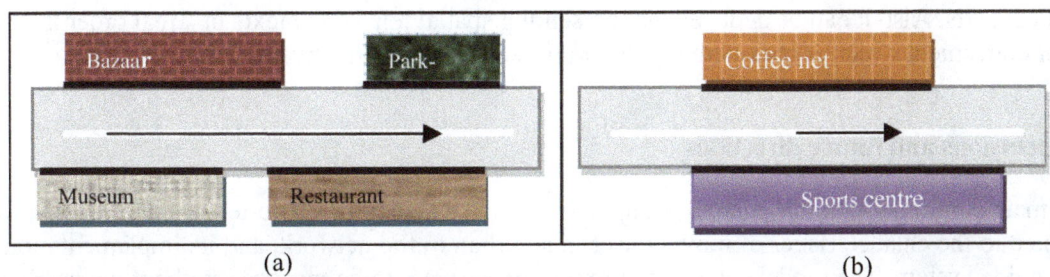

(a)　　　　　　　　　　　　　　(b)

Fig. 13. Dependency of the size of the DSI to the velocity of the user: a) at a higher speed, b) at a lower speed.

4.3). The results of the evaluation demonstrated that the algorithm is able to accurately detect the spatio-temporally relevant contexts of a moving user in an appropriate run-time and with a high level of user satisfaction with the system's performance. Also the comparison results between the current approach and previous research [33] demonstrated that using VCRQ increases the efficiency of the system by decreasing the number of false or undetected contexts. Furthermore the satisfaction of the users by using VCRQ in the first step of the proposed algorithm leas to the 2% increase rather than when we used DRNQ [33].

Another parameter which is important in context-aware system is usability of the model [13] STRM-CAS have 4 main components for modeling spatio-temporal relevancy (DSI, SI, TIU and TIC) and it is possible for every navigation applications to adapt with defining them. Indeed only the related contexts are changed based on the user preferences and goals.

While the advantages of the proposed spatio-temporal relevancy model are clearly evident, there are also some limitations to this research which should be addressed in future studies. Although the model handles spatio-temporal relevancy in urban traffic networks, it may not be suitable for a pedestrian user and may be not applicable in indoor spaces that do not have directed networks, such as a museum.

5.1. Comparison with related works

This section gives an overview of the related works that address the use of qualitative and quantitative spatio-temporal relationships to select relevant objects. The comparison is performed based on the characteristics of the STRMCAS with regard to other closely related studies and projects. One of these studies is the research of Holzmann and Ferscha [12], who defined ZOI for contexts and specified the direction, distance spatial relationships between related sensors. Gellersen et al. [19] expressed the quantitative relationships by the distance between devices and the orientation angle, and the qualitative relationships by the spatial arrangement of one device with respect to another. The Spatial Audit Policy Model (SAPM) introduced the concept of the spatial audit rule and supported the homogeneous representation of all spatial aspects involving objects and adapted information such as the user's position, with a 9-intersection topological approach [43]. The comparison of these projects with respect to the different aspects concerning spatial relationships and adaptation parameters is given in Table 7.

As seen from Table 7, using spatio-temporal relationships is a common technique for spatio-temporal relevancy modelling but in no case are all of the spatial relations such as metric, directional and topological (with all mutually ordered relationships) considered. In adapting the model to the user's movement characteristics there is no method that incorporates the velocity of the user in providing spatially relevant objects to the user. The consideration of the influence domain is an essential assumption in the recent studies, but the innovation of this paper is the definition of linear spatial intervals for a user and his/her related contexts. Also it can be deduced that presenting spatial related contexts in a real order is needed for user convenient which is considered in the proposed approach effectively.

6. Conclusions and future directions

The main contribution of this paper is the specification of a model for spatio-temporal relevancy, which is adapted to the characteristics of moving user in an urban traffic network, and its implementation in a tourist guide system. The model enables context-aware services to be managed without the user's prior knowledge of the area. Adaptation of the application to the user is based on the Voronoi Continues Range Query and Multi Interval Algebra. With customizing the spatio-temporal relationships of MIA, 72 spatial relationships between intervals of the user and related contexts are specified to detect the spatio-temporal relevant contexts.

In this research the tourist guide is equipped with a PDA or Laptop system and a tool for positioning system like GPS. The tourist could execute this program in his/her device and receive the expected context-aware service conveniently. The experimental results show that the proposed approach could detect spatio-temporal relevant contexts at the right position at the right time with a high level of satisfaction. The results of accuracy with Chi-square fitness-of-use and time performance demonstrated the implemented model. Also the evaluation of filled questionnaire form of user indicated that the proposed approach could satisfy the user in providing and introducing context-aware services.

Appendix

In this section we explain an approach of bearing computation for a directed line. Consider a line (L_{AB}) with the origin of A and destination of B. Indeed the direction of L_{AB} is from A to B. Bearing of L_{AB}(B_{AB}) is a clockwise angle from the magnetic north to the directed line (Fig. a) which is computed as Eq. (a), Eq. (b), Eqs. (c) and (d) [44]:

If $\Delta x > 0$ and $\Delta y > 0$
then $B_{AB} = \arctan(\Delta x/\Delta y)$ \hfill (a)

If $\Delta x > 0$ and $\Delta y < 0$
then $B_{AB} = 180° - |\arctan(\Delta x / \Delta y)|$ \hfill (b)

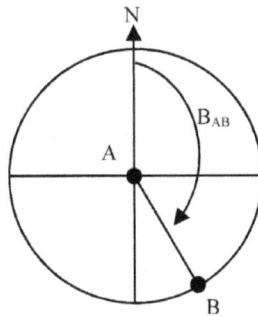

Fig. a. Bearing of the direction AB.

If $\Delta x < 0$ and $\Delta y > 0$
then $B_{AB} = 180° + |\arctan(\Delta x / \Delta y)|$ $\qquad$ (c)

If $\Delta x < 0$ and $\Delta y < 0$
then $B_{AB} = 360° - |\arctan(\Delta x / \Delta y)|$ $\qquad$ (d)

where $\Delta x = x_B - x_A$ and $\Delta y = y_A - y_B$

References

[1] A. Coronato and G. Pietro, Formal specification and verification of ubiquitous and pervasive system, *ACM Transactions on Autonomous and Adaptive Systems* **6**(1) (2011), 9.

[2] A.D. Kshemkalyani, Immediate detection of predicates in pervasive environments, *J Parallel and Distrib Comput* **72** (2012), 219-230.

[3] A.G. Cohn, B. Bennett, J. Gooday and N.M. Gotts, Representing and reasoning with qualitative spatial relations about regions, in: Spatial and Temporal Reasoning, Kluwer Academic Publishers, 1997, pp. 97-134.

[4] A. Jimenez-Molina and I.Y. Ko, Spontaneous task composition in urban computing environments based on social, spatial and temporal aspects, *J Eng Applications of Artificial Intelligence* **24** (2011), 1446–1460.

[5] A.K. Dey, Understanding and using context, *Personal and Ubiquitous Computing* **5** (2001), 4–7.

[6] A. Schmidt, Ubiquitous computing – computing in context, PhD Thesis, Lancaster University, 2002.

[7] A.U. Frank, Tiers of ontology and consistency constraints in geographical information systems, *International Journal of Geographical Information Science* **15**(7) (2001), 667–678.

[8] B.N. Schilit, N. Adams and R. Want, Context-aware computing applications. In: proceedings workshop on mobile computing systems and applications, Santa Cruz, California, USA Dec. 8–9, 1994, pp. 85–90.

[9] B.Y. Lim and A.K. Dey, Assessing demand for intelligibility in context-aware applications, UbiComp 2009, Sep 30–Oct 3, Orlando, Florida, USA, 2009, pp. 195-204.

[10] C. Becker and D. Nicklas, Where do spatial context-models end and where do ontologies start? A proposal of a combined approach, In: Proceedings of First International Workshop on Advanced Context Modelling, Reasoning and Management in conjunction with UbiComp 2004, 2004, pp. 48–53.

[11] C. Gena, Methods and techniques for the evaluation of user-adaptive systems, *The Knowledge Engineering Review* **20**(1) (2005), 1–37.

[12] C. Holzmann and A. Ferscha, A framework for utilizing qualitative spatial relations between networked embedded systems, *Journal of Pervasive and Mobile Computing* **6** (2010), 362–381.

[13] C. Mulwa, S. Lawless, M. Sharpm and V. Wade, The evaluation of adaptive and personalised information retrieval systems: A review', *Int J Knowledge and Web Intelligence* **2**(2/3) (2011), 138–156.

[14] C. Stiller, F. Ro and Ch. Ament, Integration of Spatial User-Item Relations into Recommender Systems, International Journal for Infonomics (IJI), *Infonomics Society* **3**(1) (2011), 190–196.

[15] D. Taniar, M. Safar, Q.T. Tran, W. Rahayu and J.H. Park, Spatial network RNN queries in GIS, *The Computer Journal* **54**(4) (2011), 617–627.

[16] G. Iwerk, Maintenance of spatial queries on continuously moving points. PhD Thesis, University of Maryland at College Park College Park, MD, USA, 2004.

[17]   G. Tychogiorgos and Ch. Bisdikian, Selecting relevant sensor providers for meeting "your" quality information needs,. Proc. IEEE Conference on Mobile Data Management (MDM), 2011, Lulea, Sweden.

[18]   G. Zhao, K. Xuan, D. Taniar and B. Srinivasan, Incremental k-nearest-neighbor search on road networks, *Journal of Interconnection Networks* **9**(4) (2008), 455–470.

[19]   H. Gellersen, C. Fischer, D. Guinard, R. Gostner, G. Kortuem, C. Kray, E. Rukzio and S. Streng, Supporting device discovery and spontaneous interaction with spatial references, *Journal of Personal and Ubiquitous Computing* **13**(4) (2009), 255-264.

[20]   I. Afyouni, C. Ray and Ch. Claramunt, Spatial models for indoor and context-aware navigation systems: A survey, *J Spatial Info Sci* **4**(1) (2011), 85–123.

[21]   J.F. Allen, Maintaining Knowledge about temporal intervals. Comm, *ACM* **26**(11) (1983), 832–843.

[22]   J. Hong, E.-H. Suh, J. Kiim and S. Kim, Context-aware system for proactive personalized service based on context history, *Journal of Expert Syst. with Applications* **36** (2009), 7448–7457.

[23]   J. Renz, A spatial odyssey of the interval algebra: Directed intervals, in: Proceeding of the 17th Znt 'I Joint Conference on Artificial Intelligence, B. Nebel, ed., Morgan Kaufmann Publishers Inc. San Francisco, CA, USA, August, 2001, pp. 51–56.

[24]   J. Renz and F. Schmid, Customizing qualitative spatial and temporal calculi, M.A. Orgun and J. Thornton, eds, AI 2007, LNAI 4830, pp. 293-304, Springer-Verlag Berlin Heidelberg.

[25]   K. Xuan, K. Xuan, G. Zhao, D. Taniar, W. Rahayu, M. Safar and B. Srinivasana, Voronoi-based range and continuous range query processing in mobile databases, *Journal of Computer and System Sciences* **77** (2011), 637–651.

[26]   K. Xuan, G. Zhao, D. Taniar and B. Srinivasan, Continuous Range Search Query Processing in Mobile Navigation, Proceedings of the 14th International Conference on Parallel and Distributed Systems (ICPADS 2008), IEEE, 2008, pp. 361–368.

[27]   K. Xuan, G. Zhao, D. Taniar, M. Safar and B. Srinivasan, Constrained range search query processing on road networks, Concurrency and Computation, *Practice and Experience* **23**(5) (2011), 491–504.

[28]   M.C. Golumbic and R. Shamir, Complexity and algorithms for reasoning about time: A graph theoretic approach', *J of the ACM* **40**(5) (1993), 1128–1133.

[29]   M. Grossmann, M. Bauer, N. Honle, U. Kappeler, D. Nicklas and T. Schwarz, Efficiently managing context information for large-scale scenarios, in: Proceedings of Pervasive Computing and Communications, IEEE Computer Society, 2005, pp. 331-340.

[30]   M. Safar, D. Ebrahimi and D. Taniar, Voronoi-based reverse nearest neighbor query processing on spatial networks, *Multimedia Systems* **15**(5) (2009), 295–308.

[31]   M. Safar, D. El-Amin and D. Taniar, Optimized skyline queries on road networks using nearest neighbors, *Personal and Ubiquitous Computing* **15**(8) (2011), 845–856.

[32]   N. Neisany Samany, M.R. Delavar, N. Chrisman and M.R. Malek, An ontology for spatial relevant objects in a context-aware system: case Study: A tourist guide system, World Academy of Science, *Engineering and Technology* **63** (2012), 878–884.

[33]   N. Neisany Samany, M.R. Delavar, N. Chrisman and M.R. Malek, Spatial relevancy algorithm for context-aware systems (SRACS) in urban traffic networks using dynamic range neighbor query and directed interval algebra, in press for Journal of Ambient and Smart Environments.

[34]   N. Neisany Samany, M.R. Delavar, N. Chrisman and M.R. Malek, Modeling spatio-temporal relevancy in context-aware systems using multi-interval algebra, In: Proceeding of the Joint International Conference and exhibitions on Geomatics-2011 and ISPRS Conference on Data Handling and Modeling of Geospatial Information for Management of Resources, 15–16 May 2011, National Cartographic Center of Iran, Tehran.

[35]   Q.T. Tran, D. Taniar and M. Safar, Reverse k nearest neighbor and reverse farthest neighbor search on spatial networks, *Transactions on Large-Scale Data- and Knowledge-Centered Systems* **1** (2009), 353–372.

[36]   P.M. Berry, T. Donneau-Golencer, Kh. Duong, M. Gervasio, B. Peintner and N. Yorke-Smith, Evaluating user-adaptive systems: Lessons from experiences with a personalized meeting scheduling assistant, Association for the Advancement of Artificial Intelligence, 2009, (www.aaai.org).

[37]   Sh.L. Tsang and S. Clarke, Mining user models for effective adaptation of context-aware applications, *Int J Security and its Applications* **2**(1) (2009), 53–62.

[38]   Sh. Wang, D. Liu, J. liu and X. Wang, An algebra for moving objects, In Advances in Spatio-Temporal Analysis. Taylor and Francis Group, London, 2008, pp. 111-122.

[39]   T. Olsson, T. Kakkainen, E. Lagerstam and L. Venta- Olkonen, User evaluation of mobile augmented reality scenarios, Journal of Ambient Intelligence and Smart Environments (JAISE) 4 (2012), 29–47.

[40]   T. Reichenbacher, The concept of relevance in mobile maps. Location Based Services and Tele-CartographyLecture Notes in Geo-information and Cartography, Section III, 2005, pp. 231–246.

[41]   T. Saracevic, Relevance reconsidered, In: Proceeding, The Second Conference on Conceptions of Library and Information Science (CoLIS2), Copenhagen, Denmark, Oct. 14–17, 1996, pp. 201–218.

[42] W. Pan, Z. Wang and X. Gu, Context-based adaptive personalized web search for improving information retrieval ef-
      fectiveness. In: Proceeding of IEEE International Conference on Wireless Communications, Networking and Mobile
      Computing, Shanghai, China, Oct. 8–10, 2007, pp. 5427–5430.
[43] Zh. Pingping, J. Shiguang and Ch. Weihe, A location-based secure spatial audit policy model, International Conference
      on Computer Science and Software Engineering, IEEE Computer Society, *CSSE* **4** (2008), 619–622.
[44] zn.wikipedia.org/.

M. Reza Delavar received the BSc. degree in Civil Eng. Surveying from K.N.Toosi Univ. Iran in 1989, M.E. in Civil Eng.-
Photogrammetry and Remote Sensing from Univ. of Roorkee, India, and a PhD in Geomatic Eng.-GIS from Univ. of New South
Wales, Australia in 1997. Since 1998 he has been working in Dept. of Surveying Eng., Eng. Faculty, University of Tehran as
an assistant professor. His research interest includes GIS-based artificial intelligence, agent based spatial modeling, spatial data
quality, spatio-temporal GIS, land administration, spatial data infrastructure, disaster management and cadastre. He is president
of Iranian Society of Surveying and Geomatic Eng. He is also national representative of Iran in UDMS and he is a board member of center of excellence in disaster management in Iran.

Nicholas Chrisman has working in the domain of geographic information for 40 years, developing innovative techniques to
analyze and display information about the earth and the people who inhabit it. He has some long term themes in his personal
work: addressing data quality, dealing with time and change, and examining institutional and social settings of technology. His
specialties are geographic information analysis, geospatial sciences, geomatics, data quality investigations, network of scientists, collaboration, research and innovation networks. For the past 8 years, he managed the GEOIDE Network, a network of
researchers and user communities across Canada.

M. Reza Malek is currently an Assistant at the Geodesy and Geomatics Engineering faculty of K.N.Toosi University of Technology. He has more than hundred peer reviewed journal articles, book chapters, and international conference papers. Some of
Dr. Malek's research interests include Ubiquitous and Mobile GIS, Spatial analysis and Uncertainty modeling. In May 2005, he
received an award for the best researcher in the Planning and Management Organization of Iran from the minister of Research
and science of Iran.

Najmeh Neysani Samani received her B.S. degree in Civil Eng.-Surveying engineering from industrial K.N.Toosi Univ,
Tehran, Iran in 2004, and M.S. degree in geospatial information system (GIS) engineering from the Tehran University, Tehran,
Iran in 2006, being as a GIS researcher in NCC from 2005 to 2008. She is the GIS PhD student from 2007 at Tehran University,
Tehran, Iran. Her current research interests are pervasive and ubiquitous computing, spatio-temporal modeling, wayfinding,
mobile systems, fuzzy control and approximate reasoning in GIS.

3

A weight-aware recommendation algorithm for mobile multimedia systems

Pedro M.P. Rosa[a], Joel J.P.C. Rodrigues[a,*] and Filippo Basso[b]

[a]*Instituto de Telecomunicações, University of Beira Interior, Covilhã, Portugal*
[b]*Zirak s.r.l., Italy*

Abstract. In the last years, information flood is becoming a common reality, and the general user, hit by thousands of possible interesting information, has great difficulties identifying the best ones, that can guide him in his/her daily choices, like concerts, restaurants, sport gatherings, or culture events. The current growth of mobile smartphones and tablets with embedded GPS receiver, Internet access, camera, and accelerometer offer new opportunities to mobile ubiquitous multimedia applications that helps gathering the best information out of an always growing list of possibly good ones. This paper presents a mobile recommendation system for events, based on few weighted context-awareness data-fusion algorithms to combine several multimedia sources. A demonstrative deployment were utilized relevance like location data, user habits and user sharing statistics, and data-fusion algorithms like the classical CombSUM and CombMNZ, simple, and weighted. Still, the developed methodology is generic, and can be extended to other relevance, both direct (background noise volume) and indirect (local temperature extrapolated by GPS coordinates in a Web service) and other data-fusion techniques. To experiment, demonstrate, and evaluate the performance of different algorithms, the proposed system was created and deployed into a working mobile application providing real time awareness-based information of local events and news.

Keywords: Mobile computing, ubiquitous computing, location-aware, content-aware, iphone applications, multimedia applications, mobility

1. Introduction

Utilization of mobile devices in everyday life has expanded rapidly over the past few years; consumers are changing their habits by using resources offered by the Internet, that is now always available in a click. Content-adaptation and context-awareness are more and more necessary, to be able to deal with the current information flood provided to the user [14].

The use of smartphones all over the world is growing rapidly, with a large adoption rate especially among teenagers and adolescents [35]. Mobile devices can nowadays natively support several kinds of multimedia, and mobile services are supporting the users with an exponential growth of specialized applications/services for almost every need: as a general thermometer can be used the number of deployed applications in the main mobile Markets/Stores present in Internet. The expression *"Information at your fingertips anytime, anywhere"* has been driving the mobile computing development in the past two decades. However, mobile devices do not have the same features in what concerns to conventional information processing, such as PC's and laptops [21], especially in terms of computing power,

*Corresponding author: Joel J.P.C. Rodrigues, Instituto de Telecomunicações, University of Beira Interior, Rua Marquês D'Ávila e Bolama, 6201-001 Covilhã, Portugal. E-mail: joeljr@ieee.org.

human-machine interaction (HMI) resources, general limitations of network, battery cycle, and other specific topics, becoming important limitations in mobile computing [23]. The rise of smartphones like the Blackberry, Android, and iPhone allow not only voice communication but also communications via SMS (Short Message Service)/MMS (Multimedia Messaging Service)/E-Mail/Social Networks. They also have the capacity to process intensive activities such as multimedia playback, document editing, and audio/video streaming via dedicated coprocessors. In spite of the inherent limitations, some authors are supporting the idea that conventional laptops will be replaced by smartphones and tablets soon [33].

Mobile devices provide a wide range of opportunities to a global society through online social networks, blogs and web pages, among others. New approaches to access information in mobility are being improved; over the last years the research community has studied and developed new technologies, services and applications to enable ubiquitous environments based on mobile technology [38]. The new generation of mobile devices has improved the efficiency of representation of the information, enhancing user experience [5]. Ubiquitous computing is becoming a reality with mobile computing, due to a rapid advance in wireless technologies and Internet [35].

The small screen size of the mobile phone does not help to read contents. It is difficult to present news efficiently or to display only the context of mobile news and events, capturing users' interest. Despite all the effort made to contribute to improved navigation and usability in mobile news, results should be improved constantly, not to decrease in popularity among users and result in a poor experience while navigating the mobile news [21].

Applications based on geo-location can benefit of a more stable Internet connection, enabling the user to access a wide range of server-side location-based contents, such as transport timetables, open restaurants or event calendars. Location-awareness plays an important role in the jungle of the context-awareness parameters, and everyday there are new apps using the global positioning system (GPS), its services and new features [38].

Joining ubiquitous mobile devices and web services brings into the user's hands the best of both worlds: resources and processing algorithms are server side while content-adaptation and direct context-awareness parameters are on client side [15]. The application focuses on what, where and how the users want to see and interact with the content provided. In general native applications produce better results than multi-platform solutions, contributing to a satisfactory exploration of cultural events fully utilizing the resources of the device [4].

In this scenario, recommendation systems are important to help a user to make choices, identifying the best news in an ocean of potentially good ones. For every specific application should be identified several parameters, or relevance, that can influence the filtering/ordering choices, in order to provide to the users the more important information. In some content-adaptation methods where some hundreds of hits are possible to be meaningfully shown, like a map with clustered points, it is important to maximize *relevance* to have at least 100–200 results; in other methods of data-presentation, like the list of best sold apps in AppStore, it is very important to have very high *precision* in the first 10 hits. The main issue is to identify the best parameters and the best measures to be optimized in order to them in the specific recommendation system. It is assumed that does not exist the best system, but different tunings depending on results the system is focusing on.

To provide an environment for experimenting, demonstrating, and comparing results coming from different algorithms, a dedicated application to provide news and events to a generic user was created. The results are ordered taking into account parameters coming from habits and instantaneous position. The habits chosen for this solution are the frequency of visits of a pre-defined taxonomy structure and the active sharing on social networks elements of the same taxonomy.

Nonetheless within the same technological framework several other parameterizations, both direct (background noise volume) and indirect (local temperature extrapolated by GPS coordinates) can be considered. Relevance coming from these three context-awareness parameters are then combined, or fused, with some classical data-fusion techniques, to evaluate different ways of producing an ordered list of results. The following data-fusion methods are used and experimented: *i*) CombSUM, weighted CombSUM; and *ii*) CombMNZ, weighted CombMNZ. One of the important steps leading to a successful data-fusion is the re-normalization process that was studied case by case, for each one of the three relevance cases.

The main contributions of the paper include the study and the construction of a complete generic recommendation system, with both client mobile application and Internet service, and the verification of several re-normalization and data-fusion techniques applied to the specific case of news and events recommendation system [32].

The remainder of the paper is organized as follows. Section 2 elaborates on the related work about mobile recommendation systems. Section 3 gives a mathematical overview of the used re-normalizations and data-fusion techniques. Section 4 focuses on the application demonstration and validation. Section 5 gives a performance evaluation and results of the proposed application. Finally, Section 6 concludes the paper and pinpoints directions for future work.

2. Related work

The possibility of materializing the vision of ubiquitous computing that was drawn at the beginning of the 90's is approaching [29,36], combining new features of mobile devices with the growth of short-range ad-hoc networks. In the early 90's, Marc Weiser [26,27] introduced his vision of ubiquitous computing. He presented a concept of a man-technology interaction with a complete abstraction of the user. However, Weiser's vision faced several problems, mainly lack of technological support [27]. A good example of ubiquitous technology is the Internet; users are not interested in the underlying technology behind the Internet, but only in the information and all the services provided by it [35]. Users are connected everyday to several social networks, without knowledge of protocols and network architectures used on a mobile device [28].

Mobile or portable devices such as mobile phones, personal digital assistant (PDA), and tablets are smaller and lighter, can be transported anywhere, and can easily fit in the suit pocket or briefcase. These portable devices have a good number of features, such as SMS (short message service), eMail, packet switching for access to the Internet, gaming, Bluetooth and Wi-Fi connectivity, infrared, photo camera and video recording, music player, radio and GPS antennas, memo recording, and, more importantly, make and receive phone calls. Mobile devices offer the opportunity to create a better and fast growing globally connected society, with social networks, blogs, and Web pages. These new approaches improved user-access to information through mobile communication [24,47].

The research community has been studying and developing new technologies, new services and new applications over the years to enable ubiquitous environments based on the mobile technology [38]. Mobile devices improved communication efficiency, enhancing user-experience [19]. The exponential growth of people using mobile devices leads to a constant improvement of smart communications.

The ubiquitous collaboration between mobile devices and Web services brings the best of the two worlds: the server side resources and the client-side context and location [36]. The application is centered in the user preferences (where, when, and how), providing context and content-awareness. The information and context of the events on a native application offer a better visualization to users, contributing to a satisfying physical exploration of cultural and information events [9].

Ubiquitous computing is one way to improve the computers usage [39]. The main goal of ubiquitous computing turns human-computer interaction invisible, fully integrating the computer with the actions, and behavior of its users [40]. Computer systems that surround us are proactive and are linked together, or trying to establish links between themselves constantly. Ubiquitous computing requires Internet connectivity and this feature is often used and characterized for ubiquity [43].

Geo-location applications are becoming very useful due to anytime and anywhere full connectivity. Smart phones with GPS capability are becoming more widespread. This ability of smartphones can be used as a personal navigator and a communicator device. There are various mobile navigation techniques to determine the location used on mobile multimedia applications [30]. Due to their portability and vast range of applications, mobile devices are being appointed as the future Internet navigation devices. Native applications offer a good support for multimedia contents. This access is made using Internet services in native applications [30]. By adopting these techniques and concepts it was possible to develop an intelligent mobile multimedia application.

The context-sensitive computing (Context-Aware Computing) has emerged as the field of ubiquitous computing, studying the relationship between changes in the environment and information systems [2], and thus raising new technical challenges for implementation. In the computing context-sensitive computing (Context-Aware Computing) devices try to understand and to automatically capture the context where they are inserted. Offering a better interaction between the environment and the user. This interaction can happen in terms of hardware, software or communication [1].

In recent years, many platforms have been developed for pervasive and context-aware systems in order to support rich contextual features. Many of these systems are open and available [7]. An important aspect of mobile devices is their portability and mobility. With the advent of wireless communication in mobile phones, smartphones, and tablets, ubiquitous computing has evolved as well. They take advantage of their portability to have wireless connectivity almost everywhere [24]. In mobile computing there are some technological hurdles to overcome, such as the variation on the quality of a wireless network, local access limitation, and energy constraints. These problems affect the user and also the computing experience directly [4,12].

Mobile devices with GPS (Global Positioning System) capabilities have been around for some time. Location-awareness in mobile and other devices is changing the users life [38]. Devices with such capabilities and location-aware applications will lead the mobile market [3], such as pinpointing the location on a Google Map, tracking friends, a geo-located event, giving the idea on what is going on in the user area. They can tell us what is the nearest place to eat, giving a list of all shopping or stores in the user area, tell us where to go for a party or cultural events, and other businesses [13,17,25].

Data fusion algorithms deployed (CompSUM, CompMNZ and their relative weighted version) are already found in the works of Salton and Fox [16] and improved by Lee's article three years later [25] where it is shown that in specific situations of information retrieval data-fusion it is very important the renormalization phase, not to incur in wrong evaluations of the best methodology. Several articles are proposing new methods, more complex and adapted from other branches of knowledge like neural nets, algorithms of democratic voting from the Social Choice Theory, logistic regression formulas. Other fusion techniques applied to other scientific areas shows new simple ideas as outperforming all other state-of-the-art techniques, because of new boundary conditions, and with no answer to how and which conditions act on the effectiveness of a specific technique, except generic statements [22]. Analyzing several results, it become clear that the classical CompSUM and CompMNZ are enough simple and effective, compared to other specialized formulas, and the work was focused on them. Renormalization techniques investigated were based on simple linear transformations and an exponential transformation in the case of location-based parameters. Several parameters are found in the literature, and mobile information retrieval is maybe the more creative and rich in relevance parameterizations; as an example, in *"Music Recommendation Using Content and Context Information Mining"* [39] uses context conditions such as location, time, air temperature, noise, light, humidity, and motion to make music recommendations that are sensitive to the user's mood [14]; another interesting approach can be found in [8], where a Movement Pattern Mining is determined starting from a User Movement Database (UMD), determining on a statistical base (that for data-fusion can be individual and/or collective) the patterns and distribution of future positions of the user.

Inside the Apple App Store users can find location-aware applications, such as *Loopt-Friendfinder* [18], application with virtual earth display that, allows user to share their location with the community. *NearPics* [41] is a location-aware photo browser, and *Weatherbug* [45] is a location aware weather service with predefined cities. *Appetite* [3] is a restaurant picker based on user location; *AroundMe* [42] is an application that gets your location and allows you to choose the nearest bank, bar, gas station, hospital, hotel, movie theatre, restaurant, supermarket and taxi. Every day a new item is added to this list.

An application for iPhone, called *N4MD – News for Mobile Devices*, is presented in [19]. This is an application to visualize the weekly news produced at the *Urbi et Orbi* newspaper from the University of Beira Interior, Portugal. The application *N4MD* runs natively on iPhone devices and now, it is similar to hundreds of newspapers and magazines mobile applications. The system proposed in this paper is composed by CityEvents app and an online recommendation service that is unique on its way of gathering an arbitrary set of relevance parameters, merging them into a personalized and meaningful global relevance able to filter and order events and news to fit the needs of a generic recommendation system. Several algorithms were evaluated and compared, and based on revised scientific knowledge, it is proposed a new approach that is able to investigate with little efforts other more recent algorithms and parameters.

3. Weighted scoring algorithm for mobile information applications

Considering a system with N meaningful parameters that expresses some kinds of context-awareness relative to a specific user, and where, out of a huge number of atomic informations (news, events, or more in general documents), some information should be provided to the user sorted and/or filtered in order to maximize precision. A general system require a parameters re-normalization phase, where the parameters are meaningfully mapped to the $[0, 1]$ interval, and a data-fusion phase, where the N parameters are joint into a single one, from which ordering and/or filtering can be provided, dependent to the specific content-adaptation provided by the mobile device. Weights are provided to allow user requests personalization, and real-time perception of the differences that can occur if, for example, he/she considers location to be more important than habits or vice-versa.

In this work, the following three parameters were analyzed:

1. *Location-based scoring*, giving a score relative to the Euclidean distance between the user and the position of the news/event; other types of distance can include the path distance following roads or cycle-paths, or estimated time for a specific vehicle to move from the user to the news/event position.

Table 1
Location-based scoring parameters

Transport mode	D	α
Walking	2.000	2/2.000
By bicycle	12.000	2/12.000
By car	40.000	2/40.000

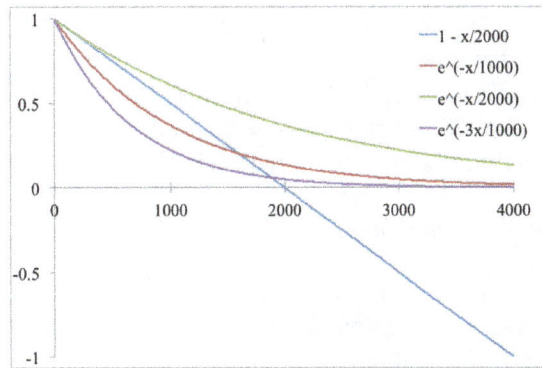

Fig. 1. Effect of different parameters in Location-based scoring.

2. *Habit-based scoring*, the frequency of visits of a pre-defined taxonomy structure; in this case a flat categorization was a sufficient approximation, but a more complex structure provided with an internal metric will benefit of the same amplified benefits.

3. *Habit-based scoring*, the active sharing on social networks of elements of the same taxonomy structure; assuming a correlation between active sharing and interest, several measures can be built to provide meaningful relevance; in this case, to simplify, it was used the frequency of article sharing, grouped by the same flat categorization of parameter 2.

3.1. Location awareness scoring

Given the Euclidean distance (d), in meters, between a device and an event, let s_d be the distance scoring; a simple scoring, useful for this approach, can be defined as follows:

$$S_{d1} \equiv \max \left(\left(1 - \frac{d}{D} \right), 0 \right) \tag{1}$$

$$S_{d2} \equiv e^{-\alpha d} \tag{2}$$

To define the constants D and α it can be noticed that they identify different exclusive scoring functions with a specific physical meaning for a user that commonly uses a bicycle (or a car, or just walks around the city), it is important to distinguish between events that are 100 m or 300 m away from the current position, while it is not so important if they are at 30 km or 100 km away; the same can be false for a user that commonly uses a car. Table 1 presents these constant values defined empirically, based on user feedback from a dozen of users.

Taking into account the scoring system used by the user that is commonly walking, Fig. 1 highlights how the factor D represents the value where s_{d1} gives zero scoring, thus considering identical all the events that are more than 2 km away from the user.

From Table 1 it can be seen that good empirical values for α can be obtained with the equation $\alpha = 2/D$, to keep scoring and its derivative not too different from the linear case in $[0, D]$ and being not too high when $S_{d1} = 0$. A factor 3 (green line, Fig. 1) should give too much importance to nearby events (being the derivative much smaller for small distance), while a factor 1 would behave like the linear function only for nearby events, but will be too high for events with distance D or bigger.

In the first deployments of the algorithms it was experimented accurately the first scoring system on Eq. (1), while after some intensive experiments was noticeable the different behavior with situations that does not use the awareness scoring system, but only the location awareness scoring. Thus, it was chosen a more valid approach following the second scoring system given by the Eq. (2). It is very simple, easy to manage, monotonically decreasing, and with not appreciable changes in computational time.

Other meaningful parameters can be used, such as time estimate instead of distance, but in the current approach only the above-described exclusive distance functions were selected.

3.2. Habit awareness scoring

While the described location awareness scoring is mainly one-dimensional (having different exclusive scoring equations for different transportation habits), the deployed context awareness scoring is multi-dimensional, giving different meaningful scorings to measure how a person can be interested in some context more than others. In this case two relevant parameters were selected. One is based on the frequency of seeing the details or sharing social networks news or events belonging to a specific taxonomy. Another parameter is based on a flat categorization, but it can be easily extended to a generic taxonomy provided with a good metric [31].

In general, assuming N different non-negative context-based relevance, $S_c = \{S_{c1}, S_{c2}, \ldots, S_{cN}\}$, it is important to keep in consideration the possible correlation between them, and remap the values of every relevance in the interval $[0, 1]$, similarly to the approach followed on the location scoring. The correlation between the relevance can be dealt with appropriate weights in an upcoming phase, when a total scoring should be defined through a data-fusion algorithm.

To re-normalize the values, the function s_{cj} as function of a discrete set of information was considered, $J = \{j_1, j_2, \ldots, j_M\}$. Thus, with limited values, remapping a simple linear mapping is shown in Eq. (3).

$$\begin{cases} s'_{cj}(0) = 0 \\ s'_{cj}(\max\{s_{cj}(I), c_{cj}\}) = 1 \end{cases} \tag{3}$$

The c_{cj} values are empirically defined as minimum values that depends on the context based relevance, and represent the minimum scorings that can be considered valuable (for example, if $s_{c1}(k)$ represents the number of times that an user shared information of the category k via a social network, c_{c1} can be set to a value around 15–20 units); in the extensive performed experiments, this lower limit helped to obtain a more precise global relevance in those cases where the specific scoring method was under-utilized.

3.3. Comparison of habit-based and location scoring

Comparing habit-based and location-based relevance, it is immediately noticeable a difference: the maximum value 1 is often present in the image of every non trivial element of s_c, if the set I is non-empty and c_{cj} properly set, while it is almost always absent in the image of s_d. Even if a simple linear remapping can be used to correct this different behavior, it was observed empirically that this difference could be positive for the global relevance. The underlying reason is that if there are no nearby informations, in general, it is more important the habit-based scoring.

In the current deployment several habit-based scoring parameters and three exclusive scoring parameters for location-based scoring (depending on main transportation method: walking, bike or car) were used. Other useful parameters that could be considered will be suggested for future work, such as the following:

– Frequency of "page-views" and/or "clicks";
– Frequency of "sharing" on a social network, grouped by social network;

- Frequency of "send-to" specific users (the user forwards the information not because is interested, but because the target user is interested in the topic);
- Frequency of "like" or other feedback systems (percentage of the details shown, if the content is presented in scrolling or tree views);
- Specific user search terms.

The parameters could be grouped as follows:

- Inside a fixed taxonomy with a metric;
- Through document similarity, using the text in title and document body, after stop-words removal and stemming;
- By information-provider, or other natural groups.

The relevant parameters should become more complex when a bigger set of information flooding is needed; for example, time or day dependent parameters can give better results (the preferences of a user should change between working hours and non-working hours, between a week-day or weekends), media-content analysis can give new ranges of context awareness scoring, and semantic Web services can be used to provide online deep analysis, giving scope to the user to define his/her own areas of interest, that will reflect similar structure of his/her "to do lists", the eMail inbox, the documents organization, as well as the informations scoring.

3.4. Combining context and location scorings

In order to produce a single global relevance, out of a set of N parameters is usually mentioned as "data-fusion". Several algorithms on data fusion topic were proposed, most of them more complex out-performing the more simple ones just in very specific conditions. It is not observed any advantage for the current proposal the use of more complex and processor intensive algorithms. Then, this work focuses in four main intuitive algorithms described in the classical work of Fox and Shaw [11]: CombSUM, Weighted CombSUM, CombMNZ, and Weighted CombMNZ.

These methods use relevance and not just their induced ordering (rankings), and they are defined as follows. For a document d, in our case an event or a news, and a set $S = \{s_1, \ldots, s_N\}$ of different relevance parameters, let n_d the number of systems that returned positive relevance for a document d, and let $W = \{w_1, \ldots, w_N\}$ a set of weights that expresses not only user preferences but also the independence of the parameters, it is commonly defined as presented in Eqs (4)–(7).

$$CombSUM_s(d) = \sum_i s_i(d) \tag{4}$$

$$CombMNZ_s(d) = n_d \sum_i s_i(d) \tag{5}$$

$$wCombSUM_s(d) = \sum_i w_i s_i(d) \tag{6}$$

$$wCombMNZ_s(d) = n_d \sum_i w_i s_i(d) \tag{7}$$

From Lee's works [14,25] it was shown that data fusion techniques could be applied to combine different relevance, but only when the relevance is appropriately re-normalized (and the process of re-normalization can influence the results of the data fusion itself). In his work, CombMNZ presented by Eq. (5) performs better than CombSUM – Eq. (4), but, in this proposal, it was noticed that both are very

Fig. 2. Home Window.

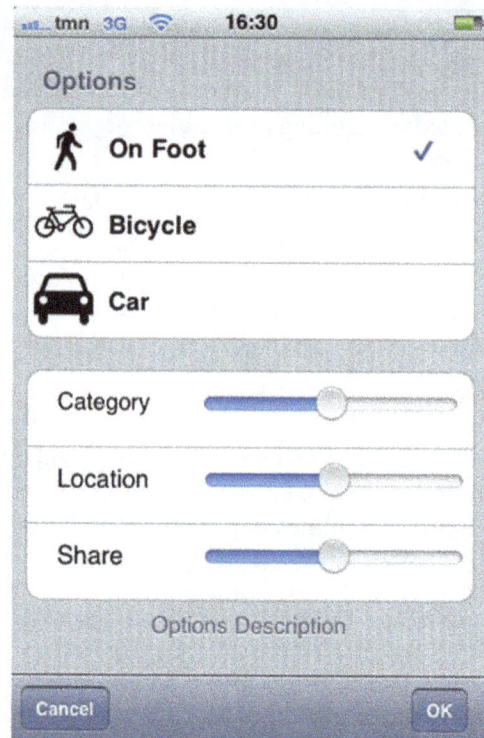

Fig. 3. User preferences screen.

similar and one does not outperform the other in all the situations, being easy to produce common cases where CombSUM, Eq. (4), or CombMNZ, Eq. (5), are better one another. CombSUM, Eq. (5), it is an algebraic mean, except a re-normalization factor $1/N$, and thus represent a stable fusion technique, quite intuitive in the results, and well behaving in our experiments.

CombMNZ, Eq. (4), it is like a class-algebraic mean where the number of non-zero parameters defines classes and their relative weights. For example, in this proposal all the documents with 1, 2, or 3 non-zero parameters were grouped.

Weighted fuses, presented in Eqs (6) and (7), can have a big impact in mobile recommendation systems because they can give personalized results, depending on parameters that user can define in real-time, and perceive the feedback of this change (while statistical measures changes have usually a more stable behavior). The user can decide that these choices should be defined more by location (for instance, in several days/hours where he/she sees that there is more traffic), while it can be defined more by interests and habits in other days/hours. Asking feedback to users was noticed that using weights in the range [0.33, 1] gives best empirical results for users that provide significant relevance in all the 3 included parameters, probably because it will keep some balance between the extremes of founded values in a neighborhood of zero and others in a neighborhood of 1.

4. CityEvents demonstration and validation

CityEvents is a simple mobile application for viewing events or news. The CityEvents system includes a user-friendly layout using the user-interaction capabilities of the iPhone. The user's interface is shown

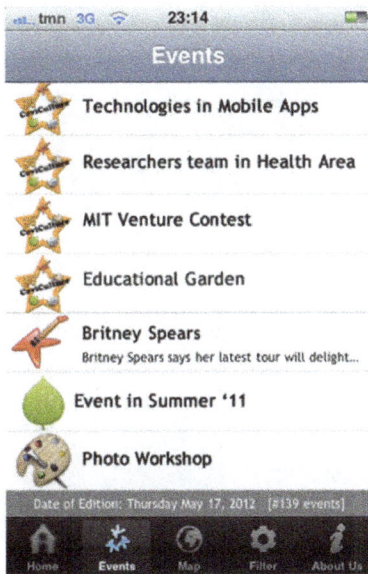

Fig. 4. List of all the available events.

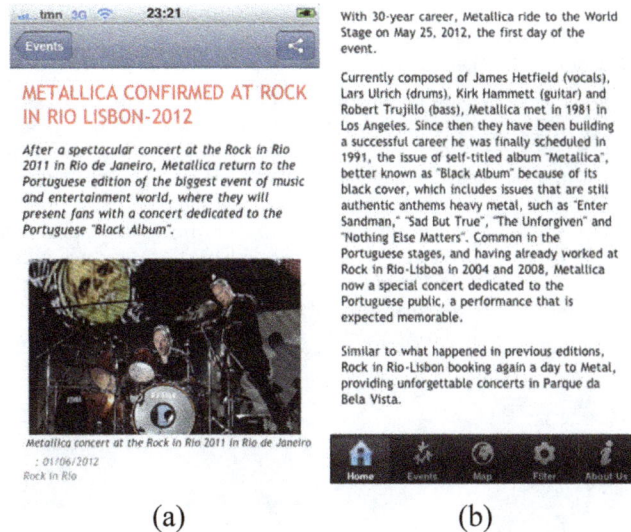

(a)   (b)

Fig. 5. Example of an event detail.

along the CityEvents system demonstration presented in this section. Figure 2 shows the *Home* window of the application. This window displays several informations (e.g. title, image, and a brief summary) about the week event, which is the most important category for the user. By clicking on the event image, the application leads the user to the detail page of the event. This feature allows users to know which event is considered the main week event, without the need to search it.

On the right top, a button for the *User Preferences* is available. Figure 3 displays all the preferences for the user, including the user's means of transportation and some criteria for being used. Those changes will affect the scoring of the events. Depending on user choices and preferences the results are shown on a map and a list of events. The map shows the location of the events. The most relevant events are displayed with colored icons while the less important have a black icon. There are icons for each category of events. This list of events is sorted from the most important to the less important ones.

At the tab bar *Events*, a list of events is shown as may be seen in Fig. 4. Displaying a list of events sorted by the user's preferences. At the bottom of this tab, the user can see the number of events available on the list. In the specific case shown in Fig. 4, the user may choose from a range of 139 events.

When an event of the list is selected, the system will get the corresponding information about it showing the details to the user (as shown in Fig. 5). As may be seen in Fig. 5(a), the details include the event title, a brief summary, an image with the corresponding caption, the event date, the publication date, and the author of the post. The description of the event can be found at the end of the view (Fig. 5(b)). When the description field is bigger than the available window area, the user is allowed to scroll the text in order to read it.

At the *Filter* tab shown in Fig. 6(a), the user may choose a specific category of events. Figure 6(a) also presents a list of the available categories to the user. Figure 6(b) shows the result when the user chooses a category. The application will fetch only the user's choice and the information of these events. With this feature, CityEvents implements a cognitively distinct group of users.

At the *Event Detail* window the user has the possibility to share an event with friends, by tapping the share button. Three options are available (Fig. 7): sharing an event in the user Facebook page, sending it by email, or sending a text message.

Table 2
Questions in CityEvents survey

Question	Description
Q1	Is the application design attractive?
Q2	Is the application easy to use?
Q3	Is the application environment user friendly and intuitive?
Q4	Are the navigation options clear and consistent?
Q5	Are fonts easy to read on the screen?
Q6	Is the feedback and response time of the application fast enough?
Q7	Is the application store with users' preference useful?
Q8	Is the events/news sorted in function of users' preferences useful?
Q9	Is the map with all the points of interest and events based on users' preferences useful and attractive?
Q10	Is the events/news scoring based on users' preferences and location useful?

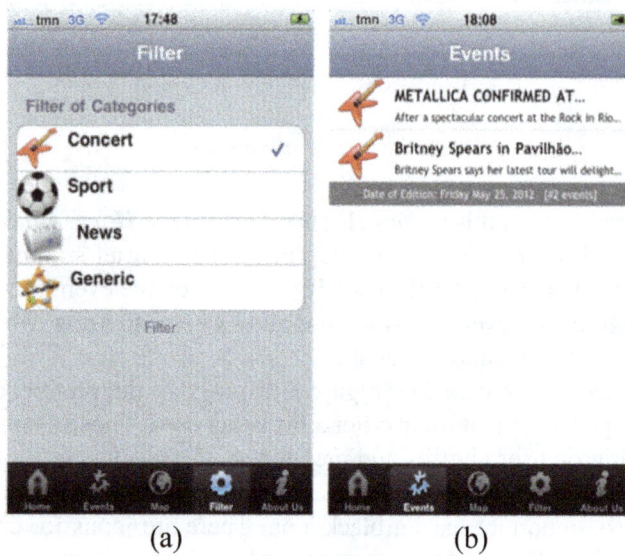

(a)   (b)

Fig. 6. List of categories and corresponding events.

Fig. 7. Options to share events.

Figure 8 shows an example of a shared event. At the left inside, it is shown the event share on Facebook while at the right inside, the event is shared by email. On the lower portion of Fig. 8, the use of SMS is illustrated. It can be noticed that with these sharing features an event will receive a bigger score on the user's profile.

Figure 9 shows the *Map* tab presenting a map with the localization of an event. It is also shown all the close available events, using pinpoints to identify the event location. The nearby event is represented by the corresponding icon category. The user location is also shown on the map. When a category is selected, the user can read the title and a brief description of the event. Figure 9(a) shows an alert of the most relevant event to the user, catching his/her attention for an event of her preferred category. Figure 9(b) presents a map with user's preferred events, based on the most important preferences and categories to him/her.

The CityEvents system was evaluated in several mobile devices (iPhone and iPad) and performed very well, as expected, and it is ready for use.

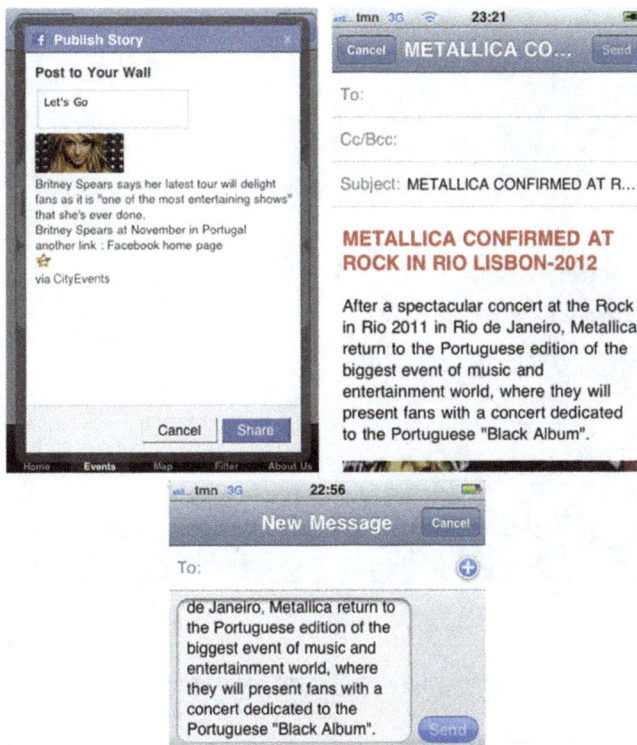

Fig. 8. Sharing events.

5. Performance evaluation and results

The proposed application should be experimented by real users in order to evaluate their experience. Then, a survey was proposed to support this study. A total of 114 users, 65% male and 45% female users from the University of Beira Interior, Portugal have answered the survey. The users have used the system for some time to become familiar with it in order to make a good experiment. After some experiments, they filled the CityEvents survey. The survey questions are available in Table 2 and the results may be found in Figs 10 and 11. It can be seen that the majority of the users agree the CityEvents application has an attractive design, it is user friendly and intuitive, presents a good navigation, and the options are clear and easy to use. The event detail is consistent and the text is easy to understand, fonts and all the event detail are easy to read on the screen. A large percentage of users also consider the application is very easy to use and helpful to bring the best events to him/her. The most part of the users agrees the application storing the users preferences and sort the events based on their preferences is good. Some users also agree that map view (with all the points of interests and events based on their preferences) is useful and attractive. Mostly users agree that events scored based on their preferences and location are useful.

Another issue is the distribution of the location-based scoring. Considering the news are usually distributed relative to an area of interest (for example, a country or a smaller area), high oscillations in the number of documents per relevance can be found rise to further optimizations. Thus, the distribution of nine sets of documents categorized by Accommodation, Leisure, Attractions, Community, Car, Food, Shopping, Supermarket, and Travel were analyzed and it was calculated the number of documents per

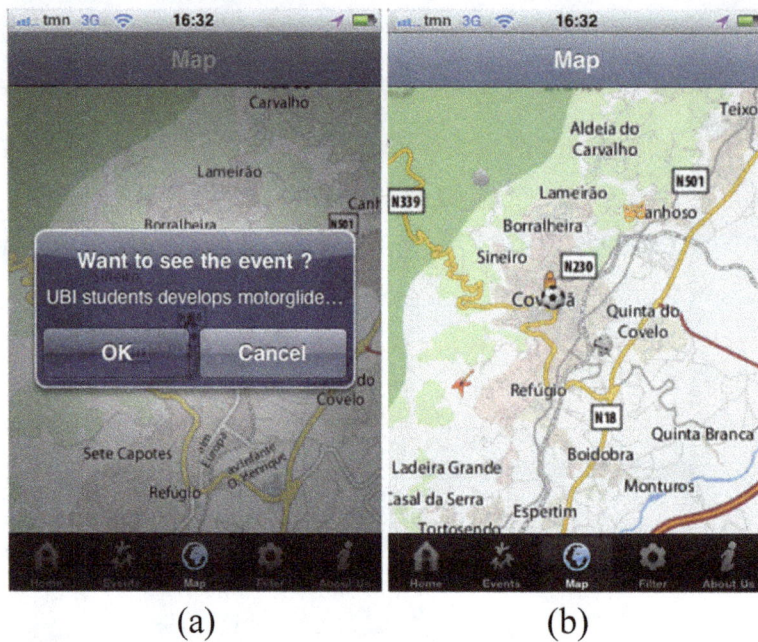

(a) (b)

Fig. 9. Maps with alert of event and scoring events.

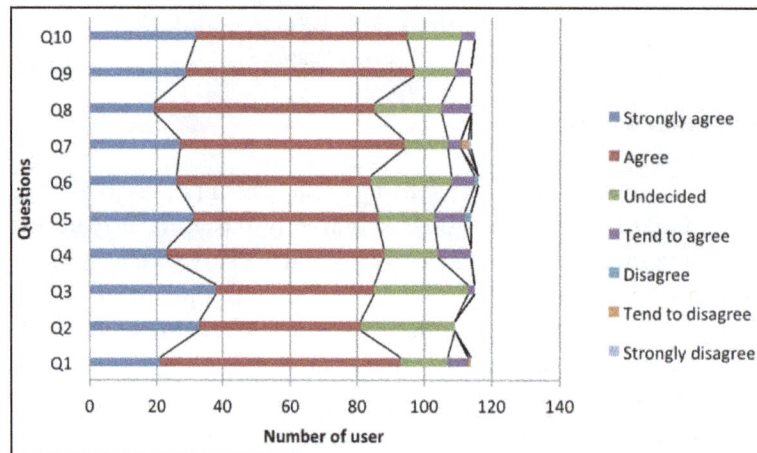

Fig. 10. Results of CityEvents users survey.

relevance windows. Figure 12 shows the results of the three categories with more documents and the average over all the nine categories, with a windowing width of 0.05, leading to a resolution of 20 intervals in the range [0, 1].

The results of Fig. 12 clearly shown they are clustered around the chosen point of reference (the center of the bigger city in the area of interest) and the distribution curve is rather smooth, keeping almost a slowly increasing value between 0.10 and 0.85. At higher resolutions (window width 0.1, with 100 intervals in the range) there are also not high oscillations, as hypothesized. Thus, the study on this direction was stopped.

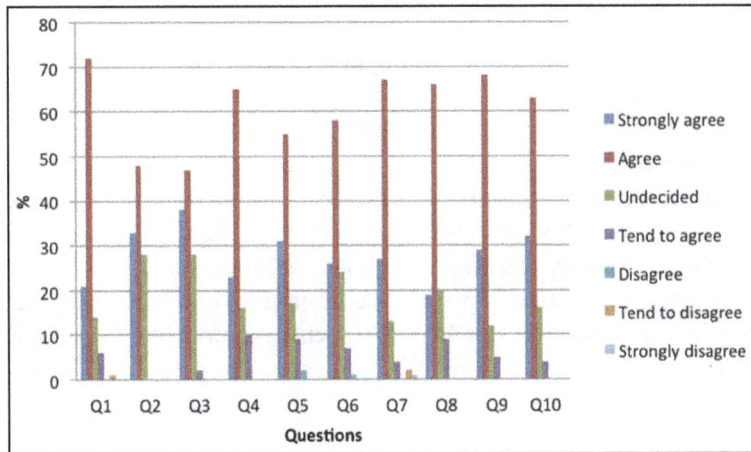

Fig. 11. Results of the users survey (percentages).

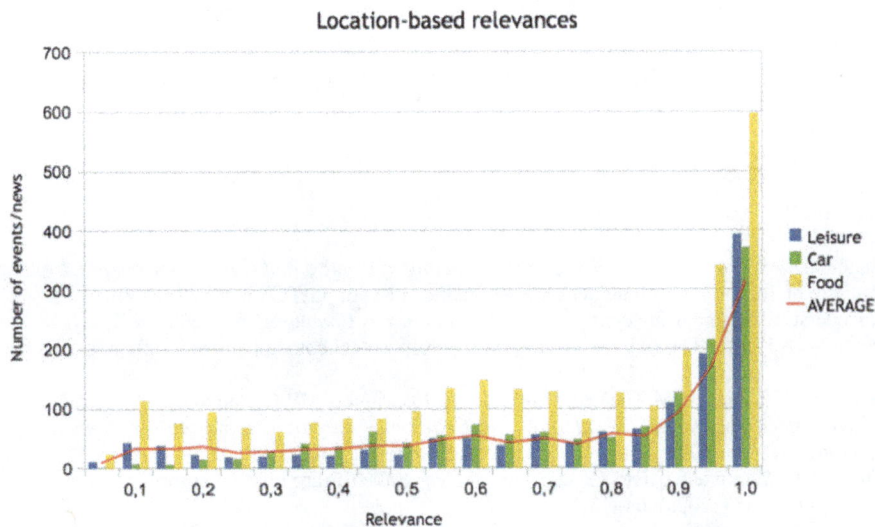

Fig. 12. Location-based relevance average.

The proposed re-normalization techniques and data-fusion equations are not intensive in terms of resources, processor or memory, and are by far less problematic than the re-sorting algorithms to be utilized in the content-adaptation layer of the application, or the display of the chosen news/events as points in a map on the mobile device.

6. Conclusion and future work

This paper proposed a ubiquitous location, habit, and context-aware mobile recommendation system including a client-side application called CityEvents and a Internet-based service providing sorted multimedia Geolocated cultural events to display on mobile devices. The user receives the best information, filtered and sorted the results based on user habits, context, location, and potentially other parameters useful to achieve sharper results, in a personalized and interactive way. Different equations for re-normalization and data-fusion were evaluated and deployed in a complete demonstrative mobile application to study users experience. Some research directions were stopped when results from experimental data were available, while the majority of improvements and ideas developed during this work are suitable for further research.

As a future work, some improvements and research directions are suggested as follows: evaluating the effect of using more advanced data-fusion techniques; evaluating the effect of more complex taxonomies endowed with proper metrics; evaluating the effect of rank-based scoring instead of relevance-based scoring; considering documents as structured entities to identify some interesting relevance parameters; studying possibilities to use generic browsing history to generate some interesting relevance parameters; and the migration of the proposed mobile solution to other mobile platforms like Android or Windows Mobile.

Acknowledgments

This work has been partially supported by the *Instituto de Telecomunicações*, Next Generation Networks and Applications Group (NetGNA), Portugal, by National Funding from the FCT – *Fundação para a Ciência e a Tecnologia* through the PEst-OE/EEI/LA0008/2011 Project, and by the company Zirak s.r.l., Italy.

References

[1]   A.K. Dey and G.D. Abowd, *Towards a Better Understanding of Context and Context-Awareness*, College of Computing, Georgia Institute of Technology, Atlanta GA USA, Technical Report GITGVU-99-22, 1999.

[2]   A.K. Dey, Understanding and using context, *Personal Ubiquitous Computing* **5**(1) 2001, 4–7.

[3]   Apps Portugal – Appetite [Online]. Available: http://appsportugal.com/app/appetite-iphone/100 [Accessed: October 2012].

[4]   B.M.C. Silva, P.A.C.S. Neves, M.K. Denko and J.J.P.C. Rodrigues, MP-Collaborator: A mobile collaboration tool in pervasive environment, in: *2009 IEEE International Conference on Wireless and Mobile Computing, Networking and Communications (WiMob 2009)*, Marrakech, Morocco, October 12–14, 2009, pp. 344–349.

[5]   C.A. da Costa, A.C. Yamin and C.F.R. Geyer, Toward a general software infrastructure for ubiquitous computing, *IEEE Pervasive Computing* **7**(1) (2008), 64–73.

[6]   C. Wolfram, Q&A: Conrad Wolfram on communicating with apps in Web 3.0 [Online]. Available: http://www.itpro.co.uk/621535/q-a-conrad-wolfram-on-communicating-with-apps-in-web-3-0 [Accessed: October 2012].

[7]   C. Dobre, Context-aware platform for integrated mobile services, in: *2011 International Conference on Emerging Intelligent Data and Web Technologies (EIDWT)*, September 7–9, 2011, pp. 198–203.

[8]   D. Taniar and J. Goh, On mining movement pattern from mobile users, *International Journal of Distributed Sensor Networks* **3**(1) (2007), 69–86.

[9]   D. Quercia, N. Lathia, F. Calabrese, G. Di Lorenzo and J. Crowcroft, Recommending social events from mobile phone location data, in: *10th International Conference on Data Mining (ICDM) 2010 IEEE*, December 13–17, 2010, pp. 971–976.

[10] D. Wang and A.A. Abouzeid, On the cost of knowledge of mobility in dynamic networks: An information-theoretic approach, *IEEE Transactions on Mobile Computing* **11**(6) (June 2012), 995–1006.

[11] E. Fox and J. Shaw, *Combination of Multiple Searches*, NIST SPECIAL PUBLICATION SP, National Institute of Standards & Techonology, 1994, pp. 243–243.

[12] E. Pitoura and G. Samaras, Locating objects in mobile computing, *IEEE Transactions on Knowledge and Data Engineering* **13**(4) (July/August 2001), 571–592.

[13] F. Chen, M.P. Johnson, Y. Alayev, A. Bar-Noy and T.F. La Porta, Who, when, where: Timeslot assignment to mobile clients, *IEEE Transactions on Mobile Computing* **11**(1) (January 2012), 73–85.

[14] F.S. Tsai, M. Etoh, X. Xie, W.C. Lee and Q. Yang, Introduction to mobile information retrieval, *Intelligent Systems, IEEE* **25**(1) (January–February 2010), 11–15.

[15] G. Ortiz and A.G. de Prado, Mobile-aware Web services, in: *3rd International Conference on Mobile Ubiquitous Computing, Systems, Services, and Technologies (UBICOMM 2009)*, Sliema, Malta, October 11–16, 2009, pp. 65–70.

[16] G. Salton, E.A. Fox and H. Wu, Extended boolean information retrieval, *Communications of the ACM* **26** (1983), 1022–1036.

[17] I. Sharp, K. Yu and T. Sathyan, Positional accuracy measurement and error modeling for mobile tracking, *Transactions on Mobile Computing, IEEE* **11**(6) (June 2012), 1021–1032.

[18] iPhone's Location – Aware Apps – Loopt's FriendFinder [Online]. Available: https://www.loopt.com/ [Accessed: October 2012].

[19] J.J.P.C. Rodrigues, M. Oliveira and B. Vaidya, New trends on ubiquitous mobile multimedia applications, *EURASIP Journal on Wireless Communications and Networking* (September 2010).

[20] J.J.P.C. Rodrigues, I.M.C. Lopes, B.M.C. Silva and I. de la Torre, A new mobile ubiquitous computing application to control obesity: SapoFit, in: *Informatics for Health and Social Care, Informa Healthcare*, New York, USA, 2012, pp. 1–17.

[21] K.S. Su, D. Chan and F.C. Chan, Navigation on mobile news, in: *2nd International Conference on Mobile Technology, Applications and Systems*, November 15–17, 2005, pp. 6–6.

[22] K.B. Ng and P.B. Kantor, An investigation of the preconditions for effective data fusion in ir: A pilot study, in: *Proceedings of the 61th Annual Meeting of the American Society for Information Science*, 1998.

[23] L. Guan, X. Ke, M. Song and J. Song, A survey of research on mobile cloud computing, in: *2011 IEEE/ACIS 10th International Conference on Computer and Information Science (ICIS 2011)*, May 16–18, 2011, pp. 387–392.

[24] L. Jessupand and D. Robey, The relevance of social issues in ubiquitous computing environments, *Communications of the ACM* **45**(12) (December 2002), 88–91.

[25] L. Joon Ho, Analyses of multiple evidence combination, in: *Proceedings of the 20th annual international ACM SIGIR conference on Research and development in information retrieval (SIGIR '97)*, ACM, New York, 1997, pp. 267–276.

[26] M. Weiser, The World is not a Desktop, *ACM Interactions Magazine* **1**(1) (January 1994), pp. 7–8.

[27] M. Weiser, The computer for the 21st century, *ACM SIGMOBILE Mobile Computing and Communications Review* **3**(3) (1999).

[28] M. Satyanarayanan, Pervasive computing: Vision and challenges, *IEEE Personal Communications* **8**(4) (August 2001), 10–17.

[29] M. Weiser and J. Seely Brown, *The coming age of calm technology*, Xerox PARC, October 5, 1996 [Online]. Available: http://www.ubiq.com/hypertext/weiser/acmfuture2endnote.htm [Accessed: October 2012].

[30] M. Weiser, Some computer science issues in ubiquitous computing, *Commun ACM* **36**(7) (1993), 75–84.

[31] N. Verma, D. Mahajan, S. Sellamanickam and V. Nair, Learning hierarchical similarity metrics, in: *IEEE Conference on Computer Vision and Pattern Recognition (CVPR) 2012*, June 16–21, 2012, pp. 2280–2287.

[32] P.M.P. Rosa, J.A. Dias, I.C. Lopes, Rodrigues, J.J.P.C. Rodrigues and K. Lin, An ubiquitous mobile multimedia system for events agenda, in: *Wireless Communications and Networking Conference (WCNC), 2012 IEEE*, April 1–4, 2012, pp. 2103–2107.

[33] Q.H. Mahmoud and P. Popowicz, A mobile application development approach to teaching introductory programming, in: *Frontiers in Education Conference (FIE), 2010 IEEE*, October 27–30, 2010, pp. T4F-1–T4F-6.

[34] R. Ling, *The Mobile Connection: The Rise of a Wireless Communications Society*, Morgan Kaufmann Publishers, USA, 2004, pp. 83–121.

[35] S.J. Hong, J.Y.L. Thong, J.Y. Moon and K.Y. Tam, Understanding the behavior of mobile data services consumers, *Information Systems Frontiers* **10** (2008), 431–445.

[36] S.K. Mostefaoui, Z. Maamar and G.M. Giaglis, *Advances in Ubiquitous Computing: Future Paradigms and Directions*, IGI Publishing, New York, 2008.

[37] S. Pohl, A. Moffat and J. Zobel, Efficient extended boolean retrieval, *IEEE Transactions on Knowledge and Data Engineering* **24**(6) (June 2012), 1014–1024.

[38] S. Quebe, J. Campbell, S. DeVilbiss and C. Taylor, Cooperative GPS navigation, *Position Location and Navigation Symposium (PLANS), 2010 IEEE/ION*, Indian Wells, CA, USA, May 4–6, 2010, pp. 834–837.

[39] S. Ja-Hwung, Y. Hsin-Ho, P.S. Yu and V.S. Tseng, Music Recommendation Using Content and Context Information Mining, *IEEE Intelligent Systems* **25**(1) (January 2010), 16–26.

[40] T. Teng Chen and D.C. Yen, Technical research themes of the mobile ubiquitous computing, in: *Eighth International Conference on Mobile Business (ICMB 2009)*, June 27–28, 2009, pp. 221–226.

[41] The Mac Box – NearPics [Online]. Available: http://themacbox.co.uk/nearpics/ [Accessed: October 2012].

[42] Tweakersoft – AroundMe [Online]. Available: http://www.aroundmeapp.com/ [Accessed: October 2012].

[43] V. Santos, Applying social paradigms in mobile context-aware computing, *6th Iberian Conference on Information Systems and Technologies (CISTI), 2011*, June 15–18, 2011, pp. 1–7.

[44] W.M.Y.W. Bejuri, M.M. Mohamad and M. Sapri, Ubiquitous positioning: A taxonomy for location determination on mobile navigation system, *International Journal Signal & Image Processing (SIPIJ)* **2**(1) (March 2011).

[45] Weather for Apple iPhone Smartphones – WeatherBug [Online]. Available: http://weather.weatherbug.com/mobile/
      weatherbug-for-iphone.html [Accessed: October 2012].
[46] Y. Bejerano, Coverage verification without location information, *IEEE Transactions on Mobile Computing* **11**(4) (April
      2012), 631–643.
[47] Y. Özgün and C. E. Rıza, iConAwa–An intelligent context-aware system, *Expert Systems with Applications* **39**(3) (2012),
      2907–2918, Elsevier.

Pedro M. Pinto Rosa received his Master degree in Informatics Engineering from University of Beira Interior, 2012 under
the supervision of Professor Joel Rodrigues. He is an MSc member of the Next Generation Networks and Applications Group
(NetGNA) at the *Instituto de Telecomunicações*, Portugal since March 2009. His main research topics include wireless sensor
networks and modeling tools, mobile and ubiquitous computing. He has authored or co-authored of several papers in international journals and conferences.

Joel J. P. C. Rodrigues is a professor at the University of Beira Interior (UBI), Covilhã, Portugal, and researcher at the *Instituto
de Telecomunicações*, Portugal. He received a PhD degree in informatics engineering, an MSc degree from the University of
Beira Interior, and a five-year BSc degree (licentiate) in informatics engineering from the University of Coimbra, Portugal.
He is the Director of the Master degree in Informatics Engineering at UBI. He is the leader of NetGNA Research Group
(http://netgna.it.ubi.pt), the Chair of the IEEE ComSoc Technical Committee on Communications Software, the Vice-Chair
of the IEEE ComSoc Technical Committee on eHealth, and Member Representative of the IEEE Communications Society on
the IEEE Biometrics Council. He is the editor-in-chief of the International Journal on E-Health and Medical Communications,
the editor-in-chief of the Recent Patents on Telecommunications, and editorial board member of several international journals.
He has been general chair and TPC Chair of many international conferences. He is a member of many international TPCs
and participated in several international conferences organization. He has authored or coauthored over 240 papers in refereed
international journals and conferences, a book, and 2 patents. He had been awarded the Outstanding Leadership Award of IEEE
GLOBECOM 2010 as CSSMA Symposium Co-Chair and several best papers awards. Prof. Rodrigues is a licensed professional
engineer (as senior member), member of the Internet Society, an IARIA fellow, and a senior member of ACM and IEEE.

Filippo Basso received his 4-year M.Sc. degree cum laude in General Mathematics in 1999, from the University of Pisa and
the Science Faculty degree cum laude in the same year from Scuola Normale Superiore di Pisa. He worked in Research and
Development since 2000 in Zirak s.r.l., Italian SME, where focused on Information Retrieval, eHealth, Mobile and Ubiquitous
Computing and generic modelling, algorithms and optimizations. His research interests comprehend theoretical frontiers topics
and their multi-disciplinary applications.

4

A P2P query algorithm for opportunistic networks utilizing betweenness centrality forwarding

Jianwei Niu[a,*], Mingzhu Liu[a] and Han-Chieh Chao[b]

[a]*State Key Laboratory of Software Development Environment, Beihang University, Beijing, China*
[b]*Institute of Computer Science and Information, National Ilan University, I-Lan, Taiwan*

Abstract. With the proliferation of high-end mobile devices that feature wireless interfaces, many promising applications are enabled in opportunistic networks. In contrary to traditional networks, opportunistic networks utilize the mobility of nodes to relay messages in a store-carry-forward paradigm. Thus, the relay process in opportunistic networks faces several practical challenges in terms of delay and delivery rate. In this paper, we propose a novel P2P Query algorithm, namely Betweenness Centrality Forwarding (PQBCF), for opportunistic networking. PQBCF adopts a forwarding metric called Betweenness Centrality (BC), which is borrowed from social network, to quantify the active degree of nodes in the networks. In PQBCF, nodes with a higher BC are preferable to serve as relays, leading to higher query success rate and lower query delay. A comparison with the state-of-the-art algorithms reveals that PQBCF can provide better performance on both the query success Ratio and query delay, and approaches the performance of Epidemic Routing (ER) with much less resource consumption.

Keywords: Opportunistic networks, P2P Query, betweenness centrality, social networks, mobile devices, epidemic routing

1. Introduction

Opportunistic networks [33] utilize the "store-carry-forward" paradigm and leverage the mobility of nodes to relay messages from the source to the destination. Thus in this sense, opportunistic network is typically a special case of Delay Tolerant Network (DTN) [9]. Recent years have witnessed a number of proof-of-concept opportunistic network applications such as ZebraNet [16], CarTel [15], Pocket Switch Networks [22], and P2P networks [5]. Information dissemination in opportunistic networks has very different features compared with traditional networks. In traditional wired/wireless networks, there are usually sustained end-to-end paths existing between the source and the destination. As a result, information can be disseminated by maintaining a certain routing topology. However, in opportunistic networks, due to the intermittent links between mobile nodes and the lack of end-to-end routing path, data can only be forwarded hop-by-hop in an opportunistic way. In this situation, the key problem is to find a routing strategy to determine whether to forward or keep the data for the owner when two nodes meet. In this paper, we focus on P2P query in opportunistic networks, more specifically the Social Opportunistic Network (SON) which is composed by mobile devices held by people. In P2P query, a query node

---

*Corresponding author: Jianwei Niu, State Key Laboratory of Software Development Environment, Beihang University, Beijing 100191, China. E-mail: niujianwei@buaa.edu.cn.

first sends an inquiry message across the network to search for another node carrying the corresponding response message. Then, the response message is forwarded to the query node. Before describing our work, we first review some related work.

Routing in delay tolerant networks have been extensively studied during the lasting few years, which focuses on how to select the next-hop data carrier during each encounter. Most work seeks to maintain a balance between message delivery rate and the cost, i.e., the number of in-network message copies. Epidemic Routing (ER) [30] is one of the most famous routing algorithms in opportunistic networks, which utilizes flooding to disseminate information throughout the network. If the resource is not the main concern, ER guarantees the lowest transmission delay and the highest data delivery rate. However, constrains in terms of node energy, bandwidth, and buffer size prevent ER from been widely used in practice. A number of protocols try to reduce the cost of flooding while still maintaining a high message delivery rate. In [12], the authors proposed a 2-hop flooding algorithm, where the source node relays messages to the first L contacts, and then these L nodes forward the messages to the destinations. Consequently, each message reaches its destination in two hops, and there are (L + 1) message copies in the network, which highly reduces the cost incurred by ER flooding. A following proposed algorithm, i.e., Spray and Wait (SW) [24] further improves ER by limiting the number of copies and avoiding performance deterioration caused by channel competition. Recently, the Spray and Focus (SF) [25] algorithm improves the SW algorithm by forwarding messages to nodes with higher probabilities to encounter destinations. Similar routing algorithms are also proposed in [17–19,35].

Several protocols are proposed for disseminating information to nodes with special properties, e.g., within a certain area. In [34], Xu et al. presented a spatial-temporal aware algorithm for data dissemination in which each message carries a time-to-live field and its location information. By this means, the algorithm effectively restricts the data dissemination range and survival time in the network, and thus the network load and bandwidth consumption are reduced. Wischhof et al. [31] proposed to use a periodically active broadcasting algorithm for message propagation and introduced dynamic adjustment of broadcasting cycles to improve message relay efficiency. Based on this idea, Eichler et al. [9] proposed to set the broadcast cycle according to urgency levels of messages. Caliskan et al. [6] introduced location-based fusion techniques into periodically active broadcasting algorithms, which further improved the message distribution efficiency.

Costa et al. [7] proposed a response message return algorithm which for the first time combined deterministic routing and random routing. In this algorithm, inquiry nodes broadcast an Inquiry Message (IM) periodically to nodes within their n hops, and then the IM can spread around the network through node movement. When the message source receives an inquiry message, the response message will be accurately routed to the inquiry node if the source node is close enough to the inquiry node and the source has reserved deterministic routing information. Otherwise, the response message will be randomly relayed by choosing some neighboring nodes until the response message's time to live (TTL) expires. Another inquiry/response propagation method is proposed in [3] which combines the content-based routing with the probability-based routing. The forwarding strategy of response message considers not only the distance between the information carrier and the inquiry node, but also the predicted encountering probabilities between nodes to select the most appropriate node as the next hop, which reduces the delivery delay.

For the same purposes, Pan et al. proposed in [23] a combined info-acquisition algorithm of inquiry message and source message. The algorithm targeted at a specific application scenario – Pocket Switched Networks (PSN) and a content routing-based algorithm is proposed for message inquiry. The algorithm assimilated the overall inquiry process to anti-permeation in physics, in which each inquiry node periodically broadcasts the inquiry messages with TTL, gradually forming concentration gradient of the IM around them. When the source receives an IM, response messages are forwarded to areas with a higher IM concentration like solvent molecules, and eventually back to the inquiry node. On this basis, Eichler et al. [9] proposed the idea of setting broadcasting cycles according to importance level. This algorithm defines several different importance levels for different messages due to limited bandwidth. Also by using the optimization theory, the algorithm preferentially broadcasts the message with the maximum priority through dynamic adjustments of broadcasting priority and frequency, so as to enhance distribution efficiency. Caliskan et al. [6] introduced GPS-based message fusion technology into the message dissemination of periodically initiative broadcasting and adjusted message dissemination rates according to nodes' information pattern, which further improved the dissemination rate.

Ouri et al. [32] proposed a Rank-Based Broadcast (RBB) inquiry algorithm in which inquiry nodes and source nodes periodically broadcast their IMs and source messages, respectively. Intermediate nodes maintain two queues for IMs and source messages, respectively. The IMs are ranked according to their request time, while the source messages are ranked according to their matching degree to IMs. When the communication channel is established, intermediate nodes will first broadcast the IM with the source message with the highest priority. This dissemination method can achieve the optimum use of the broadcast channel, and thus it maximizes the success rate of message dissemination. In [13], it is proposed to use metadata to reduce the additional expenses in RBB. Different from RBB, source nodes only broadcast the metadata instead of an entire source message. When the metadata matches a certain IM, the request message is sent to the message source, triggering it to transmit the source message. Inquiry/response is also studied in VANET (Vehicle Ad hoc NET), where Ilias et al. [20] proposed the concept of vehicle clustering to improve the inquiry process. The message source node periodically broadcasts the source message, collects IMs from neighboring vehicles, and divides inquiry vehicles into clusters according to their moving directions. The message source node selects the cluster with the maximum amount of IMs and forwards the source message to vehicles driving towards the cluster.

So far, we have made a brief survey of routing and inquiry/response algorithms in the literature. In this work, we focus on SON, where nodes are mobile devices carried by people. With the wide spreading of smart phones, we believe inquiry/response is a promising research topic in opportunistic networks. Because the devices are carried by human, the node mobility has some social characteristics, which have rarely been exploited for inquiry/response by the state of the art. We assume that all the nodes in a SON are willing to collaborate to relay messages. We introduce a concept, namely Betweenness Centrality (BC) in SONs. Then, we propose an algorithm for message inquiry/response based on nodes' BCs. Intuitively, nodes with higher BC values indicate that they are more active, and thus have more opportunities to encounter different nodes. Therefore, they should undertake more message forwarding tasks. Cai et al. in [8] found that both the node inter-contact time and nodes' contact frequencies in SON follow the power-law distribution, indicating that a few nodes play a decisive role in network connectivity and message transmission while most of the nodes are not so significant [21,26]. Literature [14,36] revealed that opportunistic networks have a typical "small world" property that on average, most messages can reach their destination through a short path (about 4–5 hops). Therefore, our design is inspired by those interesting findings.

Our major contributions in this paper are summarized as follows:

1) We propose a P2P (peer-to-peer) message inquiry scheme called PQBCF based on nodes' BC value in SON. PQBCF includes the calculation method for the number of copies of IMs and response messages.

2) We carried out extensive evaluating on the performance (message delivery ratio and delivery delay) of the PQBCF algorithm.

The remainder of this paper is organized as follows. Section 2 presents the PQBCF algorithm. In Section 3, the performance of PQBCF is evaluated by simulations. Section 4 provides some concluding remarks.

2. P2P query algorithm based betweenness centrality forwarding

2.1. Problem description and term definitions

The application scenario of our approach is: in SON, when a node (carried by a mobile phone user) intends to request some certain information (e.g., an MP3 file), it first generates an IM with descriptions of the required data. In order to reduce the expected inquiry latency, the generator produces multiple copies of the IM. Intuitively, more copies lead to shorter query delay, however, introducing higher overhead in buffer management for each node. As a result, an inquiry/response strategy should seek to find a tradeoff between query delay and overhead. To achieve this objective, PQBCF calculates the number of copies of the IM in the network according to several factors which include the expected query delay, the mobility, and density of nodes. Then, the IMs are disseminated to the network by using our PQBCF algorithm which is based on a metric called Betweeness Centrality. If the IM is passed to a node with data matching the IM, the node will generate a response message, calculate the number of copies of the response message, and send the response message back to the inquiry node. The key advantage of PQBCF is that it utilizes the betweeness centrality metric inspired from social network analysis, and as a result, PQBCF can significantly reduce the inquiry/response delay. Before describing the details of PQBCF, we first introduce several definitions used throughout this paper.

Definition 1. *Inquiry Message (IM). When a node requests certain information, it generates an Inquiry Message (IM)with format as follows:*

$$IM\left(Type, ID\_n, ID\_m, TTL, Topic, Path, Content\right)$$

where, *Type* is used to distinguish between an IM or a response message; *ID_n* and *ID_m* are the identity of the IM and the inquiry node, respectively; *TTL* indicates the survival time of this message; *Topic* is the query subject; *Path* keeps track of the nodes passed by during the message delivery process; *Content* stands for the query (e.g., requesting an MP3 file).

Definition 2. *The Response Message (RM). When a node receives an IM and has requested content, the node will generate a RM. Its format is defined as follows:*

$$RM\left(Type, ID\_n, ID\_m, TTL, ID\_d, Path, Content\right)$$

where *Type, ID_n, ID_m, TTL, Path* are exactly the same with those in the definition of IM; *ID_d* is the identity of the destination node (i.e., the node that posts the inquiry); *Content* is the content requested by the inquiry node (e.g., the requested MP3 file).

Definition 3. *Node Betweenness Centrality. The Betweenness Centrality $C_B\left(P_i\right)$ of node P_i is used to quantify the importance of the node P_i in message delivery in the whole network. To formally define $C_B\left(P_i\right)$, let g_{sd} denote the number of all the messages successfully delivered between a pair of nodes*

Table 1
Network betweenness centrality

	Sum	P_1	$\ldots$	P_i	$\ldots$	P_n
P_1	g_{1d}	0	$\ldots$	$g_{1d}(P_i)$	$\ldots$	$g_{1d}(P_n)$
$\ldots$	$\ldots$	$\ldots$	$\ldots$	$\ldots$	$\ldots$	$\ldots$
P_s	g_{sd}	$g_{sd}(P_1)$	$\ldots$	$g_{sd}(P_i)$	$\ldots$	$g_{sd}(P_n)$
$\ldots$	$\ldots$	$\ldots$	$\ldots$	$\ldots$	$\ldots$	$\ldots$
P_n	g_{nd}	$g_{nd}(P_1)$	$\ldots$	$g_{nd}(P_i)$	$\ldots$	0

P_s and P_d. $g_{sd}(P_i)$ *represents the number of the messages that pass through* P_i *during these message forwarding processes. Then,* $b_{sd}(P_i)$ *is defined as shown in Eq. (1),*

$$b_{sd}(P_i) = \frac{g_{sd}(P_i)}{g_{sd}} \tag{1}$$

where $b_{sd}(P_i)$ represents the ratio of the messages successfully delivered via P_i to all messages successful messages between P_s and P_d. $b_{sd}(P_i)$ indicates the importance of P_i in delivering messages for P_s and P_d. Suppose the number of nodes in the network is n, and then the betweeness centrality value of P_i can be calculated by Eq. (2).

$$C_B(P_i) = \frac{2}{(n-1)(n-2)} \sum_{s=1,s\neq i}^{n} \sum_{d=1,d\neq i,s\neq d}^{n} b_{sd}(P_i) \tag{2}$$

Therefore, $C_B(P_i)$ means the average of all $b_{sd}(P_i)$ for any pair of nodes P_s and P_d in the network. In SONs, $C_B(P_i)$ represents the centrality of node P_i. A high $C_B(P_i)$ indicates that node P_i has more opportunities to contact with other nodes. However, $C_B(P_i)$ is only the average for all destination nodes. In order to analyze the messages that node P_i forwards to a specific destination node P_d, we define $C_{Bd}(P_i)$ which can be calculated by Eq. (3). It looks similar with Eq. (2), and however, the destination node is fixed. $C_{Bd}(P_i)$ indicates the importance of P_i in delivering message to d.

$$C_{Bd}(P_i) = \frac{2}{(n-1)(n-2)} \sum_{s=1,s\neq i}^{n} b_{sd}(P_i) \tag{3}$$

To calculate $C_B(P_i)$ and $C_{Bd}(P_i)$, node P_d maintains a table $T(P_d)$ as shown in Table 1. The rows in Table 1 are the source nodes P_s ($s = 1, 2, \cdots, n, s \neq d$), and the columns are the intermediate nodes P_i ($i = 1, 2, \cdots, n, i \neq d$). The first column ($Sum$) is the number ($g_{sd}$) of messages successfully delivered from the source node P_s to the destination node P_d. Each of the other columns records the numbers of messages delivered from the source node P_s to the destination node P_d via the intermediate node P_i, respectively.

One should note that Table 1 can be obtained by each node in a distributed fashion. A message is forwarded from its source node to its destination via multiple intermediate nodes. Each intermediate node can thus save its ID into the Path field of the message. When the message reaches its destination node P_d, P_d updates its $T(P_d)$ table according to the *Path* field of this message.

According to the table of Betweenness Centrality, a node is able to calculate a $b_{sd}(P_i)$ value and a $C_{Bd}(P_i)$ value setting itself as the destination node. When two nodes meet, they exchange their reserved $b_{sd}(P_i)$ values and $C_{Bd}(P_i)$ values, till all reserved $b_{sd}(P_i)$ values are synchronized and all $C_{Bd}(P_i)$ values updated. After each encounter, the node consciously updates a $C_B(P_i)$ value according to all reserved $b_{sd}(P_i)$ values. Since Table 1 only maintains the local information corresponding to the node P_d, its buffer consumption is less than $O(n^2)$ even under the worst circumstances, which is desirable for current mobile devices.

2.2. *Calculate the number of inquire message copies*

In order to reduce the delay of obtaining the corresponding response for an inquiry node, it may disseminate multiple IMs into the network. However, mobile devices usually have very limited buffer size. Thus, caching IMs is the main overhead for an inquiry/response system. Intuitively, more IMs can help to reduce the inquiry delay but incur a higher cost. As a result, the system should carefully organize the number of IM copies to find a tradeoff between inquiry delay and cost. To achieve this goal, we propose to calculate the number of the IM copies (denoted as L) in the network according to the required inquiry propagation delay, node mobility model and network density. An inquiry node can increase L to have a higher probability to receive the response message in the required delay. Therefore, determining the value of L is the key challenge for our algorithm.

In [24], Spyropoulos et al. studied the average message transmission delay in intermittently connected mobile networks. Let S denote the network area. N is the number of mobile nodes. K is the transmission range. Nodes in the network perform independent random walk and *Direct Transmission* (source nodes only forward messages to their destinations). Then, the expected delay of direct transmission is

$$ED_{dt} = 0.5S \left(0.34 \log_2 S - \frac{2^{K+1} - K - 2}{2^K - 1} \right) \tag{4}$$

The expected delay of the optimal algorithm is

$$ED_{opt} = \frac{H_{N-1}}{N-1} ED_{dt} \tag{5}$$

It is proved in [24] that ED_{opt} is the low bound of the expected transmission delay of any algorithms, where H_n is the nth harmonic number, i.e,

$$H_n = \sum_{i=1}^{n} \frac{1}{i} \tag{6}$$

The expected IM propagation delay (ED) should be larger than ED_{opt}, i.e., ED meets the following condition where a satisfies $a \geqslant 1$

$$ED = aED_{opt} \tag{7}$$

Then the relationship between *ED*, L and the total number of nodes N in the network can be described as follows.

$$(H_N^3 - 1.2)L^3 + \left(H_N^2 - \frac{\pi^2}{6} \right) L^2 + \left(a + \frac{2N-1}{N(N-1)} \right) L = \frac{N}{N-1} \tag{8}$$

Equation (8) shows the relationship between L, *ED* and N. However, the number of nodes in the network is global information that can only be obtained in a centralized way. Moreover, due to the frequent node movements, it may be dynamically changing over time. As a result, the value of L should be adaptively tuned according to the number of nodes in current areas.

We use a heuristic method to estimate the number of nodes in the network. Assume that the nodes move independently and are identically distributed. It is proved in [24] that the delay (denoted as T_1) of node i in meeting with another node within the area follows an exponential distribution D_{dt} with mean of $D_d/(N-1)$. Recall that the expected delay of direct transmission is ED_{dt}. Therefore, we have

$$T_i = \frac{ED_{dt}}{N-1} \tag{9}$$

The delay (denoted as T_2) of node i in meeting with two different nodes follows an exponential distribution with mean of $D_{dt}(1/(N-1)+1/(N-2))$, i.e.,

$$T_2 = ED_{dt}\left(\frac{1}{N-1}+\frac{1}{N-2}\right) \tag{10}$$

We can derive an estimation of the number of nodes N by combining Eqs (9) and (10),

$$N = \frac{2T_2 - 3T_1}{T_2 - 2T_1} \tag{11}$$

One should note that the node is able to measure T_1 and T_2 through locally. Thus, the total number of nodes N in the network can be estimated according to Eq. (11) in a distributed way. After obtaining N, we can determine the number L of copies of distributed messages according to Eq. (8).

2.3. Propagation algorithm for inquiry messages

An inquiry node needs to disseminate the generated L IM copies in the network. In this paper, we propose the Spread distribution algorithm based on improved binary tree, which chooses nodes with greater values of BC as transmission nodes, making it possible to forward messages as quickly as possible, with specific procedures as follows:

1. Inquire node P_s generates inquiry messages, determines the ratio a of expected transmission delay ED and the theoretically optimal transmission delay expected ED_{opt} according to the degree of urgency of inquiry contents; calculates the total number of nodes N according to the value of T_1, T_2, confirming the number L of distributed copies of inquire messages, records the node distribution task with mark $token(P_s) = L$, turn to step 2;
2. Node P_m carrying inquiry messages decides whether the TTL value of IMs is 0, if positive, delete the message, turn to step 2; otherwise, to determine whether the label of distributed task token is 1, if positive, turn to step 3; if negative, turn to step 5;
3. Node P_m carrying IMs meets certain node P_i to exchange the value of $T(P_m)$, $T(P_i)$ of their BC tables, and calculates the updated value of BC $C_B(P_m)$, $C_B(P_i)$. Node P_m asks whether node P_i has been carrying the IM, if positive, repeat steps 3, P_m continues to meet with other nodes; otherwise, turn to step 4;
4. Nodes P_m, P_i determine whether $C_B(P_m) < C_B(P_i)$, if established, first of all P_m minus TTL value of the inquire messages by 1, and then forward the message to P_i, meanwhile deleting its own copies of the IM, set the of node P_i as 1, turn to step 2; If negative, directly turn to step 2;
5. Node P_m carrying IMs meets certain node P_i to exchange the value of $T(P_m)$, $T(P_i)$ of their BC tables, and calculate the updated value of BC $C_B(P_m)$, $C_B(P_i)$. Node P_m asks whether node P_i has been carrying the IM, if positive, turn to step 2; if not, to determine whether $C_B(P_m) < C_B(P_i)$; if established, turn to step 6; If negative, turn to step 2;

6. First of all, P_m minus TTL value of the IMs by 1, and then forward the message to P_i, meanwhile allocating half of the distributed task to node P_i, then goes on to the remaining other half of the task, namely $token(P_i) = \lceil token(P_m)/2 \rceil$, $token(P_i) = \lfloor token(P_m)/2 \rfloor$, turn to step 2.

In this algorithm, two nodes update their $T(P_m)$ and $T(P_i)$ value when they encounter each other, and conduct a comparison operation, therefore the algorithm complexity is $O(E)$, wherein E stands for the number of times that the node carrying IMs meets with other nodes. $E(= ax^{-k}, k > 1)$ obeys the power law distribution [8], which means that most nodes in the network have low computational complexity, and that the algorithm manifests good scalability.

2.4. Back forwarding algorithm for response messages

When a certain node of the network receives an IM, it first checks the *Topic* field of the IM to determine whether it is able to respond to such inquiries, if positive, then it performs the following Back Forwarding algorithm; if negative, the node will perform as a transmission node to implement Spread distribution algorithm above. The Back Forwarding algorithm is based on the probability of meeting with the target node and selecting forwarding nodes. The specific procedures are as follows.

1. Response node P_r generates a response message, views the source node ID of ID_n field of the inquire message, and fills in the source node ID (the message's destination node ID, without loss of generality, assumed as P_d), turn to step 2;

2. Response node P_r checks the source node's expected inquiry delay *ED* preserved in the *Control* field of the IM, determines to back-forward the ratio of expected delay of the response message and ED_{opt}; measures the values of T_1, T_2 of the response node P_r, calculates the total number of nodes N, determines the number of forwarding copies L of the response message, records the forwarding task mark of response messages $token\_r(P_r)$, turn to Step 3;

3. Node P_{rm} carrying IMs decides whether the TTL value of IMs is 0, if positive, delete the message, turn to step 3; otherwise, determines whether the forwarding task label $token\_r(P_{rm})$ is 1, if positive, turn to step 4; if negative, turn to step 6;

4. Node P_{rm} carrying IMs meets certain node P_i to exchange the value of $T(P_{rm})$, $T(P_i)$ of their BC tables, and calculate the updated value of BC $C_{Bd}(P_{rm})$, $C_{Bd}(P_i)$. Node determines whether node is the destination node of the IM, if positive, forwards the message, 'end of algorithm'; if negative, P_{rm} asks whether P_i carries the response message; if positive, repeat step 4; If negative, turn to step 5;

5. Nodes P_{rm}, P_i determine whether $C_{Bd}(P_{rm}) < C_{Bd}(P_i)$, if established, P_{rm} will first minus the TTL value of the response message by 1, and then forwards the message to P_i, meanwhile deleting its own copies of the response message, sets $token\_r(P_i) = 1$, turn to step 3; If negative, directly turn to Step 3;

6. Node P_{rm} carrying IMs meets certain node P_i to exchange the value of $T(P_{rm})$, $T(P_i)$ of their BC tables, and calculates the updated value of BC $C_{Bd}(P_{rm})$, $C_{Bd}(P_i)$. Node P_{rm} determines whether P_i is the destination node of the IM; if positive, forwards the message, end of calculation; if negative, node P_{rm} asks whether node P_i has been carrying the IM, if positive, turn to step 3, if negative, determines whether $C_{Bd}(P_{rm}) < C_{Bd}(P_i)$; if established, turn to step 7; If negative, turn to step 3;

7. Node P_{rm} first minus TTL value of the IMs by 1, and then forward the message to P_i, meanwhile allocating half of the distributed task to node P_i, then goes on to the remaining t other half of the task, namely $token\_r(P_i) = \lceil token\_r(P_m)/2 \rceil$, $token\_r(P_m) = \lfloor token\_r(P_m)/2 \rfloor$, turn to step 3.

Table 2
Main simulation parameters

Simulation parameter	values
Simulation time (hour)	12 (ONE internal timer)
PQBCF learning time (hour)	3
Simulation of the regional area (width, height: m)	4500, 3400
The number of mobile nodes	300, 500, 1000
Node cache (Mbyte)	5
Message size (byte)	512
Minimum node speed (m / s)	0.5
Node maximum speed (m / s)	3
Transmission distance (m)	10, 30, 50, 75, 100
Transfer rate (kbyte / s)	250
FIFO buffer queue model	FIFO
Message generation cycle (minute)	15
A	1, 2, 3, 4, 5, 6, 7, 8, 9, 10
TTL (hop)	1, 2, 3, 4, 5, 6, 7, 8, 9, 10

In this algorithm, two nodes update value $T(P_m)$, $T(P_i)$ on meeting each other, and conduct a comparison operation, therefore the algorithm complexity is $O(E)$, where E stands for the number of times that the node carrying inquire messages meets with other nodes. $E(= ax^{-k}, k > 1)$ obeys the power law distribution [8], which means that most nodes in the network have low computational complexity and that the algorithm manifests good scalability.

3. Simulation

3.1. Simulation environment

We use the Opportunistic Network Environment (ONE) [29] to evaluate the proposed PQBCF algorithm. The simulation area is a 4500×3400 rectangle, where nodes follow the Random Walk (RW) mobility Model. Employing RM model is due to its preferable robustness. The moving speed of each node is within [0.5, 3] m/s. The whole simulation lasts 12 hours. Each node has a buffer size of 5 Mbytes and the size of each message is 512 bytes. The buffer at each node is managed in the FIFO fashion. When the buffer overflows, the first entry would be deleted. We set the sending rate of each node to 250 Kbyte/s. Note that each message has a TTL. When the TTL decreases to 0, the message is automatically discarded. We will evaluate the setting of TTL in our simulation. The detailed simulation settings are shown in Table 2.

At the beginning, nodes are evenly distributed in the network. During the initialization period, each node calculates its BC $C_B(P_i)$ using local encounter information. In our simulation, we set the length of the initialization period to 3 hours. IMs are generated by the Message Event Generator which comes with ONE. An IM is generated every 15 minutes. The query and the corresponding data are generated in a random fashion across all nodes. The queries from the same node vary from time to time.

In PQBCF, each node needs to do online estimation for parameters such as N and I. Consequently, each node calculates the values of T_1, T_2 periodically (every 15 minutes) in order to estimate the amount of nodes N_{new} as shown in Eq. (11). The new estimation is then used to obtain a new value of N via a moving average with N_{old} (the old estimation of the number of nodes). This new N is then substituted into Eq. (8) to update the value of L.

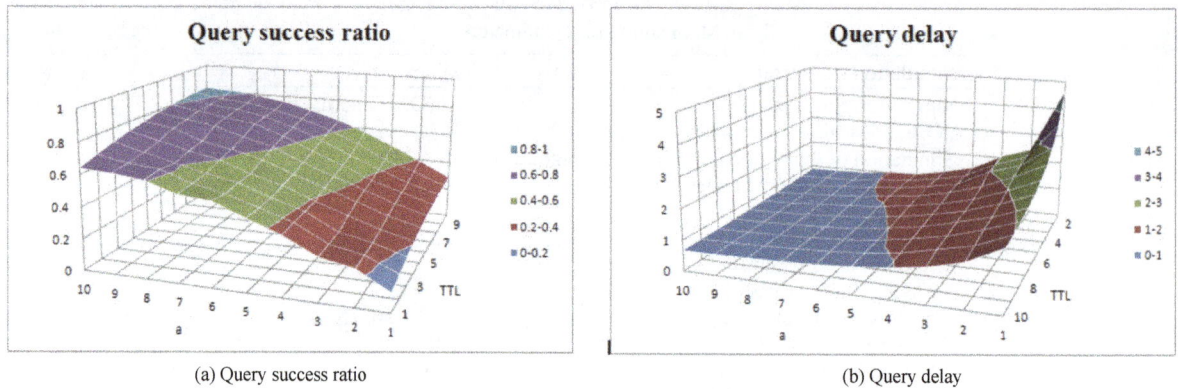

(a) Query success ratio

(b) Query delay

Fig. 1. Query success ratio and delay under different settings of a and TTL.

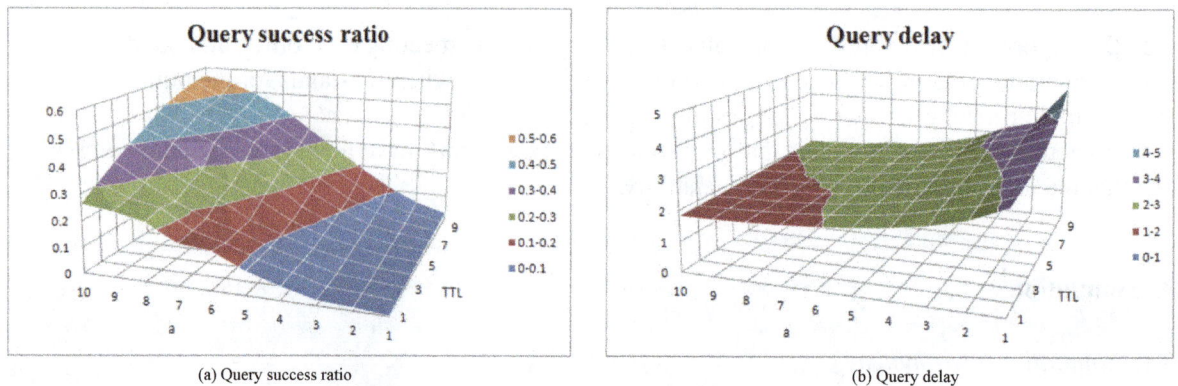

(a) Query success ratio

(b) Query delay

Fig. 2. Impact of a and TTL on PQBCF's performance (range $=$ 30).

3.2. Simulation results

We evaluate the performance of our PQBCF algorithm with two metrics, i.e., inquiry success rate and inquiry delay. Inquiry success rate is the ratio of IMs which are successfully responded to all generated IMs. Note that each IM carries a TTL field. When the TTL of an IM decreases to 0, the IM is dropped which means that the IM fails. Inquiry delay means the duration between generating an IM and receiving the first correct response message. As for failed inquiries, the inquiry delay is set as 1.5 times of the simulation time, namely 18 hours. The inquiry delay in this experiment is the averaged across all the inquire delays.

3.2.1. The impact of expected delay/TTL/range

As discussed previously, the expected delay ED is set to a times of the theoretic value ED_{opt}, where a is called the expected delay coefficient. In this set of experiments, a increases from 1 to 10. A larger a means the inquiry node expects to receive the response in a longer delay. The TTL value of IMs is selected among 1 to 10 hops. A TTL of 10 means that if an IM fails to meet nodes being able to respond after 10 hops, this IM would be deleted.

Figure 1 shows the query success ratio under different expected delay coefficient a and TTL. The transmission range of each node is set to 100 meters. In Fig. 1, we can see that the query success ratio

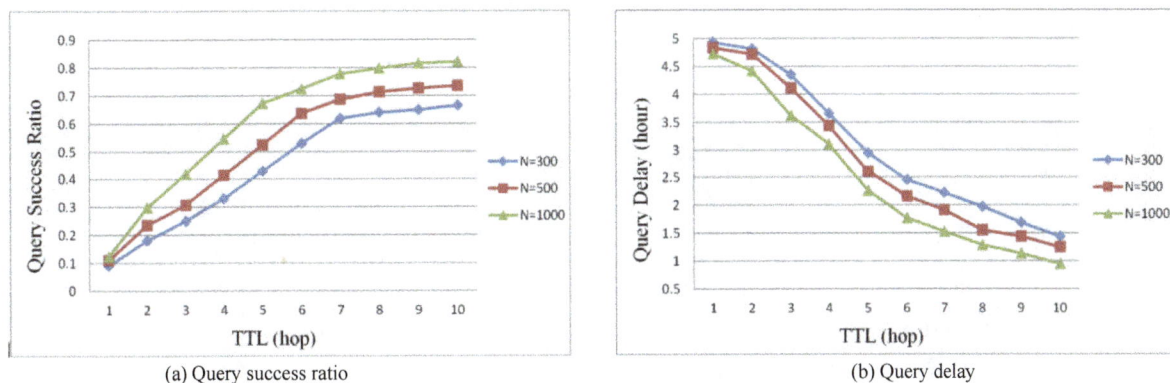

(a) Query success ratio   (b) Query delay

Fig. 3. Impact of the number of nodes on PQBCF' performance.

grows with the increase of a and also TTL. With $a = 10$ and TTL $= 10$, we can achieve a query success ratio of 82.6%. In practice, the value of a and TTL may vary from one application to another. We can set a small a for emergency scenarios where we only care responses returned in a short duration. This is especially useful when the data has to meet certain timeliness. A long delay will make the received data at the inquiry node out of date. On the other hand, a large a means the query is not in urgency and the inquiry node can wait for a slightly long time to obtain the response. The TTL is used to control the number of IMs in the whole network. A larger TTL makes each IM live longer. As a result, there will be more IMs, which is the main cost of PQBCF. In contrast, if TTL is set too small, IM copies might be deleted soon after generation, resulting in low query success ratio. Figure 2 shows the impact of a and TTL on the performance of inquiry when the transmission range is set to 30 m. In comparison with Fig. 1, we can see a lower query success ratio with higher query delay, which is reasonable since nodes have less chance to communicate with each other. However, they yield the similar trend as that shown in Fig. 1.

3.2.2. The impact of node density and node's moving speed

Figure 3 shows the impact of the node density in the network on the performance of PQBCF. In this set of simulations, we set $N = 300$, $N = 500$, and, $N = 1000$ respectively. a is set to 6 and the transmission range is set to 100 meters.

Figures 3(a) and (b) show that a higher node density yields a higher query success ratio and a lower query delay, respectively. This is because, with the increase of the total number of nodes, the number of nodes being able to respond to IMs also increases according to the message generation model in our experiment. Another observation in Fig. 3 is that the performance improved by increasing the density of nodes is not significant in comparison with that by increasing the TTL.

Figure 4 indicates the influence of node's moving speed on the performance of PQBCF algorithm. In this experiment, we examined its performance selectively on $v = 0.5$ m/s, $v = 1$ m/s, $v = 3$ m/s. In this series of experiment, $a = 6$, and the total quantity of nodes is 500 with communication radius of Wi-Fi signal as 100 m.

Figure 4 indicates that with the increase of nodes' moving speed, the inquiry success rate of PQBCF improves accordingly, and vice versa. The higher inquire success rate and decrease in transmission delay result from the following process. With the improvement of nodes' moving speed, the encounter frequency will increase, thus the inquiry message is relayed in higher efficiency based on PQBCF forwarding strategy. Compared to the effectiveness of increasing node number, advancing moving speed

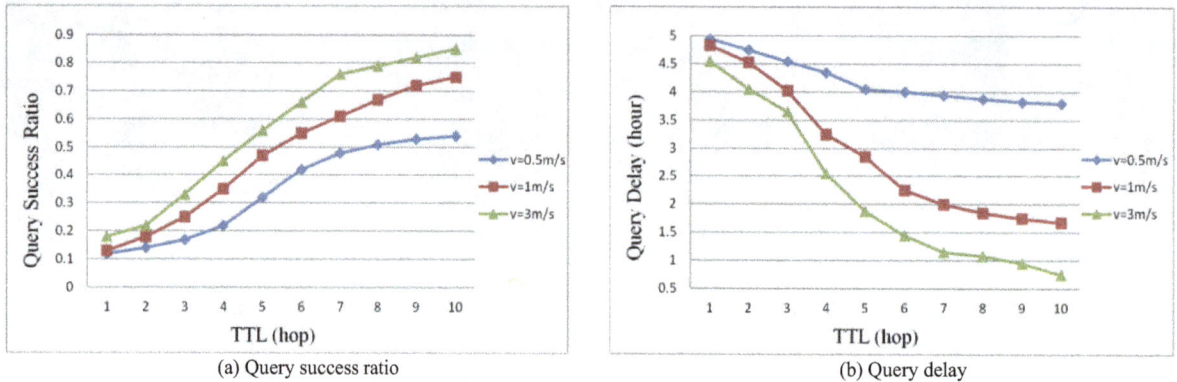

(a) Query success ratio   (b) Query delay

Fig. 4. Impact of the nodes' moving speed on PQBCF' performance.

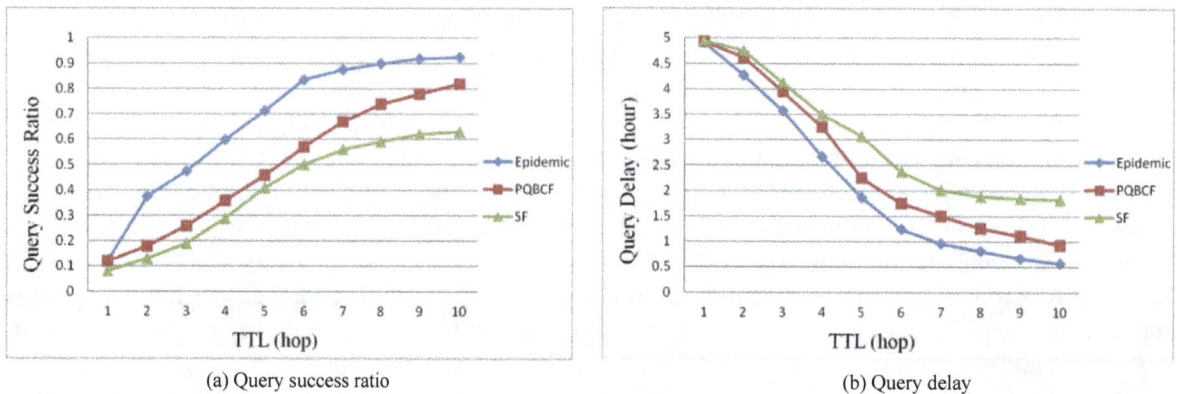

(a) Query success ratio   (b) Query delay

Fig. 5. Impact of the TTL on algorithms' performance.

promotes the performance of PQBCF in a larger degree. Since the core strategy of PQBCF is based on message relay utility rather than redundancy in quantity, therefore the increase in speed and further increase in relaying efficiency is more effective for the improvement of performance of PQBCF. Yet similar to increase of node quantity, the increase in nodes' mobility is equally unobvious compared with enhancing TTL.

3.2.3. Performance comparision

We compare the performance between PQBCF and several other routing algorithms for opportunistic networks, including Epidemic routing and SF. Epidemic algorithm [4] is a redundant based flooding algorithm where nodes keep copying query messages to all contacts. This algorithm has the highest success rate if no the bandwidth and buffer constrains. However, it consumes too much network resource, resulting in an impractical method in practical scenarios. Spray and Focus algorithm (SF) [28] is a hybrid forwarding algorithm. According to SF, messages are forwarded continuously from nodes with low utility value to higher ones until reaching the destination.

The metric used here are query success ratio and query delay. If nodes in the network have infinite bandwidth and memory size, the Epidemic algorithm absolutely outperforms all others. In our evaluation, Epidemic routing serves as the performance upper bound. SF algorithm first calculates the number of copies L, and then forwards L copies of the same IM to another node that may meet the destination nodes with a higher frequency. Note that the value of the L in PQBCF algorithm and SF algorithm is the same, so the comparison between two algorithms is fair. Our PQBCF algorithm works similar with the SF algorithm except that PQBCF uses the BC of nodes as the utility of forwarding. Figures 5(a) and (b) show the inquiry success rate and inquiry delay for the three different algorithms under different TTL values.

In this set of experiments, $a = 6$, the node transmission range is 100 meters. Figure 5(a) shows that the query success rates of all the three algorithms grow with the increase of TTL. This is because, the greater the TTL is the higher the probability that the IM is forwarded to nodes that can make a successful response to the IM. Because the network resources in our experiment settings are not the bottleneck, Epidemic routing achieves the highest success rate. However, the gap between the Epidemic and our PQBCF is not quite small. The SF algorithm performs the worst among the three. The PQBCF has a significantly higher success ratio than that of the SF algorithm when the TTL is greater than 6.

Figure 5(b) shows the average query delay for the three algorithms with different TTL values. When the TTL is small, the IMs may be discarded before reaching a node with corresponding information leading to a fail query. In Fig. 5(b) we can see our PQBCF outperforms SF and yield a close performance with Epidemic routing.

Figure 6 shows the performance of the three algorithms under different transmission range. In this set of experiments, $a = 6$ and TTL $= 10$. It can be seen from Fig. 6(a) that, with the increase of transmission range, the inquiry success rate of all three algorithms increased gradually. This is because the increasing transmission range improves the network connectivity, and thus, the IMs can reach the nodes being able to respond with a higher probability. Therefore, the inquiry success rate of the algorithm will be improved accordingly.

In Fig. 6(b), we can see with the increase of range radii of nodes' wireless signal, the inquiry delay of each algorithm has reduced significantly. When the range radiuses approaches 100 meters, the inquiry delay of PQBCF is close to that of the Epidemic algorithm. Compared with the SF algorithm, there is a more significant drop of PQBCF's inquiry delay. This is because, as the range radiuses increases, the network connectivity becomes better, and the nodes forward IMs faster, also forward response messages back to the inquire node in time, therefore, the inquiry delay of the PQBCF algorithm shows a more obvious fall.

4. Conclusion

In this paper, we focus on message acquisition scenarios for opportunistic networks consisting of smart phones. We proposed an efficient P2P query algorithm based on Betweenness Centrality Forwarding (PQBCF) for opportunistic networking. BC is introduced from social networks to indicate the forwarding roles of nodes in data transmission. The PQBCF algorithm selects the nodes with major impacts on data transmission and greater values of BC to forward IMs and response messages. We also design the algorithms to calculate the optimum number of IM and response message copies in the network to meet a certain delay constraint. The simulation experiments show that compared to other algorithms such as the SF algorithm, PQBCF effectively increases inquiry success rate and reduces inquiry delay.

Next, we will implement a prototype system of the PQBCF algorithm and deploy a pilot application on our campus so that it can be evaluated in a real environment.

Acknowledgments

This work was supported by the Research Fund of the State Key Laboratory of Software Development

Environment under Grant No. BUAA SKLSDE-2012ZX-17, the National Natural Science Foundation of China under Grant No. 61170296 and 61190120, the Program for New Century Excellent Talents in University under Grant No. NECT-09-0028.

References

[1]   A Community Resource for archiving wireless data, http://crawdad.cs.dartmouth.edu/.

[2]   L.A. Adamic and A. Eytan, Friends and Neighbors on the Web, *Social Networks* **25**(3) (2003), 211–230.

[3]   R. Baldoni, R. Beraldi and M. Migliavacca, Content-based routing in highly dynamic mobile ad hoc networks, *International Journal of Pervasive Computing and Communications* **1**(4) (2005), 277–288.

[4]   V.D. Becker, Epidemic routing for partially connected ad hoc networks, Technique Report, CS-2000-06, Department of Computer Science, Duke University, Durham, NC, 2000.

[5]   C.-F. Lai, Y.-M. Huang and H.-C. Chao, DLNA-based Multimedia Sharing System over OSGI Framework with Extension to P2P Network, *IEEE Systems Journal* **4**(2) (June 2010), 262–270.

[6]   M. Caliskan and D. Graupner, Decentralized discovery of free parking places, Proc. of the 3rd international workshop on Vehicular ad hoc networks, Los Angeles, ACM Press, 2006, pp. 30–39.

[7]   P. Costa and G.P. Picco, Semi-Probabilistic Content-Based Publish-Subscribe in Distributed Computing Systems, *Proc. of the 25th IEEE International Conference ICDCS 2005*, Washington, IEEE Press, 2005, pp. 575–585.

[8]   Q.S. Cai and J.W. Niu, Properties of message delivery path in opportunistic networks, *Proc. of the 5th International Conference on Future Information Technology*, Busan, IEEE Press, 2010, pp. 1–6.

[9]   S. Eichler, C. Schroth, T. Kosch and M. Strassberger, Strategies for context-adaptive message dissemination in vehicular ad hoc networks, *Proc. of third Annual International Conference on Mobile and Ubiquitous Systems*, California, IEEE Press, 2006, pp. 1–9.

[10] K. Fall, A delay-tolerant network architecture for challenged Internets, *Proc. of 2003 Conf. on Application, Technologies, Architectures, and Protocols for Computer Communications*, Karlsruhe, ACM Press, 2003, pp. 27–34.

[11] L.C. Freeman, Centrality in social networks: Conceptual clarification, *Social Networks* **1**(3) (1978), 215–239, Gossglause.

[12] M. Gossglauser and D. Tse, Mobility increases the capacity of ad hoc wireless networks, *IEEE/ACM Trans. on Networking* **10**(4) (2002), 477–486.

[13] F. Guidec and Y. Maheo, Opportunistic Content-Based Dissemination in Disconnected Mobile Ad Hoc Networks, *Proc. of UBICOMM '07 International Conference*, Papeete, French, 2007, pp. 49–54.

[14] J. Ghosh, S. Philip and C. Qiao, Sociological orbit aware location approximation and routing (solar) in DTN, Technique Report, CSE-2005-12, Department of Computer Science and Engineering, State University of New York, Buffalo, NY, 2005.

[15] B. Hull, V. Bychkovsky and Y. Zhang, CarTel: A distributed mobile sensor computing system, *Proc. of the 4th Int'l Conf. on Embedded Networked Sensor Systems*, Boulder, ACM Press, 2006, pp. 125–138.

[16] P. Juang, H. Oki and Y. Wang, Energy-efficient computing for wildlife tracking: Design tradeoffs and early experiences with ZebraNet, *Proc. of the 10th ASPLOS*, New York, ACM Press, 2002, pp. 96–107.

[17] A. Lindgren, A. Doria and O. Schelén, Probabilistic routing in intermittently connected networks, *ACM SIGMOBILE Mobile Computing and Communications Review* **7**(3) (2003), 19–20.

[18] C. Liu and J. Wu, Routing in a cyclic mobispace, *Proc. of ACM MobiHoc* New York, ACM Press, 2008, pp. 351–360.

[19] L. Zhou, H.-C. Chao and A.V. Vasilakos, Joint Forensics-Scheduling Strategy for Delay-Sensitive Multimedia Applications over Heterogeneous Networks, *IEEE JSAC* **29**(7) (August 2011), 1358–1367.

[20] I. Leontiadis and C. Mascolo, Opportunistic spatio-temporal dissemination system for vehicular networks, *Proc. of MobiOpp'07*, San Juan, ACM Press, 2007, pp. 39–46.

[21] C. Marta, A. César and B. Albert, Understanding individual human mobility patterns, *Nature* **453** (2008), 779–782.

[22] H. Pan, A. Chaintreau and J. Scott, Pocket switched networks and human mobility in conference environments, *Proc. of the 2005 ACM SIGCOMM workshop on Delay-tolerant networking*, Philadelphia, ACM Press, 2005, pp. 244–251.

[23] H. Pan, J. Leguay and J. Crowcroft, Osmosis in Pocket Switched Networks, *Proc. of the First International Conference on Communications and Networking in China (CHINACOM 2006)*, Beijing, IEEE Press, 2006, pp. 1–6.

[24] T. Spyropoulos, K. Psounis and C.S. Raghavendra, Spray and wait: An Efficient routing scheme for intermittently connected mobile networks, *Proc. of ACM SIGCOMM workshop on Delay Tolerant Networking (WDTN)*, Philadelphia, ACM Press, 2005, pp. 252–259.

[25] T. Spyropoulos, K. Psounis and C.S. Raghavendra, Spray and focus: Efficient mobility-assisted routing for heterogeneous and correlated mobility, *Proc. of the IEEE PerCom Workshop on Intermittently Connected Mobile Ad Hoc Networks*, New York, IEEE press, 2007, pp. 79–85.

[26] C.M. Song, Z.H. Qu, N. Blumm and A. Barabási, Limits of Predictability in Human Mobility, *Science Magazine* **327** (2010), 1018–1021.

[27] L.D. Sailer, Structural Equivalence: Meaning and definition, computation and applications, *Social Network* **1**(1) (1978), 73–90.

[28] T. Spyropoulos, K. Psounic and C.S. Raghavendra, Spray and focus: Efficient mobility-assisted routing for heterogeneous and correlated mobility, *Proc. Of the IEEE PerCom Workshop on Intermittently Connected Mobile Ad hoc Network*, 2007.

[29] The Opportunistic Network Environment simulator, http://www.netlab.tkk.fi/tutkimus/dtn/theone/.

[30] A. Vahdat and D. Becker, *Epidemic Routing for Partially-Connected Ad Hoc Networks*, Technique Report, CS-2000-06, Department of Computer Science, Duke University, Durham, NC, 2000.

[31] L. Wischhof, A. Ebner and H. Rohling, Information dissemination in self-organizing intervehicle networks, *IEEE Transactions on Intelligent Transportation Systems* **6**(1) (2005), 90–101.

[32] O. Wolfson, B. Xu and H.B. Yin, Search-and-Discover in Mobile P2P Network Databases, *Proc. of ICDCS 26th IEEE International Conference*, Lisboa, IEEE Press, 2006, pp. 65–67.

[33] Y.P. Xiong, L.M. Sun, J.W. Niu and Y. Liu, Opportunistic networks, *Journal of Software* **20**(1) (2009), 124–137.

[34] B. Xu, A. Ouksel and O. Wolfson, Opportunistic resource exchange in inter-vehicle ad-hoc networks, *Proc. of Mobile Data Management IEEE International Conference*, California, IEEE Press, 2004, pp. 4–12.

[35] Q. Yuan, L. Cardei and J. Wu, Predict and relay: an efficient routing in disruption tolerant networks, *Proc. of ACM MobiHoc'09*, Louisiana, ACM Press, 2009, pp. 95–104.

[36] J. Yoon, B. Noble, M. Liu et al., Building realistic mobility models from coarse-grained traces, *Proc. of ACM MobiSys 2006*, Uppsala, ACM Press, 2006, pp. 177–193.

Jianwei Niu received his Ph.D. degrees in 2002 in computer science from Beijing University of Aeronautics and Astronautics (BUAA). He was a visiting scholar at School of Computer Science, Carnegie Mellon University, USA from Jan. 2010 to Feb. 2011. He is a professor in the School of Computer Science and Engineering, BUAA. He has published more than 90 referred papers on such as INFOCOM, TECS, JPDC, and etc., and filed more than 30 patents in mobile and pervasive computing. He served as the Program Chair of IEEE SEC 2008, TPC members of ICC, WCNC, Globecom, IWCMC, and etc. He has served as associate EiC of Int. J. of Ad Hoc and Ubiquitous Computing, associate EiC of Journal of Internet Technology, editors of Journal of Network and Computer Applications (Elsevier). He received New Century Excellent Researcher Award from Ministry of Education of China 2009, Innovation Award from Nokia Research Center, and won the best paper award in CWSN 2012 and the 2010 IEEE International Conference on Green Computing and Communications (GreenCom 2010). His current research interests include mobile and pervasive computing, mobile video analysis.

Mingzhu Liu received his B.S. at Beijing Institute of Petrochemical Technology, Beijing in 2007. Now he is a master graduate at Beihang University. His current research interests are DTN and opportunistic networking.

Han-chieh Chao received his MS and PhD at School of Electrical Engineering, Purdue University in 1989 and 1993, respectively. He is a Professor of Department of Electrical Engineering, National Dong Hwa University and Joint Appointed Professor, Department of Electronic Engineering, Institute of Computer Science & Information Engineering, National Ilan University. He is Fellow of IET (IEE) and Fellow & Chartered IT Professional of British Computer Society (BCS). Prof. Chao's research interest include: wireless and mobile computing, high speed networks, IPv6 and digital arts technology.

5

A safe exit algorithm for continuous nearest neighbor monitoring in road networks

Hyung-Ju Cho*, Se Jin Kwon and Tae-Sun Chung
Department of Information & Computer Engineering, Ajou University Woncheon-dong, Suwon Si Yeongtong-gu, Gyeonggi-Do, South Korea

Abstract. Query processing in road networks has been studied extensively in recent years. However, the processing of moving queries in road networks has received little attention. In this paper, we introduce a new algorithm called the Safe Exit Algorithm (SEA), which can efficiently compute the safe exit points of a moving nearest neighbor (NN) query on road networks. The safe region of a query is an area where the query result remains unchanged, provided that the query remains inside the safe region At each safe exit point, the safe region of a query and its non-safe region meet so that a set of safe exit points represents the border of the safe region. Before reaching a safe exit point, the client (query object) does not have to request the server to re-evaluate the query This significantly reduces the server processing costs and the communication costs between the server and moving clients. Extensive experimental results show that SEA outperforms a conventional algorithm by up to two orders of magnitude in terms of communication costs and computation costs

Keywords: Continuous monitoring, nearest neighbor query, safe exit algorithm, road network

1. Introduction

The points of interest (POIs; e.g., accommodation, restaurants, and gas stations) marked on web mapping services such as Google Maps, Bing Maps, and Yahoo Maps are located in a road network and their proximity is measured as their shortest path distances [7,8,13,14]. A k nearest neighbor (kNN) query is one of the most important query types in location-based services [2,16], where a user submits a kNN query to the service provider for the k-nearest objects (i.e., POIs) to his/her current location. With the advances in map software, location detection devices, wireless communication, and database systems, the kNN query has been extended from the Euclidean space to the road network environment [2, 5,6,15,16,18,29], where the user can submit the kNN query to request his/her k-nearest objects on the basis of the network distance. In many cases, the kNN query result based on the network distance is more relevant to the user than that based on the Euclidean distance because the user's movement is typically restricted by an underlying road network.

Recently, the kNN query has been further extended via continuous monitoring. However, continuous monitoring algorithms for moving queries related to POIs in road networks have received little attention [26]. We can consider an example of a monitoring query, i.e., "find the three closest restaurants to my current location for the next 10 minutes". The main idea of continuous nearest neighbor monitoring

---

*Corresponding author: Hyung-Ju Cho, Department of Information & Computer Engineering, Ajou University Woncheon-dong, Suwon Si Yeongtong-gu, Gyeonggi-Do, 443-749, South Korea. E-mail: hjcho@ajou.ac.kr.

is to find the k-nearest objects of interest to a user's current location while he/she moves freely. This study assumes that the path is not known in advance and that the user moves arbitrarily in road networks.

The key problem with continuous monitoring algorithms is maintaining the freshness of the query answer when the query point moves freely and arbitrarily. A simple approach involves the client q (i.e., query point) requesting the server to re-evaluate the query periodically (e.g., every second). However, such a periodic monitoring approach cannot solve the problem because the query answer may still become stale in between each call to the server. This approach also places an excessive computation burden on the server side as well as the high communication frequency burden imposed on the communication channel.

To overcome the problems (i.e., excessive computation and communication costs) of periodic monitoring, safe region-based algorithms have been introduced [4,9,26,27]. The safe region of a query is the region where a query answer remains unchanged, provided that the query point is within the safe region. The safe region approach allows a client to get fresh query answers without excessive overheads on the server side or communication channel. However, the network bandwidth required to provide the client with a safe region (which may consist of complex road segments) is more than that required to provide a set of safe exit points [26] that represent the boundary of the corresponding safe region.

We address this issue by proposing a new Safe Exit Algorithm (SEA), which efficiently computes the safe exit points for moving NN queries in road networks. A safe exit point denotes a point where the safe region of q and its non-safe region meet. Safe exit points satisfy the following four constraints which are introduced in [1] for the design of the safe region in the Euclidean space: (1) lightweight construction, (2) compact representation, (3) fast containment check and (4) device heterogeneity. In particular, in road networks, the set of safe exit points is more concise than the safe region (which may consist of complex road segments); hence, it incurs a low communication cost between clients and the server. Until a client q reaches a safe exit point, he/she is guaranteed to remain in the safe region and thus the query answer is valid. Upon traveling beyond the safe exit point, the client requests the server to evaluate the query in order to refresh the query answer and its safe exit points.

The following are the distinguishing features of our study: (1) SEA focuses on the computation of safe exit points for moving NN queries in road networks, whereas Yung et al. [26] focus on the computation of safe exit points for moving range queries. (2) SEA does not require the path for the computation of safe exit points, whereas existing works (e.g. [6,18]) focus solely on the computation of split points for a given path. The split points indicate the locations in the path, at which the k NNs of a moving query object change. The contributions of this paper are summarized as follows:

- We propose a Safe Exit Algorithm (SEA) for the continuous monitoring of moving NN queries over static objects by assuming that the path is not known in advance and that query points move arbitrarily in road networks
- We present both lemmas and mathematical analyses that make it possible to effectively determine the safe exit points of a moving NN query.
- A thorough experimental study confirms that SEA clearly outperforms a traditional approach that evaluates queries periodically, in terms of both communication and computation costs.

The remainder of the paper is structured as follows: Section 2 reviews existing work on spatial queries in road networks. Section 3 presents the specific background and formulates the preliminaries of the problem. Section 4 elaborates on the proposed SEA for computing the safe exit points of moving NN queries in road networks. We present the performance analysis in Section 5, followed by the conclusion in Section 6.

2. Related work

Significant research attention has been given to developing techniques for spatial queries in road networks. NN queries [5,15,16,20,28–30] and range queries [4,11,16,19,20,22–26] are among the most studied spatial queries in road networks.

The related research can be classified into the following four categories, according to the mobility of the queries and the objects: (1) static queries for static objects, (2) static queries for moving objects, (3) moving queries for static objects, and (4) moving queries for moving objects. Section 2.1 reviews previous studies that have dealt with static queries for static objects or moving objects. Sections 2.2 and 2.3 review previous studies of moving queries for static objects and moving queries for moving objects, respectively.

2.1. Static queries for static/moving objects

Papadias et al. [16] proposed a framework to support nearest neighbor queries, closest pairs queries, range queries and distance joins in a road network. However, they assumed that queries and objects have fixed positions in a spatial network. Wang et al. [20] proposed an infrastructure known as MOVNET that utilizes dual index structures for location-based services with moving objects in road networks. Based on the infrastructure, they presented two algorithms for processing snapshot range queries and kNN queries. Pesti et al. [17] proposed a road network-based query-aware location update framework known as RoadTrack to reduce the communication costs of moving objects. RoadTrack partitions the road network into precincts and identifies the relevant range queries for each precinct. A moving object reports its latest location to the server only when it enters/leaves a precinct or the range of some query.

2.2. Moving queries for static objects

Our work belongs to this category, in which queries move freely while data objects are static. Chen et al. [5] studied path k-NN queries that return kNNs with respect to the shortest path that connects the destination to the user's current location. Bao et al. [2] introduced a new type of query known as a k-range nearest neighbor (kRNN) query to find the k closest objects for every point on road segments inside a given query region on the basis of the network distance. Cheema et al. [4] proposed a safe region-based approach to the continuous monitoring of range queries for static objects in Euclidean space and in road networks. They devised pruning rules and a unique access order in order to efficiently compute the safe region. Xuan et al. proposed several Voronoi-based algorithms for continuous range queries [23] and continuous kNN queries [29] in road networks. Wang et al. extended the functionality of MOVNET [20] to support continuous range query processing [21]. Yung et al. [26] proposed an algorithm to compute the safe exit points of a moving range query for static objects in road networks. Safar et al. [18] and Cho et al. [6] proposed eDAR algorithm and UNICONS algorithm, respectively, both of which effectively compute split points within a given path in order to support continuous NN queries in road networks.

SEA is characterized by the following features First, unlike the previous work [11], SEA makes very few assumptions on moving clients. These assumptions are as follows: (1) No computational capabilities, (2) no storage capabilities, and (3) no velocity assumptions. Second, existing algorithms (e.g. [6,18]) compute split points for a given path, whereas SEA computes safe exit points for a given query point. In other words, SEA considers a more practical situation where the path is not known apriori as well as the moving speed and direction of the client are arbitrary.

Finally, as in the case of most continuous monitoring algorithms [4,15,26], this study assumes (1) a client server model and (2) main-memory query evaluation on the server side. In other words, the client (i.e., a moving query) can communicate with the server via a wireless communication infrastructure (e.g., cellular services and Wi-Fi) and the server's main memory (e.g., 4 GB memory) is, in general, sufficiently large to accommodate the entire dataset.

2.3. Moving queries for moving objects

Mouratidis et al. [15] addressed the issue of processing continuous k-NN queries by proposing two algorithms (i.e., IMA and GMA) that handle arbitrary object and query movement patterns on road networks. Liu et al. [11] developed a distributed processing technique to solve the moving range queries for moving objects. Their strategy leverages the computational power of the moving objects and each moving object reports to the server when it affects the results of one or more queries. Kriegel et al. [10] studied the problem of proximity monitoring in road networks. Given a proximity threshold and a set of moving objects, a server responsible for proximity monitoring continuously reports the pairs of objects that are within a specific distance of each other. Stojanovic et al. [19] proposed a technique for the continuous monitoring of range queries for moving objects where the mobility pattern (e.g., speed, direction, and route) of the moving objects is utilized. However, the assumption of being able to predict the trajectories of moving objects is not always acceptable. This approach cannot be used if the prediction of an object's movements fails (e.g., a shopper randomly strolling in a shopping mall).

3. Preliminaries

Section 3.1 defines terms and notations used in the paper, while Section 3.2 formulates our continuous monitoring problem using an example of a road network.

3.1. Definition of terms and notations

Road network A road network is represented by a weighted undirected graph $G = (NEW)$ where $N = \{n_1, n_2, \ldots, n_{|N|}\}$ is a set of nodes, $E = \{e_1, e_2, \ldots, e_{|E|}\}$ is a set of edges (i.e., road segments) each of which connects two distinct nodes, and $W(e)$ is a function that returns a weight for an edge e in G. Each edge is given the length of its corresponding road segment as a weight. Note that $|S|$ denotes the cardinality of a set S.

Sequence A sequence $SQ_{(n_s, n_{s+1}, \ldots, n_e)}$ is a path between two nodes, n_s and n_e, such that the degrees of n_s and n_e are not equal to 2 and all intermediate nodes $n_{s+1}, \ldots, n_{e-1}$ on the path have a degree of 2. In particular, the sequence where a query q remains is called the *active sequence* of q. The two end nodes, n_s and n_e, of the sequence are referred to as the boundary nodes. The sequence length is the total weight of the edges in the sequence. A boundary node with a smaller node id is referred to as the base node of the sequence. This study assumes that $n_s \leqslant n_e$. Hence, n_s is the base node.

To simplify the presentation, Table 1 summarizes the notations used in this paper.

3.2. Problem formulation

To provide a clear explanation, we use the example road network shown in Fig. 1 where there are five POIs (i.e., a, b, c, d, and e) and a NN query q that requests three NNs. The number on each edge

Table 1
Notations used in the paper

Notation	Definition
$G = (N, E, W)$	The graph model of road network
$A_q = \{c, d, e\}$	The length of the shortest path between objects a and b
n_i	A node in the road network
$e_i = (n_j, n_k)$	An edge in the edge set E
$W(e_i) = d(n_j, n_k)$	The weight of the edge $e_i = (n_j, n_k)$
$SQ_{(n_s, n_{s+1}, \dots, n_e)}$	A sequence where n_s (or n_e) is the start (or end) boundary node and the others $A_q = \{c, d, e\}$ are intermediate nodes. Note that $O_{(n_2, q)} = \{c\}$, $SQ_{(n_2, n_5)}$, and $SQ_{(n_2, n_5)}$ where $A_{n_5} = \{c, d, e\}$ indicates the degree of a node n_i, which is the number of edges incident to n_i
q	A query point in the road network
k	The number of requested NNs
Ap	The set of k NNs at a point p
$O_{(n_s, n_{s+1}, \dots, n_e)}$	The set of objects on the sequence $SQ_{(n_5, n_6)}$
p_{se}	A safe exit point at which the safe region of q and its non-safe region meet
p_{anchor}	Anchor point which becomes the start point of expansion in the sequence such that $A_{n_6} = \{c, d, e\}$
$o^-_{nearest}$	The nearest non-answer object to a point $p \in G$ such that $d(p, o^-_{nearest}) = MIN(d(p, o^-_1), d(p, o^-_2), \dots, d(p, o^-_{|O-|}))$ where $O^- = \{o^-_1, o^-_2, \dots, o^-_{|O-|}\}$ indicates the set of non-answer objects
$o^+_{farthest}$	The farthest answer object to a point $p \in G$ such that $d(p, o^+_{farthest}) = MAX(d(p, o^+_1), d(p, o^+_2), \dots, d(p, o^+_{|O+|}))$ where $O^+ = \{o^+_1, o^+_2, \dots, o^+_{|O+|}\}$ indicates the set of answer objects

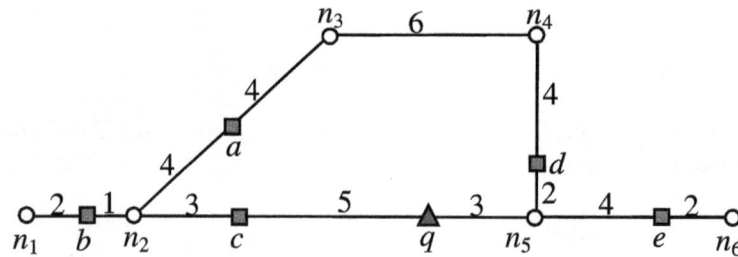

Fig. 1. Example of a road network and a 3-NN query q.

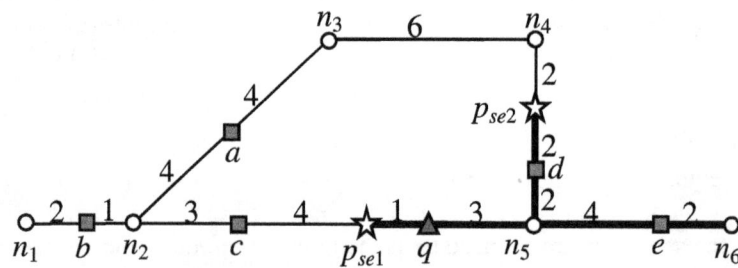

Fig. 2. Safe region of q and its safe exit points p_{se1} and p_{se2}.

indicates the network distance between two adjacent objects. For instance, $d(n_1, b) = 2$, $d(a, n_2) = 4$, etc. For simplicity, we assume that there is only one query q in the system. The active sequence of q is $SQ_{(n_2, n_5)}$ and its length is 11.

Figure 2 shows the safe region of q and its safe exit points of the example in Fig. 1. Note that the example has four sequences, i.e., $SQ_{(n_1, n_2)}$, $SQ_{(n_2, n_5)}$, $SQ_{(n_2, n_3, n_4, n_5)}$, and $SQ_{(n_5, n_6)}$. The query answer A_q is $A_q = \{c, d, e\}$ since $d(q, a) = 12$, $d(q, b) = 9$, $d(q, c) = 5$, $d(q, d) = 5$, and $d(q, e) = 7$. The safe region and safe exit points are labeled with bold lines and five-pointed stars, respectively. The safe region is the union of the sub-safe region for each sequence. The sub-safe region of a sequence can be expressed as a sequence id and the segment within the sequence. Each end point of the segment can be represented as the distance from the base node to the point.

In the example road network, the safe region (i.e., $q.SR$) of q can be encoded as follows: $q.SR = \{(SQ_{(n_2,n_5)}, [7, 11]), (SQ_{(n_2,n_3,n_4,n_5)}, [16, 20]), (SQ_{(n_5,n_6)}, [0, 6])\}$. Similarly, the set of safe exit points (i.e., P_{SE}) of q can be encoded as follows: $P_{SE} = \{(SQ_{(n_2,n_5)}, 7), (SQ_{(n_2,n_3,n_4,n_5)}, 16)\}$ where $(SQ_{(n_2,n_5)}, 7)$ and $(SQ_{(n_2,n_3,n_4,n_5)}, 16)$ correspond to safe exit points p_{se1} and p_{se2}, respectively. As shown in Fig. 2, until q reaches either p_{se1} or p_{se2}, the three NNs of q are $\{c, d, e\}$. When q passes through either p_{se1} or p_{se2}, the query is re-evaluated based on the updated location of q in order to refresh the query answer and safe exit points.

4. Safe exit algorithm for moving NN query

Section 4.1 elaborates on SEA, which determines the safe exit points of a moving NN query in road networks. Section 4.2 then discusses the computation of the safe exit points of a query in the example road network.

4.1. Safe exit algorithm

First, we formally define a set of safe exit points for a moving NN query in the road network. Let P_{SE} be the set of safe exit points for a k-NN query point q and $O = \{o_1, o_2, \ldots, o_{|O|}\}$ be the set of objects of interest to q. Assume that the answer set (i.e., A_q^+) of q and its non-answer set (i.e., $A_q^- = O - A_q^+$) are $A_q^+ = \{o_1^+, o_2^+, \ldots, o_k^+\}$ and $A_q^- = \{o_{k+1}^-, o_{k+2}^-, \ldots, o_{|O|}^-\}$, respectively. Then, it holds that $d(q, o^+) \leqslant d(q, o^-)$ for an answer object $o^+ \in A_q^+$ and a non-answer object $o^- \in A_q^-$. In addition, $A_q^+ \cap A_q^- = \phi$ and $A_q^+ \cup A_q^- = O$. Finally, P_{SE} is defined as follows:

$$P_{SE} = \{p_{se} \in G | MAX(d(p_{se}, o_1^+), d(p_{se}, o_2^+), \ldots, d(p_{se}, o_k^+))$$
$$= MIN(d(p_{se}, o_{k+1}^-), d(p_{se}, o_{k+2}^-), \ldots, d(p_{se}, o_{|O|}^-))\}$$

where $MIN()$ and $MAX()$ return the minimum and maximum values of the input array, respectively. In other words, a safe exit point p_{se} is the midpoint (i.e., $MAX(d(p_{se}, o_1^+), \ldots, d(p_{se}, o_k^+)) = MIN(d(p_{se}, o_{k+1}^-), \ldots, d(p_{se}, o_{|O|}^-)))$ between the farthest answer object and the nearest non-answer object.

Algorithm 1 depicts the skeleton for SEA, which identifies the safe exit points for a moving NN query. SEA begins with the exploration of the active sequence The traversal of sequences is terminated if no more sequences can be explored in the queue Each entry in the queue takes the form (*sequence, anchor point*) where the anchor point corresponds to a point in the sequence. More precisely, if the sequence to be explored is the active sequence, a query point q becomes the anchor point. Otherwise, either of end nodes (i.e., n_s and n_e) of the sequence becomes the anchor point. Thus, $p_{anchor} \in \{n_s, n_e, q\}$ for a sequence $SQ_{(n_s, n_{s+1}, \ldots, n_e)}$. Unless there is a safe exit point in the segment between n_s (or n_e) and the anchor point, the adjacent sequences of n_s (or n_e) are explored.

Finally, the evaluate_query(p, k) function retrieves k NNs from a given point p in the graph. According to our performance evaluation of SEA, the evaluate_query function typically accounts for most of the CPU

Algorithm 1: find_safe_exit_points (q,k)

Input: q: query point, k: the number of requested NNs
Output: A_q: a set of k NNs of q, P_{SE}: a set of safe exit points of q

1:      $queue \leftarrow \phi$ /* $queue$ is a FIFO queue */
2:      $P_{SE} \leftarrow \phi$ /* P_{SE} is initialized to the empty set */
3:      $A_q \leftarrow$ evaluate_query (q,k) /* A_q is the set of k NNs of q */
4:      $queue$.push(seq_{active},q)   /* seq_{active} is the active sequence of q*/
5:      **while** $queue$ is not empty **do**
6:          $(seq, p_{anchor}) \leftarrow queue$.pop()
7:          **if** seq has not been explored before **then**
8:              Mark seq as explored
9:              $P_{SE\_seq} \leftarrow$ find_safe_exit_pt_in_sequence (seq, p_{anchor})
10:            $P_{SE} \leftarrow P_{SE} \cup P_{SE\_seq}$
11:            **if** there is no safe exit point in $[n_s, p_{anchor}]$ **then**
12:              $queue$.push(each adjacent sequence of n_s, n_s)
13:            **if** there is no safe exit point in $[p_{anchor}, n_e]$ **then**
14:              $queue$.push(each adjacent sequence of n_e, n_e)
15:      **return** (A_q, P_{SE}) /* query answer A_q and safe exit points P_{SE} are returned to the query issuer */

---

time particularly when the object density is very low. Therefore, to avoid redundant query evaluation, we use the shared execution paradigm [2,12]. In other words, the query condition (i.e., query location and the number of requested NNs) and its query answer (i.e., the set of NNs from the query location) are stored in the memory and re-cycled.

Algorithm 2: find_safe_exit_pt_in_sequence $(SQ_{(n_s, n_{s+1}, \ldots, n_e)}, p_{anchor})$

Input: $SQ_{(n_s, n_{s+1}, \ldots, n_e)}$: sequence to be examined, p_{anchor}: anchor point
Output: P_{SE_seq}: a set of safe exit points in $SQ_{(n_s, n_{s+1}, \ldots, n_e)}$

1:      $P_{SE\_seq} \leftarrow \phi$ /* P_{SE_seq} is initialized to the empty set. */
2:      $A_{p_{anchor}} \leftarrow$ evaluate_query (p_{anchor}, k) /* Recall that $A_q = A_{p_{anchor}}$. */
3:      **if** $n_s \neq p_{anchor}$ **then** $A_{n_s} \leftarrow$ evaluate_query (n_s, k)
4:      **if** $n_e \neq p_{anchor}$ **then** $A_{n_e} \leftarrow$ evaluate_query (n_e, k)
5:      $O_{(n_s, p_{anchor})} \leftarrow$ scan_sequence(n_s, p_{anchor})/* Let $O_{(n_s, p_{anchor})}$ be the set of objects in segment between n_s and p_{anchor} */
6:      $O_{(p_{anchor}, n_e)} \leftarrow$ scan_sequence(p_{anchor}, n_e)/* Let $O_{(p_{anchor}, n_e)}$ be the set of objects in segment between p_{anchor} and n_e */
7:      **if** $n_s \neq p_{anchor}$ and $A_{n_s} \cup O_{(n_s, p_{anchor})} \neq A_{p_{anchor}}$ **then**   /* Refer to Lemma 3 */
8:          $P_{SE\_seq} \leftarrow P_{SE\_seq} \cup$ find_safe_exit_pt($A_{n_s}, A_{p_{anchor}}, O_{(n_s, p_{anchor})}$)
9:      **if** $n_e \neq p_{anchor}$ and $A_{n_e} \cup O_{(n_e, p_{anchor})} \neq A_{p_{anchor}}$ **then**   /* Refer to Lemma 3 */
10:        $P_{SE\_seq} \leftarrow P_{SE\_seq} \cup$ find_safe_exit_pt $(A_{n_e}, A_{p_{anchor}}, O_{(p_{anchor}, n_e)})$
11:      **return** P_{SE_seq}

Algorithm 2 retrieves the safe exit points in the sequence using A_{n_s}, A_{n_e}, and $O_{(n_s, n_{s+1}, \ldots, n_e)}$ Without loss of generality, lemmas 1 to 4 assume that $n_b \neq p_{anchor}$ where $n_b = \{n_s, n_e\}$ is a boundary node and $p_{anchor} = \{n_s, n_e, q\}$ is the anchor point. Naturally, if $n_b = p_{anchor}$, there is no safe exit point between n_b and p_{anchor} Lemmas 1 and 2 are introduced to prove Lemma 3.

Lemma 1. Let $A_{(n_s, n_{s+1}, \ldots, n_e)}$ be the union of a set of k NNs at every point in $SQ_{(n_s, n_{s+1}, \ldots, n_e)}$. Then, the following formula is satisfied.

$$A_{(n_s, n_{s+1}, \ldots, n_e)} = A_{n_s} \cup A_{n_e} \cup O_{(n_s, n_{s+1}, \ldots, n_e)}$$

Proof) Lemma 1 is well-known [2,6,18]; hence, the proof is omitted   □

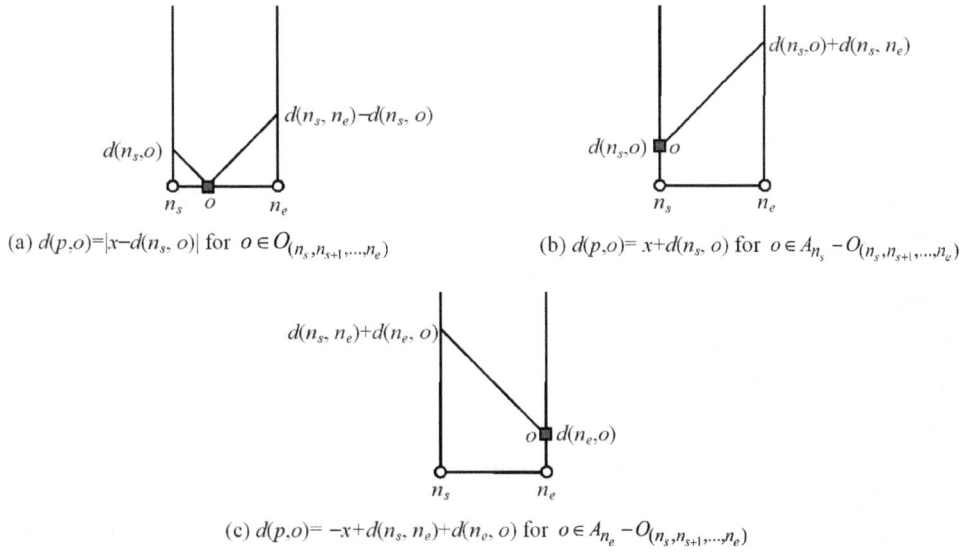

(a) $d(p,o)=|x-d(n_s,o)|$ for $o \in O_{(n_s,n_{s+1},...,n_e)}$

(b) $d(p,o)=x+d(n_s,o)$ for $o \in A_{n_s} - O_{(n_s,n_{s+1},...,n_e)}$

(c) $d(p,o)= -x+d(n_s,n_e)+d(n_e,o)$ for $o \in A_{n_e} - O_{(n_s,n_{s+1},...,n_e)}$

Fig. 3. Determination of $d(po)$ when $o \in O_{(n_s,n_{s+1},...,n_e)} \cup A_{n_s} \cup A_{n_e}$, $p \in SQ_{(n_s,n_{s+1},...,n_e)}$, and $x = d(n_s,p)$ (a) $d(po) = |x - d(n_s,o)|$ for $o \in O_{(n_s,n_{s+1},...,n_e)}$; (b) $d(po) = x + d(n_s,o)$ for $o \in A_{n_s} - O_{(n_s,n_{s+1},...,n_e)}$; (c) $d(po) = -x + d(n_s n_e) + d(n_e,o)$ for $o \in A_{n_e} - O_{(n_s,n_{s+1},...,n_e)}$.

Lemma 2. Let $[n_b, p_{anchor}]$ be the segment between n_b and p_{anchor}, and $A_{(n_b,p_{anchor})}$ be the union of a set of k NNs at every point in $[n_b, p_{anchor}]$. Then, the following formula is satisfied.

$$A_{(n_b,p_{anchor})} = A_{n_b} \cup A_{p_{anchor}} \cup O_{(n_b,p_{anchor})}$$

Proof) Lemma 2 is easily extended from Lemma 1 and the proof is omitted. □

Lemma 3. If $A_{n_b} \cup O_{(n_b,p_{anchor})} \neq A_{p_{anchor}}$, there is a safe exit point p_{se} in the segment.

Proof) This lemma can be proved using the contradiction method. To do so, let us assume that there is no safe exit point in the segment. This means that for each point $p \in [n_b, p_{anchor}]$, $A_p = A_{p_{anchor}}$, which results in $A_{(n_b,p_{anchor})} = A_{p_{anchor}}$. According to Lemma 2, $A_{(n_b,p_{anchor})} = A_{n_b} \cup A_{p_{anchor}} \cup O_{(n_b,p_{anchor})}$. This means that $A_{n_b} = A_{p_{anchor}}$, $O_{(n_b,p_{anchor})} \subseteq A_{p_{anchor}}$, and $A_{n_b} \cup O_{(n_b,p_{anchor})} = A_{p_{anchor}}$. However, this leads to a contradiction to the given condition (i.e., $A_{n_b} \cup O_{(n_b,p_{anchor})} \neq A_{p_{anchor}}$). Therefore, there is a safe exit point p_{se} in the segment. □

Lemma 4. If $A_{n_b} \cup O_{(n_b,p_{anchor})} = A_{p_{anchor}}$, there is no safe exit point in the segment.

Proof) According to Lemma 2, $A_{(n_b,p_{anchor})} = A_{n_b} \cup A_{p_{anchor}} \cup O_{(n_b,p_{anchor})}$. Because $A_{n_b} \cup O_{(n_b,p_{anchor})} = A_{p_{anchor}}$ by the given condition, $A_{(n_b,p_{anchor})} = A_{p_{anchor}}$. This leads to $A_p = A_{p_{anchor}}$ for every point $p \in [n_b, p_{anchor}]$. Therefore, there is no safe exit point in the segment. □

If a sequence satisfies Lemma 3, SEA starts to compute the location of a safe exit point as shown in lines 7 and 9 of Algorithm 2. For this purpose, we introduce $o^-_{nearest}$ and $o^+_{farthest}$. For simplicity, let us assume that $A_{n_b} \cup O_{(n_b,p_{anchor})} - A_{p_{anchor}}$ corresponds to $O^- = \{o^-_1, o^-_2, \ldots, o^-_{|O-|}\}$ and $A_{p_{anchor}}$ corresponds to $O^+ = \{o^+_1, o^+_2, \ldots, o^+_{|O+|}\}$. Then, at a point $p \in [n_b, p_{anchor}]$, $o^-_{nearest}$ is referred to as the nearest non-answer object to p such that $d(p, o^-_{nearest}) = MIN(d(p, o^-_1), d(p, o^-_2), \ldots, d(p, o^-_{|O-|}))$. Similarly, at a point $p \in [n_b, p_{anchor}]$, $o^+_{farthest}$ is referred to as the farthest answer object to p such

Table 2
Summary of $d(p,o)$ for $o \in O_{(n_s, n_{s+1}, ..., n_e)} \cup A_{n_s} \cup A_{n_e}$ and $x = d(n_s, p)$

Condition	$d(p,o)$
$o \in O_{(n_s, n_{s+1}, ..., n_e)} - (A_{n_s} \cup A_{n_e})$	$d(p,o) = |x - d(n_s, o)|$
$o \in A_{n_s} - (A_{n_e} \cup O_{(n_s, n_{s+1}, ..., n_e)})$	$d(p,o) = x + d(n_s, o)$
$o \in A_{n_e} - (A_{n_s} \cup O_{(n_s, n_{s+1}, ..., n_e)})$	$d(p,o) = -x + d(n_s, n_e) + d(n_e, o)$
$o \in A_{n_s} \cap A_{n_e} - O_{(n_s, n_{s+1}, ..., n_e)}$	$d(p,o) = MIN(x + d(n_s, o), -x + d(n_s, n_e) + d(n_e, o))$
$o \in O_{(n_s, n_{s+1}, ..., n_e)} \cap A_{n_s} - A_{n_e}$	$d(p,o) = MIN(|x - d(n_s, o)|, x + d(n_s, o))$
$o \in O_{(n_s, n_{s+1}, ..., n_e)} \cap A_{n_e} - A_{n_s}$	$d(p,o) = MIN(|x - d(n_s, o)|, -x + d(n_s, n_e) + d(n_e, o))$
$o \in O_{(n_s, n_{s+1}, ..., n_e)} \cap A_{n_s} \cap A_{n_e}$	$d(p,o) = MIN(|x - d(n_s, o)|, x + d(n_s, o), -x + d(n_s, n_e) + d(n_e, o))$

that $d(p, o^+_{farthest}) = MAX(d(p, o^+_1), d(p, o^+_2), ..., d(p, o^+_{|O+|}))$. The midpoint between $o^-_{nearest}$ and $o^+_{farthest}$ becomes a safe exit point p_{se}. That is, $d(p_{se}, o^-_{nearest}) = d(p_{se}, o^+_{farthest})$. Next, we elaborate on how we determine $d(po)$ where $o \in A_{n_s} \cup O_{(n_s, n_{s+1}, ..., n_e)} \cup A_e$ and $p \in SQ_{(n_s, n_{s+1}, ..., n_e)}$.

Figure 3 shows the determination of $d(p,o)$ for an object $o \in O_{(n_s, n_{s+1}, ..., n_e)} \cup A_{n_s} \cup A_{n_e}$ and a point $p \in SQ_{(n_s, n_{s+1}, ..., n_e)}$ according to the location of object o (i.e., $o \in O_{(n_s, n_{s+1}, ..., n_e)}$, $o \in A_{n_s} - O_{(n_s, n_{s+1}, ..., n_e)}$ and $o \in A_{n_e} - O_{(n_s, n_{s+1}, ..., n_e)}$). In the figure, the x-axis represents $d(n_s, p)$, while the y-axis represents $d(p,o)$ We note that $d(p,o)$ can be represented as a function of $x = d(n_s, p)$ for $0 \leqslant x \leqslant d(n_s, n_e)$ As shown in Fig. 3(a), if $o \in O_{(n_s, n_{s+1}, ..., n_e)}$ $d(p,o) = |x - d(n_s, o)|$ As shown in Fig. 3(b), if $o \in A_{n_s} - O_{(n_s, n_{s+1}, ..., n_e)}$, $d(p,o) = x + d(n_s, o)$. Finally, as shown in Fig. 3(c), if $o \in A_{n_e} - O_{(n_s, n_{s+1}, ..., n_e)}$, $d(p,o) = -x + d(n_s, n_e) + d(n_e, o)$

Table 2 summarizes the determination of the $d(p,o)$ value for $p \in SQ_{(n_s, n_{s+1}, ..., n_e)}$ and $o \in O_{(n_s, n_{s+1}, ..., n_e)} \cup A_{n_s} \cup A_{n_e}$ If an object belongs to more than two sets (i.e., $o \in A_{n_s} \cap A_{n_e} - O_{(n_s, n_{s+1}, ..., n_e)}$, $o \in O_{(n_s, n_{s+1}, ..., n_e)} \cap A_{n_s} - A_{n_e}$, $o \in O_{(n_s, n_{s+1}, ..., n_e)} \cap A_{n_e} - A_{n_s}$, and $o \in O_{(n_s, n_{s+1}, ..., n_e)} \cap A_{n_s} \cap A_{n_e}$), the distance is the length of the shortest path connecting two objects. For example, given that $o \in A_{n_s} \cap A_{n_e} - O_{(n_s, n_{s+1}, ..., n_e)}$, $d(n_s, o) = 3$, $d(n_e, o) = 5$, $d(n_s, n_e) = 20$, and $d(n_s, p) = 6$ $d(p,o) = MIN(x + d(n_s, o), -x + d(n_s, n_e) + d(n_e, o)) = MIN(9, 19) = 9$ where $x = d(n_s, p) = 6$

Algorithm 3: find_safe_exit_pt($A_{n_b}, A_{p_{anchor}}, O_{(n_b, p_{anchor})}$)

Input: Arguments have the same meanings as before
Output: p_{se}: a safe exit point in the segment $[n_b, p_{anchor}]$

1:   $O^-_{nearest} \leftarrow \{\langle p, o^-_{nearest} \rangle |$ for each point $p \in [n_b, p_{anchor}]$, $o^-_{nearest}$ such that $d(p, o^-_{nearest}) = $ MIN($d(p, o^-_1), ..., d(p, o^-_{|O-|})$)}

2:   $O^+_{farthest} \leftarrow \{\langle p, o^+_{farthest} \rangle |$ for each point $p \in [n_b, p_{anchor}]$, $o^+_{farthest}$ such that $d(p, o^+_{farthest}) = $ MAX($d(p, o^+_1), ..., d(p, o^+_{|O+|})$)}

3:   /* Note that p_{se} is the midpoint between $o^-_{nearest}$ and $o^+_{farthest}$ */

4:   Find the closest point p_{se} to p_{anchor}, such that $d(p_{se}, o^-_{nearest}) = d(p_{se}, o^+_{farthest})$ for $O^-_{nearest}$ and $O^+_{farthest}$

5:   /* If more than two points satisfy $d(p_{se}, o^-_{nearest}) = d(p_{se}, o^+_{farthest})$, the closest point p_{se} to p_{anchor} is chosen.*/

6:   **return** p_{se}

Algorithm 3 finds a safe exit point that is located between n_b and p_{anchor}. Let $O^-_{nearest}$ be the set of objects, each of which becomes $o^-_{nearest}$ at a point $p \in [n_b, p_{anchor}]$ and $O^+_{farthest}$ be the set of objects, each of which becomes $o^+_{farthest}$ at the point p. First, we determine $O^-_{nearest}$ and $O^+_{farthest}$. Recall that $o^-_{nearest}$ is the nearest non-answer object to p, whereas $o^+_{farthest}$ is the farthest answer object to p. Since the safe exit point p_{se} is the midpoint between $o^-_{nearest}$ and $o^+_{farthest}$, p_{se} is the solution to the following

Table 3
Computation of the safe exit points for the example road network

Sequence (or segment)	p_{anchor}	A_{n_b}	$A_{p_{anchor}}$	$O_{(n_b, p_{anchor})}$	Safe exit point
$[n_2, q]$ in $SQ_{(n_2, n_5)}$	q	$A_{n_2} = \{a, b, c\}$	$A_q = \{c, d, e\}$	$O_{(n_2, q)} = \{c\}$	$p_{se1} = (SQ_{(n_2, n_5)}, 7)$
$[q, n_5]$ in $SQ_{(n_2, n_5)}$	q	$A_{n_5} = \{c, d, e\}$	$A_q = \{c, d, e\}$	$O_{(q, n_5)} = \phi$	none
$SQ_{(n_5, n_6)}$	n_5	$A_{n_6} = \{c, d, e\}$	$A_{n_5} = \{c, d, e\}$	$O_{(n_5, n_6)} = \{e\}$	none
$SQ_{(n_2, n_3, n_4, n_5)}$	n_5	$A_{n_2} = \{a, b, c\}$	$A_{n_5} = \{c, d, e\}$	$O_{(n_2, n_3, n_4, n_5)} = \{a\}$	$p_{se2} = (SQ_{(n_2, n_3, n_4, n_5)}, 16)$

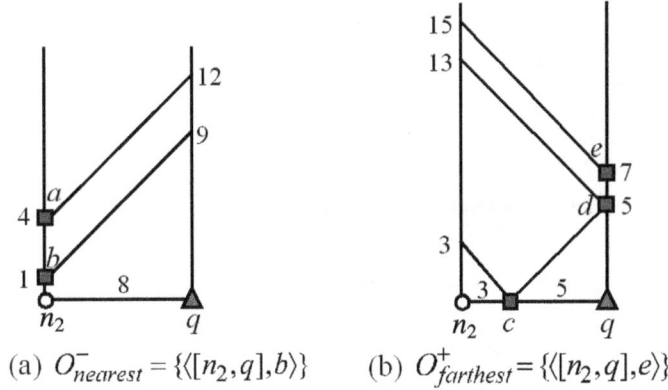

(a) $O_{nearest}^{-} = \{\langle [n_2, q], b \rangle\}$ (b) $O_{farthest}^{+} = \{\langle [n_2, q], e \rangle\}$

Fig. 4. Determination of $O_{nearest}^{-}$ and $O_{farthest}^{+}$ for $(SQ_{(n_2, n_5)}, [n_2, q])$. (a) $O_{nearest}^{-} = \{\langle [n_2, q], b \rangle\}$; (b) $O_{farthest}^{+} = \{\langle [n_2, q], e \rangle\}$.

equation: $d(p_{se}, o_{nearest}^{-}) = d(p_{se}, o_{farthest}^{+})$ If there are found more than two points in $[n_b, p_{anchor}]$, each of which satisfies $d(p_{se}, o_{nearest}^{-}) = d(p_{se}, o_{farthest}^{+})$, the closest point to the anchor point is chosen as the safe exit point.

4.2. Computation of the safe exit points for the example

We now discuss the computation of the safe exit points for the query q in the example road network shown in Fig. 1. Table 3 summarizes the computation of the safe exit points of query q for the example road network in Fig. 1. Note that the number (i.e., k) of NNs requested by q is $k = 3$ and the query answer is $A_q = \{c, d, e\}$

As shown in Algorithm 1, SEA first explores the active sequence $SQ_{(n_2, n_5)}$ where q remains. Since $SQ_{(n_2, n_5)}$ is the active sequence, the location of q is the anchor point. Each of the two segments $[n_2, q]$ and $[q, n_5]$ within $SQ_{(n_2, n_5)}$ is explored individually. For $[n_2, q]$, $p_{anchor} = q$, $A_q = \{c, d, e\}$, $A_{n_2} = \{a, b, c\}$, and $O_{(n_2, q)} = \{c\}$. By Lemma 3 (i.e., $A_{n_2} \cup O_{(n_2, q)} \neq A_q$), there exists a safe exit point in the segment $[n_2, q]$. For each point $p \in [n_2, q]$, $o_{nearest}^{-}$ is selected from the non-answer objects in $A_{n_2} \cup O_{(n_2, q)} - A_q = \{a, b\}$ while $o_{farthest}^{+}$ is selected from the answer objects in $A_q = \{c, d, e\}$. As shown in Fig. 4(a), $O_{nearest}^{-} = \{\langle [n_2, q], b \rangle\}$ because $d(p, b) \leqslant d(p, a)$ for every point $p \in [n_2, q]$. Similarly, as shown in Fig. 4(b), $O_{farthest}^{+} = \{\langle [n_2, q], e \rangle\}$ because $d(p, c) \leqslant d(p, d) \leqslant d(p, e)$ for every point $p \in [n_2, q]$

For $O_{nearest}^{-} = \{\langle [n_2, q], b \rangle\}$ and $O_{farthest}^{+} = \{\langle [n_2, q], e \rangle\}$, the midpoint (i.e., $d(o_{nearest}^{-}, p) = d(o_{farthest}^{+}, p)$) between $o_{nearest}^{-}$ and $o_{farthest}^{+}$ at a specific point p becomes a safe exit point. Thus, the safe exit point p_{se1} is determined as shown in Fig. 5. That is, $d(p_{se1}, b) = d(p_{se1}, e)$ where

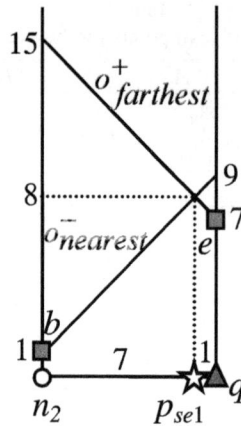

Fig. 5. Determination of location of safe exit point p_{se1}.

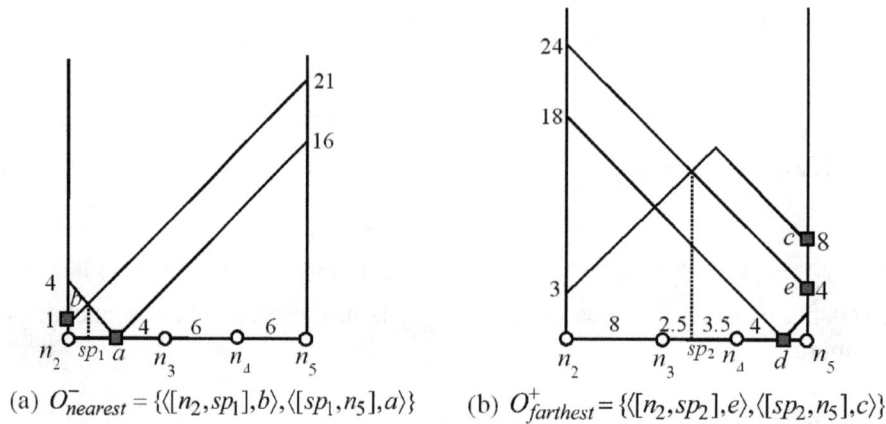

(a) $O_{nearest}^- = \{\langle [n_2, sp_1], b\rangle, \langle [sp_1, n_5], a\rangle\}$ (b) $O_{farthest}^+ = \{\langle [n_2, sp_2], e\rangle, \langle [sp_2, n_5], c\rangle\}$

Fig. 6. Determination of $O_{nearest}^-$ and $O_{farthest}^+$ for $SQ_{(n_2, n_3, n_4, n_5)}$. (a) $O_{nearest}^- = \{\langle [n_2, sp_1], b\rangle, \langle [sp_1, n_5], a\rangle\}$; (b) $O_{farthest}^+ = \{\langle [n_2, sp_2], e\rangle, \langle [sp_2, n_5], c\rangle\}$.

$d(p_{se1}, b) = x + 1$ and $d(p_{se1}, e) = -x + 15$ for $0 \leqslant x \leqslant 8$. Consequently, $x = 7$. This means that the distance from n_2 to p_{se1} is 7.

According to Lemma 4, there is no safe exit point within the segment $[q, n_5]$ because $A_{n_5} \cup O_{(q, n_5)} = A_q$, as shown in Table 3. Therefore, sequences (i.e., $SQ_{(n_2, n_3, n_4, n_5)}$ and $SQ_{(n_5, n_6)}$) adjacent to n_5 are explored using $p_{anchor} = n_5$. According to Lemma 4, there is also no safe exit point in $SQ_{(n_5, n_6)}$ because $A_{n_6} \cup O_{(n_5, n_6)} = A_{n_5}$, as shown in Table 3.

Finally, we determine a safe exit point in the sequence $SQ_{(n_2, n_3, n_4, n_5)}$. By Lemma 3 (i.e., $A_{n_2} \cup O_{(n_2, n_3, n_4, n_5)} \neq A_{n_5}$ as shown in Table 3), a safe exit point exists in the sequence. For each point $p \in SQ_{(n_2, n_3, n_4, n_5)}$, $o_{nearest}^-$ is selected from the non-answer objects in $A_{n_2} \cup O_{(n_2, n_3, n_4, n_5)} - A_{n_5} = \{a, b\}$ while $o_{farthest}^+$ is selected from the answer objects in $A_{n_5} = \{c, d, e\}$. As shown in Fig. 6(a), $O_{nearest}^- = \{\langle [n_2, sp_1], b\rangle, \langle [sp_1, n_5], a\rangle\}$ where a split point $sp_1 = (SQ_{(n_2, n_3, n_4, n_5)}, 1.5)$ is the cross point between $d(p, a) = |x - 4|$ and $d(p, b) = x + 1$ Similarly, as shown in Fig. 6(b) $O_{farthest}^+ = \{\langle [n_2, sp_2], e\rangle, \langle [sp_2, n_5], c\rangle\}$ where a split point $sp_2 = (SQ_{(n_2, n_3, n_4, n_5)}, 10.5)$ is the cross point between $d(p, c) = MIN(x + 3, -x + 28)$ and $d(p, e) = -x + 24$ Recall that $p_{anchor} = n_5$ for $SQ_{(n_2, n_3, n_4, n_5)}$

Table 4
Experimental parameter settings

Parameter	Default value	Range
Number of POIs (N_{POI})	50k	1, 5, 50, 70, 100 (k)
Number of queries (N_{qry})	5k	1, 3, 5, 7, 10 (k)
Number of requested NNs (k)	32	8, 16, 32, 64, 128
Distribution of POIs (D_{POI})	uniform	(U)niform,(S)kewed
Distribution of queries (D_{qry})	skewed	(U)niform,(S)kewed
Query speed (V_{qry})	60 km/h	20, 40, 60, 80, 100 (km/h)

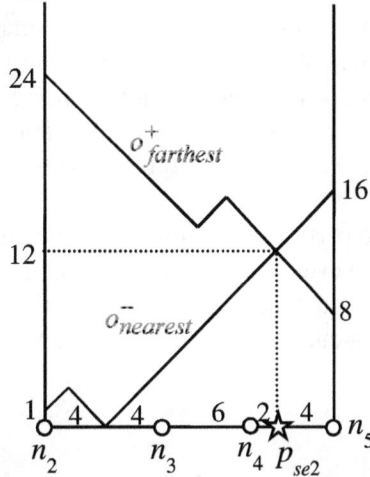

Fig. 7. Determination of location of safe exit point p_{se2}.

and that object c belongs to $A_{n_2} \cap A_{n_5} - O_{(n_2,n_3,n_4,n_5)}$ Hence, for $c \in A_{n_2}$, $d(p,c) = x + 3$, whereas for $c \in A_{n_5}$, $d(p,c) = -x + 28$ Thus, $d(p,c) = MIN(x + 3, -x + 28)$ for $p \in SQ_{(n_2,n_3,n_4,n_5)}$.

The safe exit point p_{se2} is determined as shown in Fig. 7. That is, $d(p_{se2}, a) = d(p_{se2}, c)$ where $d(p_{se2}, a) = |x - 4|$ and $d(p_{se2}, c) = -x + 28$. Consequently, $x = 16$. This means that the distance from n_2 to p_{se2} is 16. Note that object e (i.e., $d(p, e) = -x + 24$) is $o^+_{farthest}$ for $0 \leqslant x \leqslant 10.5$, whereas object c (i.e., $d(p, c) = MIN(x + 3, -x + 28)$) is $o^+_{farthest}$ for $10.5 \leqslant x \leqslant 20$. Similarly, object b (i.e., $d(p, b) = x + 1$) is $o^-_{nearest}$ for $0 \leqslant x \leqslant 1.5$ whereas object a (i.e., $d(p, a) = |x - 4|$) is $o^-_{nearest}$ for $1.5 \leqslant x \leqslant 20$.

5. Performance evaluation

In this section, we describe the performance evaluation of our proposed algorithm, SEA, by comparing it with a conventional method that evaluates each query at every timestamp. Section 5.1 describes our experimental settings while Section 5.2 presents our experimental results.

5.1. Experimental settings

In the experiments, we use a real-life road map that consists of the main roads of North America (175,813 nodes and 179179 edges) obtained from [31]. Note that the road map has been used for performance evaluation in many papers (e.g. [4,26]). Table 4 summarizes the parameters that are investigated. In each experiment, we vary a single parameter in the range shown while fixing all other parameters to the default values shown in Table 4. A moving client performs a random walk in the network and covers a fixed distance of V_{qry}, where V_{qry} is the query speed. Whenever a moving client reaches a node, one of its adjacent nodes is selected randomly as a destination and the client continues traveling. We simulate the movement of each client using the network-based moving objects generator [3].

In the performance study, we evaluate the performance of our method using the following measures: (1) the total communication cost as the total number of points (including both POIs in the query answers and safe exit points) sent by the server per timestamp, (2) the total communication frequency as the number of messages sent from the client to the server per timestamp, and (3) the total server CPU time per timestamp. Battery power and bandwidth consumption of mobile devices typically increase with the size of data transferred between the server and clients [2,26]. Thus, the size of the transferred data is used as a metric for the communication cost. Note that the CPU time at each client is negligible. Hence, we measure only CPU time at the server.

For comparison, we also report the performance of the periodic querying approach (PERIOD), which issues a new NN query to the server at every timestamp. All queries are continuously monitored for 100 timestamps. Note that the figures show the measured values per timestamp. We implement all solutions in C++ and all the experiments are conducted on a Windows machine with a Pentium 2.8 GHz CPU and 4 GB memory.

5.2. Experimental results

Figure 8 shows the comparison between SEA and PERIOD in terms of the CPU time at the server. The numbers shown in the parentheses in Figs 8 and 9 indicate the total communication frequencies, which are equal to the numbers of evaluated queries per timestamp. Note that the vertical axis in every chart in Fig. 8 is a log scale of the CPU time.

Figure 8(a) shows the effect of POI cardinality N_{POI} on the query processing time of SEA and PERIOD. Note that $N_{POI} = 1$ k corresponds to a low density of POIs while $N_{POI} = 100$ k corresponds to a high density. The query processing time of SEA is several orders of magnitude less than that of PERIOD. This is expected because about 1% of all moving clients in SEA ask the server to renew the query answers and safe exit points, whereas every client in PERIOD requests the server to refresh the query answer periodically. An interesting observation is that as opposed to PERIOD, SEA does not suffer from performance degradation when the density of POIs is very low (i.e., $N_{POI} = 1$ k). It benefits from the shared execution of multiple queries and a small number of clients that require the updated answers.

Figure 8(b) shows the performance of SEA and PERIOD as a function of query cardinality N_{qry} between 1 k and 10 k. For both SEA and PERIOD, the query processing time increases with the N_{qry} value. SEA achieves a much lower CPU time than PERIOD because most of the clients in SEA remain in their safe region, hence, they do not request query evaluations from the server. Figure 8(c) shows the query processing times of SEA and PERIOD as a function of the number k of NNs requested by clients. SEA is superior to PERIOD in all cases.

Figure 8(d) shows the effect of query speed V_{qry} on the query processing time of SEA and PERIOD. PERIOD incurs a constant processing cost because it issues queries to the server periodically, regardless of the client moving speed. However, PERIOD cannot guarantee the correctness of the result at any time;

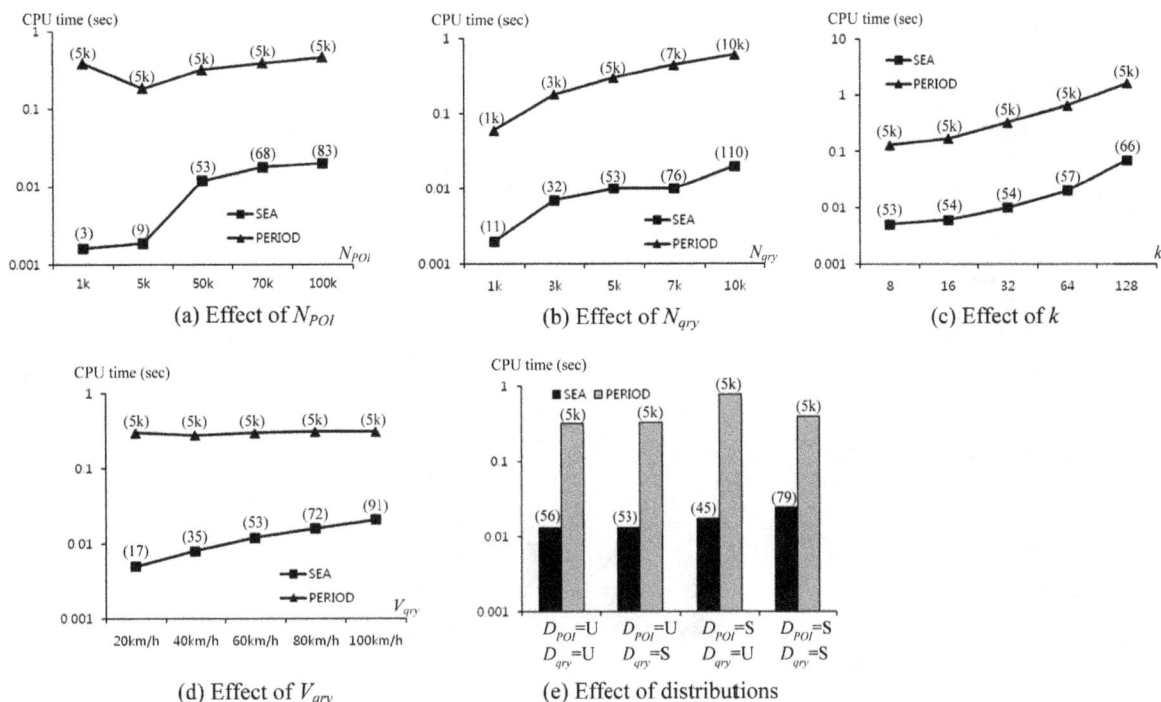

(a) Effect of N_{POI}

(b) Effect of N_{qry}

(c) Effect of k

(d) Effect of V_{qry}

(e) Effect of distributions

Fig. 8. SEA vs. PERIOD in terms of query processing costs.

hence, it suffers from stale query answers when the clients move at a high speed (e.g., 100 km/h). In SEA, the client reaches a safe exit point sooner when it moves faster and the communication frequency increases.

In all the previous experiments, the initial positions of the POIs follow a uniform distribution in the network while the clients follow a skewed distribution. In Fig. 8(e), we show the query processing time for different combinations of distributions of POIs and queries while setting the remaining parameters to the default values. The U and S values shown in the figure indicate the uniform and skewed distributions, respectively. SEA provides better performance than PERIOD in all cases.

Figure 9 shows a comparison between SEA and PERIOD in terms of communication costs given the same conditions as shown in Fig. 8. Note that the vertical axis of every chart in Fig. 9 is a log scale of the number of transferred points.

Figure 9(a) shows the effect of the N_{POI} value on the communication costs of SEA and PERIOD. We can see that PERIOD incurs constant communication costs and constant communication frequency, regardless of different densities of POIs. In contrast, the communication cost of SEA increases with the N_{POI} value. This is expected because the safe region becomes smaller as the density of POIs in the road map increases, which requires a high communication frequency. For a very low density of POIs (i.e., $N_{POI} = 1$ k and $N_{POI} = 5$ k), SEA markedly outperforms PERIOD because in SEA, a few clients ask the server to update their query results and safe exit points.

Figure 9(b) shows the communication costs of SEA and PERIOD with respect to the N_{qry} value. Again, SEA achieves much lower communication cost and communication frequency than PERIOD. For SEA and PERIOD, the communication costs increase in proportion to the N_{qry} value.

Figure 9(c) shows the effect of the number k of requested NNs on the communication costs of SEA and PERIOD. Even if PERIOD has a constant communication frequency, the number of result points

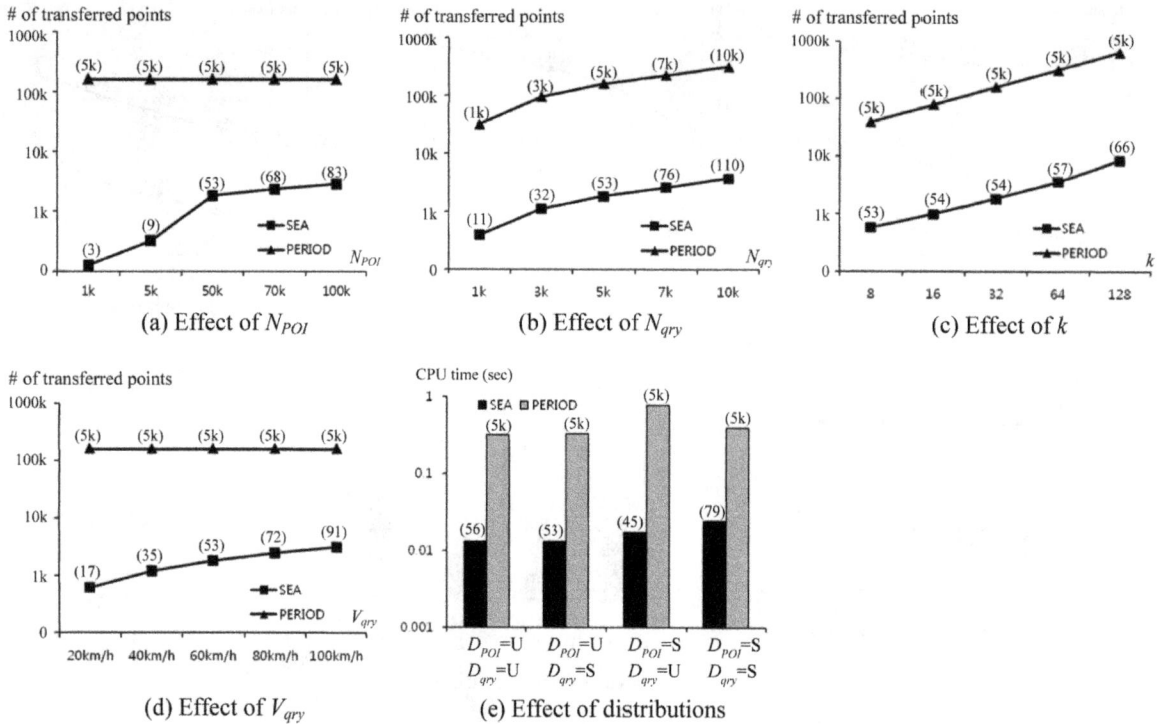

(a) Effect of N_{POI}

(b) Effect of N_{qry}

(c) Effect of k

(d) Effect of V_{qry}

(e) Effect of distributions

Fig. 9. SEA vs. PERIOD in terms of communication costs.

transferred from the server to clients increases with the k value; hence, the total communication costs also increase. In case of SEA, both the communication frequency and the number of POIs in the query answer increase with the k value. SEA significantly outperforms PERIOD, regardless of the k value. This is because only about 1% of all the clients in SEA request the updated query answers and safe exit points from the server.

Figure 9(d) shows the total communication costs and total communication frequency of SEA and PERIOD with respect to different client moving speeds (ranging from 20 km/h to 100 km/h). PERIOD incurs constant communication costs and frequency. This is because each client in PERIOD asks for periodically updated results (regardless of the client moving speed) for the precision desired by the client. However, PERIOD suffers from outdated results at high speed. For SEA, the client reaches a safe exit point sooner when the speed is increased; hence, both the communication costs and communication frequency increase. SEA outperforms PERIOD by an order of magnitude in terms of the communication costs and frequency. Figure 9(e) shows the effect of the initial distributions of POIs and moving clients on the communication costs. SEA outperforms PERIOD in all cases and achieves near-optimal communication costs.

6. Conclusion

We proposed a new algorithm called SEA which can efficiently compute the safe exit points for moving NN queries in road networks. The performance evaluation using a real-life road network shows that the communication costs of SEA are one or two orders of magnitude lower than those of a traditional method. In addition, SEA outperforms the traditional method significantly in terms of CPU time at the server. Consequently, SEA could be highly beneficial in real-life scenarios where mobile devices have limited network bandwidth and where the server demands a high throughput.

Acknowledgments

The authors would like to thank the anonymous reviewers for their valuable comments and suggestions to improve the quality of the paper. This research was supported by Basic Science Research Program through the National Research Foundation of Korea (NRF) funded by the Ministry of Education, Science and Technology (2010-0013487).

References

[1]   B. Bamba, L. Liu, A. Iyengar and P. Yu, Distributed processing of spatial alarms: A safe region-based approach, ICDCS, 2009, pp. 207–214.

[2]   J. Bao, C. Chow, M. Mokbel and W. Ku, Efficient evaluation of k-Range nearest neighbor queries in road networks, Mobile Data Management, 2010, pp. 115–124.

[3]   T. Brinkhoff, A framework for generating network-based moving objects, *GeoInformatica* **6**(2) (2002), 153–180.

[4]   M. Cheema, L. Brankovic, X. Lin, W. Zhang and W. Wang, Continuous monitoring of distance-based range queries, *IEEE Trans Knowl Data Eng* **23**(8) (2011), 1182–1199.

[5]   Z. Chen, H. Shen, X. Zhou and J. Yu, Monitoring path nearest neighbor in road networks, SIGMOD Conference, 2009, pp. 591–602.

[6]   H. Cho and C. Chung, An efficient and scalable approach to cnn queries in a road network, VLDB, 2005, pp. 865–876.

[7]   T. Delot, S. Ilarri, N. Cenerario and T. Hien, Event sharing in vehicular networks using geographic vectors and maps, *Mobile Information Systems* **7**(1) (2011), 21–44.

[8]   N. Ghadiri, A. Dastjerdi, N. Aghaee and M. Nematbakhsh, Optimizing the performance and robustness of type-2 fuzzy group nearest-neighbor queries, *Mobile Information Systems* **7**(2) (2011), 123–145.

[9]   H. Hu, J. Xu and D. Lee, PAM: An Efficient and privacy-aware monitoring framework for continuously moving objects, *IEEE Trans Knowl Data Eng* **22**(3) (2010), 404–419.

[10] H. Kriegel, P. Kröger and M. Renz, Continuous proximity monitoring in road networks, GIS, 2008, p. 12.

[11] F. Liu, T. Do and K. Hua, Dynamic range query in spatial network environments, DEXA, 2006, pp. 254–265.

[12] M. Mokbel, X. Xiong and W. Aref, SINA: Scalable incremental processing of continuous queries in spatio-temporal databases, SIGMOD Conference, 2004, pp. 623–634.

[13] D. Moon, B. Park, Y. Chung and J. Park, Recovery of flash memories for reliable mobile storages, *Mobile Information Systems* **6**(2) (2010), 177–191.

[14] F. Morvan and A. Hameurlain, A mobile relational algebra, *Mobile Information Systems* **7**(1) (2011), 1–20.

[15] K. Mouratidis, M. Yiu, D. Papadias and N. Mamoulis, Continuous Nearest Neighbor Monitoring on road Networks, VLDB, 2006, pp. 43–54.

[16] D. Papadias, J. Zhang, N. Mamoulis and Y. Tao, Query processing in spatial network databases, VLDB, 2003, pp. 802–813.

[17] P. Pesti, L. Liu, B. Bamba, A. Iyengar and M. Weber, RoadTrack: Scaling location updates for mobile clients on road networks with query awareness, *PVLDB* **3**(2) (2010), 1493–1504.

[18] M. Safar and D. Ebrahimi, eDAR algorithm for continuous knn queries based on pine, *IJITWE* **1**(4) (2006), 1–21.

[19] D. Stojanovic, A. Papadopoulos, B. Predic, S. Kajan and A. Nanopoulos, Continuous range monitoring of mobile objects in road networks, *Data Knowl Eng* **64**(1) (2008), 77–100.

[20] H. Wang and R. Zimmermann, A novel dual-index design to efficiently support snapshot location-based query processing in mobile environments, *IEEE Trans Mob Comput* **9**(9) (2010), 1280–1292.

[21] H. Wang and R. Zimmermann, Processing of continuous location-based range queries on moving objects in road networks, *IEEE Trans Knowl Data Eng* **23**(7) (2011), 1065–1078.

[22] K. Xuan, G. Zhao, D. Taniar and B. Srinivasan, Continuous range search query processing in mobile navigation, ICPADS, 2008, pp. 361–368.

[23] K. Xuan, G. Zhao, D. Taniar, J. Rahayu, M. Safar and B. Srinivasan, Voronoi-based range and continuous range query processing in mobile databases, *J Comput Syst Sci* **77**(4) (2011), 637–651.

[24] K. Xuan, G. Zhao, D. Taniar, M. Safar and B. Srinivasan, Constrained range search query processing on road networks, *Concurrency and Computation: Practice and Experience* **23**(5) (2011), 491–504.

[25] K. Xuan, G. Zhao, D. Taniar, M. Safar and B. Srinivasan, Voronoi-based multi-level range search in mobile navigation, *Multimedia Tools Appl* **53**(2) (2011), 459–479.

[26] D. Yung, M. Yiu and E. Lo, A safe-exit approach for efficient network-based moving range queries, *Data Knowl Eng Vol* **72** (2012), 126–147.

[27] J. Zhang, M. Zhu, D. Papadias, Y. Tao and D. Lee, Location-based Spatial Queries, SIGMOD Conference, 2003, pp. 443–454.

[28] G. Zhao, K. Xuan, D. Taniar and B. Srinivasan, LookAhead continuous KNN mobile query processing, *Comput Syst Sci Eng* **25**(3) (2010), 205–217.

[29] G. Zhao, K. Xuan, W. Rahayu, D. Taniar, M. Safar, M. Gavrilova and B. Srinivasan, Voronoi-Based continuous k Nearest NeighborSearch in mobile navigation, *IEEE Trans. on Industrial Electronics* **58**(6) (2011), 2247–2257.

[30] G. Zhao, K. Xuan and D. Taniar, Path kNN query processingin mobile systems, IEEE Trans. on Industrial Electronics, to appear (DOI: 10.1109/TIE.2011.2167113).

[31] Real Datasets for Spatial Databases, http://www.cs.fsu.edu/~lifeifei/SpatialDataset.htm, 2009.

Hyung-Ju Cho received his B.S. and M.S. degrees in Computer Engineering from Seoul National University in February 1997 and February 1999, respectively, and his Ph.D. degree in Computer Science from KAIST in August 2005. He is currently a research assistant professor at the department of information & computer engineering, Ajou University, South Korea. His current research interests include moving object databases and query processing in mobile peer-to-peer networks.

Se Jin Kwon received the B.S. and M.S. degrees in Computer Engineering from Ajou University, Korea, in 2006 and in 2008, respectively. He is currently PH.D. candidate (expected in August 2012) in Computer Engineering from Ajou University, Korea. His current interests include flash memory and large database systems.

Tae-Sun Chung received his B.S. degree in Computer Science from KAIST in February 1995 and his M.S. and Ph.D. degrees in Computer Science from Seoul National University in February 1997 and August 2002, respectively. He is currently an associate professor at the School of Information and Computer Engineering, Ajou University. His current research interests include flash memory storage, XML databases, and database systems.

6

Robust video communication over an urban VANET

N. Qadri, M. Altaf, M. Fleury* and M. Ghanbari

School of Computer Science and Electronic Engineering, University of Essex, Wivenhoe Park, Colchester, CO4 3SQ, UK

Abstract. Video communication within a Vehicular Ad Hoc Network (VANET) has the potential to be of considerable benefit in an urban emergency, as it allows emergency vehicles approaching the scene to better understand the nature of the emergency. However, the lack of centralized routing and network resource management within a VANET is an impediment to video streaming. To overcome these problems the paper pioneers source-coding techniques for VANET video streaming. The paper firstly investigates two practical multiple-path schemes, Video Redundancy Coding (VRC) and the H.264/AVC codec's redundant frames. The VRC scheme is reinforced by gradual decoder refresh to improve the delivered video quality. Evaluation shows that multiple-path 'redundant frames' achieves acceptable video quality at some destinations, whereas VRC is insufficient. The paper also demonstrates a third source coding scheme, single-path streaming with Flexible Macroblock Ordering, which is also capable of delivery of reasonable quality video. Therefore, video communication between vehicles is indeed shown to be feasible in an urban emergency if the suitable source coding techniques are selected.

Keywords: Error resilience, IEEE 802.11p, multiple path delivery, redundant frames, VANET, video communication

1. Introduction

This paper considers how to support robust video communication across multi-hop networks between vehicles when an urban emergency occurs. Real-time video communication allows early responders approaching an incident [28] to better understand the nature of the problem at the scene of an emergency but the lack of centralized routing and network resource management is challenging. Crash scenes, views of fleeing vehicles or burning buildings are some illustrative applications, while there is also now a strategic incentive [28] to provide coverage during a more serious, general emergency. In all these scenarios, it is the other personnel in the emergency vehicle or passengers in the vehicle that view the arriving video stream and not the driver.

Vehicular Ad Hoc Networks (VANETs) bring several advantages to video streaming within an ad hoc network. Battery power is no longer a problem if built-in transceivers are employed, implying that larger buffers (with passive and active energy consumption) can now serve to absorb any latency arising from multi-hop routing. We consider urban VANETs. Within a city, because of traffic congestion, high speeds do not generally arise. Therefore, connections are on average longer and Doppler effects are limited. Vehicle motion is indeed restricted by the road geometry but compared to a highway VANET vehicle motion is no longer linear.

---

*Corresponding author. E-mail: fleum@essex.ac.uk.

We examine three alternative video practical source-coding schemes for emergency video streaming, with one of the schemes applied in two different ways. The source-coding techniques applied exist in the context of IP networking [39] but, as far as the authors are aware, they have not been applied elsewhere in the way described within a VANET context. The first scheme examined is a variant of Multiple Description Coding (MDC) [36] in which two or more versions or descriptions of the same video stream are sent over different, preferably disjoint, routes across a network. Either description can serve to reconstruct the video but enhanced quality is produced by combining both descriptions. If adverse conditions occur on one of the paths then the packetized encoded bitstream from the other path can compensate. Video Redundancy Coding (VRC) [40] is the simplified MDC scheme employed by us that in the event of packet loss does not require additional decoder reconciliation between the two descriptions. Additionally, the VRC scheme was also trialed using distributed intra-coded macroblocks, that is H.264/AVC (Advanced Video Coding)'s [39] Gradual Decoder Refresh (GDR) [30], to avoid the reliance on prior reference frames.

This paper also proposes a second MDC scheme, employing H.264/AVC redundant frames, which when combined with multiple-path video transfer will result in higher-quality delivered video at a cost in higher data traffic. However, this cost may well be justified in an emergency. Redundant frames [32,42] (or strictly redundant slices[1] making up a frame) are coarsely quantized frames that can avoid sudden drops in quality marked by freeze frame effects if a complete frame (or slice) is lost. Again assuming even and odd frames are sent separately in two streams, then redundant frames are predicted from previous frames in the same stream but do not act as a reference to later frames.

In a third scheme, our paper proposes Flexible Macroblock Ordering (FMO) [38] with Checkerboard FMO pattern for single-path video stream transfer as an alternative to multiple-path methods. Error resilience [31] is applied at a source encoder to counter potential packet loss. FMO is an error-resilient technique newly included in the H.264/AVC codec that is suitable for error-prone channels. In good channel conditions, the overhead from sending the FMO mapping is a disadvantage but this is unlikely to be a problem when multi-hop routing occurs. Through source-encoder-independent error concealment at the decoder, FMO can aid the reconstruction of frames that have lost some of their constituent packets.

To the best of the authors' knowledge, though investigation of the concept of video streaming within a VANET has occurred, source-coding techniques have not been applied to any extent to VANETs. Some of the literature that exists on this subject so far in comparison to our work is examined in Section 2. The University of California, Los Angles (UCLA) research group under the leadership of Mario Gerla has produced a range of creative ideas on VANETs, for example [12,25,28]. However, their focus is on wireless aspects and at the 2009 Wireless Days Conference they have confirmed their interest in using a variant of our FMO scheme within live vehicle convoys as a security measure. We are flattered by their interest. Finally, we should add that another apparently unique feature of this work is that we group the emergency vehicles into a multicast group to receive the video. But we also use the other vehicles in the vicinity whether emergency or not to relay the video. In this way, the efficiency of the transfer is greatly improved. This latter feature is described in Section 3 but firstly this paper reviews other investigations of video over VANETs.

---

[1]A slice is headed by a decoder resynchronisation marker and may include reversible Variable Length Decoding, aimed at countering propagation of errors arising from the sequential dependencies of entropic encoding. Consequently, a slice is a self-contained decoding unit.

2. Related research

Earlier work on video communication over highway (not urban) VANETs [15] considered the problem of triggering remote video sources in the event of forward traffic congestion. The main problems in triggering [28] are how to reduce the number of messages reaching the remote camera(s) and how to reduce the latency in reaching those cameras (by reducing the number of hops), which is principally an issue of protocol design. In an emergency scenario as opposed to obtaining forward views of traffic congestion, it may be that video sources can be locally generated. Then an entirely different problem arises: how best to deal with heavy packet losses in the harsh urban environment. Video quality is strongly influenced by the impact of packet loss. Because successive video frames are broadly similar (except at scene cuts and changes of camera shots), to increase coding efficiency only the difference between successive frames is encoded. Consequently, at the frame level, removing temporal redundancy introduces a dependency on previously transmitted data that implies lost packets from reference frames will have an impact on future frames until a successful delivery of the next spatially-encoded anchor frame, when the decoder can be reset.

In [15], multiple vehicle video sources were modeled traveling on a 4–5 lane highway in Atlanta. Video was collected by sending from a car approaching a destination a request trigger to a camera on a remote vehicle passing that destination region. Video transport back to the requestor was by a store-carry-and-forward sub-system, though the method was not detailed in [15]. The main analysis in [15] was of delay characteristics, presumably because on a highway there should be sufficient time for the approaching vehicle to take evasive action if the forward view shows congestion or an accident.

Research in [25], extending the work in [15], simulated a two-ray wireless propagation model and imposed an application-layer Forward Error Control (FEC)-based solution through network coding. Though network coding of FEC and in particular rateless error coding is an effective means of limiting the impact of packet erasures upon streamed video, it depends on action by intervening nodes. When these nodes are not possible destinations and consequently may not be expected to make special provision for video data, then network coding is not feasible. Therefore, our paper considers alternative video protection methods that act in an end-to-end fashion, without the need for processing by intervening vehicles. Error resilience (in our schemes) is able to complement physical layer FEC, whereas when applying higher layer forms of FEC, it is better to do so in such a way that the channel code acts as an inner code to the PHY coding.

The feasibility of H.264/AVC video communication between two vehicles with IEEE 802.11b transceivers in a live setting was examined in [5]. With speeds between the two vehicles on average 15 mph (6.71 m/s) in a city setting (in Japan) it was reported that 'link availability' was 97.78%, as opposed to on a highway at an average speed of 55 mps (24.59 m/s) it was only 33.98%. Average SNR was somewhat worse in the urban setting, 19.14 dB, as opposed to 22.49 dB on the highway. Relative video quality was good (around 30–35 dB Peak Signal to Noise Ratio (PSNR)) and better in the city scenario with (slow-scan) rates ranging from 15–20 Hz and for Quarter Common Interchange Format (QCIF) (176×144 pixels/frame) to CIF-resolution (352×288 pixels/frame). The test clip was the well-known 'Foreman' clip at QCIF resolution (employed in this paper also) with medium coding complexity, though 'Paris' with less temporal complexity was employed at the larger CIF resolution. The study [5] established that, for vehicles traveling in proximity to each other, video exchange is entirely feasible albeit at slow-scan rates and resolution depending on coding complexity (which is a reflection of spatial activity (within frames) and temporal activity (between frames)). Of course, these results do not necessarily translate to multi-hop video transfer.

In a general context, the dissemination of multimedia information is a subject of active investigation within mobile systems. For example, in [21] progressive transmission of multi-resolution documents occurs so that the viewer can first view the relevance of the information before continuing with full transfer. To reduce the impact of low bandwidth capacity and delay over multi-hop connections, caching of data in the vicinity of an ad hoc node [6] is a promising approach. To support such systems in an ad hoc network, it is important that the routing protocols are optimized, for example [16] by tuning the route request flooding mechanism.

3. VANET system

3.1. Emergency application

Our system usage is captured in Fig. 1, showing an encircled crash scene. In such an urban emergency, it is envisaged that a scene is captured by one vehicle (the first emergency vehicle at the scene) that acts as the video source. Emergency vehicles now commonly carry video cameras, which in the case of the police act as a source of evidence in traffic offences. Therefore, there may be no need to capture the scene from cameras mounted on roadside masts, though these can be triggered locally as alternative video sources. Thus, either the scene can be captured manually by an emergency worker operating a vehicle mounted video camera, as already occurs when traffic police gather evidence or it can be captured through vehicle to roadside communication (vrc) or a roadside camera could be controlled by an emergency vehicle through remote communication. As how the video is captured is not central to this paper, we refer the reader to discussion of the feasibility of vrc such as in [8].

The video is distributed via WLAN-enabled vehicles to a multicast group of patrol cars, fire engines, ambulances or the like, acting as early responders to a crisis. If the multicast group consisted only of emergency vehicles (assuming less than ten responders for any one incident) then the ad hoc network size or density would be an impediment to communication. By routing the video stream over other intermediate car wireless transceivers, even though these cars are not destinations for the video stream (only the emergency vehicles are) multi-hop packet routing is more effective. Notice that though the destinations form a multicast group within the larger set of VANET-enabled vehicles, to improve robustness in all scenarios considered point-to-point communication is employed rather than a multicast protocol.

One emergency vehicle, acting as the video source, transfers the captured video by sending individual copies to each destination forming a virtual multicast group. When an MDC schemes is used each stream is split over two paths, refer to Fig. 1. Separate threads of control are able to generate these descriptions, possibly utilizing multi-core processors. It is assumed that available destinations, corresponding to other emergency vehicles in a group, are known through another emergency channel. In tests, six destinations were employed and it was found that, depending on choice of scheme, reasonable to good quality video was possible, though not for every destinations. Nevertheless, sufficient emergency vehicles would have a view of the emergency to allow preparations to be made as vehicles approached the scene.

3.2. VANET communication

Direct inter-vehicle communication can be an aid both to passenger comfort and to road safety [43]. Compared to a cellular network, a VANET may be toll free, avoids the delay in setting up a long communication circuit, and on a highway will operate where there are coverage gaps in a cellular

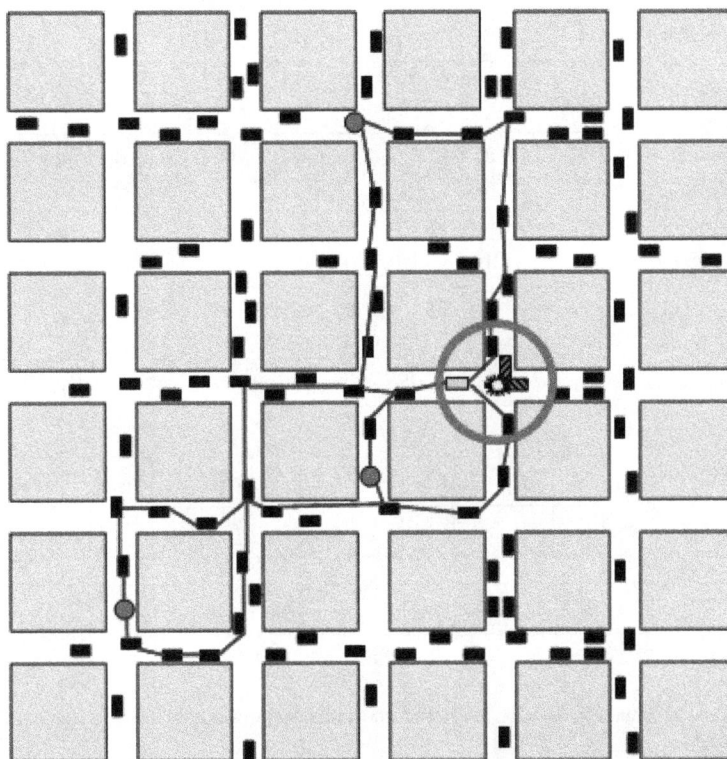

Fig. 1. VANET operating in city blocks with collision (encircled) videoed by light colored patrol car. Black rectangles are other vehicles acting as relays. Small circles are other emergency vehicles. Connecting lines show possible multi-paths for transmission of video.

network. There are strong pressures pushing car manufacturers towards equipping cars with WLAN capability, if they have not already done so. The IEEE 802.11p standard [17] will take advantage of 75 MHz of spectrum allocated both in Europe and the USA in the 5.9 GHz range with 10 MHz channels operating at up to 27 Mbps depending on modulation mode. The increased safety [3] that may arise from wireless provision is under active investigation. As well as safety alerts through wayside access points, the possibility of advertising localized services provides an additional commercial incentive to wireless take-up. It is thought that early adoption will result in around 20% of WLAN-enabled cars [4] in the near term. Therefore, at least 20% of the available cars in a city are likely to be available as relays to aid in video communication in an emergency.

If a VANET is to present an alternative to private cellular radio such as the Terrestrial Trunk Radio (TETRA) system [9] then it should provide similar services. Video communication over TETRA was explored in [7] and TETRA-2 was provided with extra bandwidth in support of multimedia communication. As an example, HW Communications Ltd. recently presented T-Serv for slow-scan video communication over TETRA, with in-vehicle video communication through IEEE 802.15.1 (Bluetooth). Compared to TETRA's cellular system, a VANET system can additionally make use of vehicles other than the emergency vehicles themselves, thus increasing coverage. (TETRA has an ad hoc mode but this is obviously confined to emergency vehicles, thus restricting the size of the ad hoc network.) Ad hoc radio is also potentially not limited by the urban 'canyons' caused by high buildings (if a base antenna is employed).

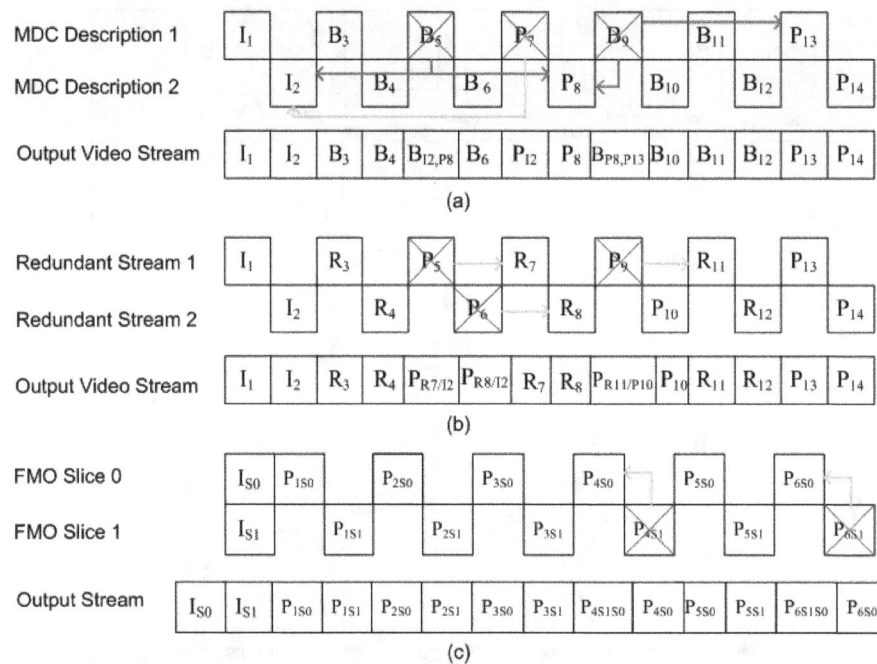

Fig. 2. Different path diversity schemes: a) VRC with odd and even descriptions b) Two streams with redundant frames, c) FMO slice replacement scheme.

If the source and destination are both assumed to be emergency vehicles, as in the scheme presented in this paper, then, when the video is routed via non-emergency vehicles, its confidentiality can be preserved through a stream cipher or alternatively through selective encryption of compressed video [44] to reduce the computation load of full encryption. (For example, only motion vectors in the compressed steam need be encrypted, as without these it is difficult to reconstruct a video). Another approach to security [22], which requires the cooperation of intermediate nodes to perform network coding, takes advantage of disjoint paths. However, though multiple paths may be available in a VANET, it is difficult and unnecessary to ensure they are node disjoint. A further approach [12] is to allow partial disclosure of some data after a time limit has expired. However, though this scheme may reduce dissemination latency, it is not clear that it allows streaming, as it requires an all-or-nothing transform, i.e. receipt of all data, before the data can be reconstructed. Partial disclosure of content is also permitted.

3.3. Video transfer schemes

Figure 2 illustrates the schemes tested in this paper. The frame numbers indicate the raw video frame from which a coded frame is constructed. Frames are decoded with motion compensation from reference frames in the same stream. By separately decoding from each stream, the problem of MDC decoder complexity is avoided. Figure 2a shows a number of frames have been dropped (marked by crossing out). Lost frames in one description can be reconstructed by reference to other correctly received packets in either description, with arrowed lines in Fig. 2a indicating the reconstruction route. For example, B_5 has been lost in description 1 and is reconstructed from either I_2 and P_8. The final row of frames in Fig. 2a shows the frame display sequence arising after substitution of reconstructed frames. Not shown in Fig. 2a is the VRC variant in which IDR-frames are no longer included in the sequence (refer to Section 4).

In Fig. 2b, showing the redundant frame scheme, the absence of B-frames allows use of the computationally efficient H.264/AVC Baseline profile. Redundant frames are sent in each stream, at a potential cost in latency but a potential gain in delivered video quality. There is only one initial IDR-frame in each sequence, which can be replaced by an intra-coded redundant frame. All other redundant frames are normally encoded in inter-mode (with the same reference as the frame that they back up) with normally a coarser quantization setting than the frame they back-up. Again example packet losses are shown and the result of reconstructing the sequences appears in the final row of Fig. 2b. For example, P_5 has been lost and reconstructed from R_7 but if R_7 did not survive it could also have been reconstructed from I_2. Figure 2c shows the FMO method of error resilience in which each frame has been split into two slice groups. That is each frame is divided into two slices and sent in different packets. The packets are multiplexed onto a single stream with slice 0's packet preceding slice 1's packet (though the order is not important). When a packet bearing one of a frame's slices is lost then the corresponding slice is normally employed to reconstruct it through the non-normative error concealment procedure. If both slices are lost then previous frame replacement is reverted to. The final row now shows the frame receiving order from a single stream.

4. Proposed schemes

The first scheme proposed for use in the VANET emergency is VRC [40], which avoids the need for decoder reconciliation in the event of packet loss. This is because two *independent* streams are formed from separately encoded odd and even frame sequences. Some lack of coding efficiency occurs as the motion between frames in any one description is likely to be greater than if the frames were coded in their original order. Consequently, the residual or difference data, which is actually coded by predictive coding, has a larger dynamic range requiring more bits to code. By insertion of IDR-frames in both descriptions (streams), the descriptions can be resynchronized even if one of the IDR frames is lost, at a cost in increased data redundancy compared to sending a single set of IDR-frames in a single stream. The macroblocks of IDR frames are completely spatially- (intra-) coded without removal of temporal redundancy and, therefore, do not reference any other frames. Consequently, they act as anchor frames for predictively- (inter-) coded subsequent frames. (In a hybrid video codec, each frame is split into macroblocks for processing purposes. Further details can be found in a textbook such as [13].)

In compensation, for the reduction in coding efficiency resulting from employing IDR-frames, bi-predictive B-frames are included within each Group of Pictures (GOP) within a VRC description to improve efficiency through multiple references (with a 10% bit-rate reduction in H.264/AVC [29]). (Notice that in the H.264/Advanced Video Codec (AVC) [41], IDR frames prevent predictive reference across GOP boundaries, whereas H.264/AVC I-frames do not (unlike their usage in earlier codecs)). Either an IDR- or an I-frame is inserted after every 12 or 15 frames making up a GOP.

IDR frames cause periodic increases in the data rate and consequently introduce additional buffering delay. Therefore, as an alternative form of VRC, we also distributed an equivalent number of intra-coded macroblocks [30] to those contained in the IDR-frames across the two VRC descriptions. Though another function of IDR-frame insertion is to provide a random access facility (supporting video player functions), this is unlikely to be required in an emergency scenario. In this variant of VRC apart from initial IDR-frames in each description only predictively-coded P-frames occur. This results in some loss of coding efficiency but improves computational efficiency (as the need to conduct more than one predictive search for B-frames is no longer needed)[2]. Along with Constant Bit Rate (CBR) encoding, an all P-frame sequence reduces delay for real-time applications. The risk of continued error propagation from the loss of any one IDR-frame-bearing packet is also reduced by distributing intra-macroblocks across all frames, H.264/AVC's GDR [39].

In general, MDC is computationally complex and requires specialist codecs [36], because synchronization between encoder and decoder is necessary to reduce motion estimation error drift. In a two stream MDC scheme, synchronization normally requires a third decoder in addition to the decoders that produce the reduced quality streams from single descriptions [14]. VRC is a simplified version of MDC with only two decoders, the output of which is interleaved before display. As mentioned previously, VRC normally requires the inclusion of IDR frames to allow decoder reset in the event of packet loss. If the packets happen to belong to an IDR frame then an IDR frame in the other sequence can serve as an anchor.

To avoid the need to send IDR frames, in Multiple State Video Coding (MSVC) [1] lost frames in one description are reconstructed from temporally adjacent frames in the other description. In this solution, all frames apart from the first IDR frame in each description are P-frames and reconstruction may also occur with the aid of past and future P-frames. However, reconstruction with P-frames from a different description reintroduces the risk of picture drift from the lack of synchronization between an encoder and decoder. For that reason MSVC is not tested in this paper, though MSVC can be credited as the basis of several later practical MDC schemes.

To overcome picture drift, redundant frames intended for error resilience in H.264/AVC [32,42], can serve to better reconstruct P-frames received in error. Though redundant frames were originally intended for Internet video streaming, applying redundant frames to multiple path streaming over general ad hoc network video was independently investigated in [26,33]. However, in [26] redundant pictures in one stream were encoded based on primary frames (frames for which redundant representations are generated) in a second stream, which requires modification of the operation of the H.264 codec. Another multiple path version [33] combines slicing with redundant data. In this version, a frame is split between alternate slices (formed from macroblock rows). Each slice is either a primary slice or a redundant slice for the matching primary slice in the other description. The need to generate this alternating pattern of slice types in each description prevents independent stream generation. That is, it is no longer possible to generate each description or decode it within its own control thread. Therefore, in this paper we apply redundant frames in a more direct manner that does not involve the need for a customized codec operation but does allow independently generated descriptions.

As also introduced in Section 1, FMO [19,38] is a promising form of video error resilience, which we use for single-path video transfer, adopting the type one checkerboard FMO pattern. By default in H.264/AVC, each frame forms a single slice group and macroblocks within that group are decoded in raster scan order. However, within a frame up to eight slice groups are possible. There are also seven different types of mapping between macroblock and slice group. Because type six supports arbitrary slice group mappings, its overhead is the greatest, as in addition to the slice group header a mapping must also be transmitted to the decoder (in a parameter set). Types three to five allow the size of each group to evolve over time, though macroblocks within a group remain geometrically contiguous. Only type one allows the assignment of geometrically dispersed macroblocks within a frame to form a slice group. The assignment is made through a mathematically function that in the two slice case results in a checkerboard

---

[2]It may also be possible to employ intra-coded macroblocks within B-frames in H.264/AVC, because unlike MPEG-4, this is now supported in H.264/AVC.

pattern. Therefore, the overhead is lighter for this type. To reduce overhead, it is also preferable to choose the option in H.264/AVC that prevents reference outside the slice group, though at some cost in coding efficiency. A detailed analysis of overhead, which depends on encoder configuration, can be found in [19].

Significantly, the type one FMO checkerboard pattern is the only H.264 predefined mapping function that supports error concealment by interpolation of data from adjacent macroblocks in order to reconstruct missing macroblocks (if one of the two checkerboard slice packets were to be lost). Error concealment in H.264/AVC is a non-normative feature [35] in which the motion vectors of correctly received slices are computed if the average motion activity is sufficient (more than a quarter pixel). The recommendation [35] gives details of which motion vector to select to give the smoothest block transition. It is also possible to select the intra-coded frame method of spatial interpolation. In our FMO experiments, though experience favors a motion vector-based method, we employed both methods and selected the superior result in terms of average PSNR across the video sequence. In a live situation, it is possible to choose the method that best reduces 'blockiness' at macroblock boundaries. Notice that in non-FMO experiments, previous frame replacement was employed at the decoder to reconstruct a frame, as this is the normal form of error concealment for comparison purposes in such tests.

5. Simulation model

Simulation is the main tool for research on VANETs [34], because it is difficult to find an analytical solution due to the large number of variables involved such as vehicle density, speeds, and mobility patterns. It is also difficult to conduct repeated live experiments.

5.1. Simulating IEEE 802.11p

The Global Mobile System Simulator (GloMoSim) [45] simulation library was employed to generate our results. GloMoSim was developed based on a layered approach similar to the OSI seven-layer network architecture. Total simulation time was 900 s, with the emergency video distribution starting after 100 s. We employed IP framing with UDP transport, as TCP transport can introduce unbounded delay, which is not suitable for delay-intolerant video streaming. We simulated a multi-path variety of the Ad Hoc On demand Distance Vector (AODV) protocol but without strict enforcement of disjoint paths (either path or node disjoint) for the grid of Fig. 1. This allowed two paths to be selected for the MDC schemes. Though we are aware of considerable research in the field of multi-path protocol design, e.g. see [23] for multi-path video transport with a random waypoint mobility model, the intention of the current paper is to concentrate on video transport aspects. The BonnMotion mobility generator (http://web.informatik.uni-bonn.de/IV/Mitarbeiter/dewaal/BonnMotion/ accessed Sept. 2009) was chosen. Though BonnMotion does not model driver behaviour in the way that a mobility model such as VanetMobiSim [10] does, we considered it sufficient for generic simulations mainly intended to check the application's behavior. Besides, the driving behavior of emergency vehicles will be quite different from normal drivers.

A two-ray propagation model with an omni-directional antenna height of 1.5 m at receiver and transmitter was selected for which the reflection coefficient was −0.7, which is the same as that of asphalt on tarmaced roads. The plane earth path loss exponent was set to 4.0, with the direct path exponent set for free space propagation (2.0). As in IEEE 802.11p, transmission was at 5.9 GHz with a bandwidth of 10 MHz. The transmission power was 33 dBm (2W). The receiver power threshold was set to −93 dBm, a normal value. Lastly, IEEE 802.11p's robust Binary Phase Shift Keying (BPSK)

Table 1
Default simulation settings for Manhattan grid mobility model

Parameter	Value
Terrain dimension	$1000 \times 1000 \text{ m}^2$
No. of vehicles	100
Size of multicast group	6
Number of x-, y- blocks	10, 10
Turn probability	0.5
Speed change probability	0.2
Minimum speed	0.5 m/s
Average speed	10.0 m/s (22 mph)
Speed standard deviation[1]	0.2 m/s
Speed update distance	10 m
Pause probability[2]	0
Transmission range	150 m
Routing protocol	AODV
Wireless technology	IEEE 802.11p
Channel model	Two-ray

[1]For normally distributed speeds. [2] If no change of speed.

modulation mode was simulated, introducing a packet length dependency through Bit Error Rate (BER) modeling in a Additive Gaussian White Noise (AGWN) channel. Accordingly, the data-rate was set to 3 Mbps.

5.2. Urban mobility model

The essential features of an urban scenario are captured by mobility models. Two approaches are possible: either the detailed microcellular approach [34] or generic models [2]. The microcellular approach has the advantage that it includes the effect of obstacles such as lane closures, uphill gradients, and potholes. Though the generic models lack the detail of the microcellular approach, these models do allow systematic investigation and easily interpretable results. For the purposes of assessing how video streams can be effectively transferred, we have used a generic model that captures the essential feature of an urban scenario, the restricted mobility patterns imposed by the presence of city blocks.

In [2], two generic models relevant to vehicular mobility are described, namely Freeway Mobility and Manhattan Mobility. The Freeway model limits vehicles to 1-D motion in either direction. Vehicles are tied to one of several lanes; the speed is dependent on a vehicle's previous speed; and in the 'car-following' restriction, a following vehicle cannot exceed the speed of a preceding vehicle to avoid approaching within a safety distance. The Manhattan model, an extension of the Freeway model, restricts the number of lanes in either direction to just one, but introduces a turning probability to give greater mobility. Both Freeway and Manhattan are related in that they should result in high spatial and temporal dependency.

Default simulation settings for the Manhattan grid mobility model are given in Table 1, while individual simulations varied from the given defaults. In many urban settings it is likely that wireless-enabled cars would be restricted to average speeds of around 10 m/s by congestion and traffic regulation, though vehicles responding to an emergency may go at faster speeds. The city block dimensions are chiefly related to the wireless range. Consequently, it is the relationship between range and block dimension that is important rather than the absolute settings.

5.3. Video configuration

The reference Foreman video clip was encoded at QCIF resolution with 4:2:0 sampling. Foreman, intended for judging communication between mobile devices, exhibits the typical features of a hand-held camera and, because of camera pans, exhibits high to medium coding complexity. Each frame was generally coded as a single slice and encapsulated in an H.264/AVC Network Abstraction Layer unit (NALU) [39] before being placed in a single packet. The combination of RTP/UDP/IP headers results in a further 40 B of overhead. IDR-frames, however, were split into two slices, which reduces the peak data rate. If one of the IDR-frame packets arrives before another [37] partial decoding can still take place while the other packet arrives. The encoder was set to output in CBR mode, with initial quantization parameter of 32. The frame rate of the video stream was set at a slow scan rate of 15 Hz to avoid injecting too great a data-rate into the network. Consequently, for each stream in the MDC schemes the data rate was approximately 60 kbps.

For VRC, the skip frame(s) facility of H.264/AVC allowed the creation of even and odd frame sequences. For each sequence when using IDR frames, the Main profile of H.264 allowed B-frames to be included. The GOP size was 15 frames with the usual repeating pattern of two B- and one P-frame until the next IDR-frame. In the Main Profile, Context-Adaptive Binary Arithmetic Coding (CABAC) results in a 9–14% bit saving at a small cost in computational complexity [24]. In the VRC variant and the redundant frame scheme, B-frames were no longer used. Intra-refresh macroblocks now provide the coding anchor points previously provided by IDR-frames. Intra-coded macroblocks were randomly selected (from a Uniform distribution) of H.264 macroblocks and were embedded with P-frames to the equivalent number needed for a QCIF IDR-frame. That is seven macroblocks per frame results in 105 macroblocks within a GOP, as opposed to 99 8×8 macroblocks for a single IDR-frame. Again the ability to generate random intra-coded macroblocks is a facility of H.264/AVC (H.264 JM Ref. Software, http://iphome.hhi.de/suehring/tml/download/, accessed Sept. 2009).

For FMO experiments, the Baseline Profile of H.264/AVC was selected with a GOP structure of IPP... In this Profile, intended for mobile devices and consequently with a smaller code footprint, Context Adaptive Variable Length Codes (dynamic Huffman entropic coding) is employed for simplicity (rather than CABAC), with some reduction in latency.

6. Evaluation

6.1. Preliminary tests

Figure 3 is a comparison between the luminance PSNR resulting from different error-resilient techniques upon Foreman, as the packet loss rate was varied. 100 simulation runs with different starting seeds were averaged to ensure convergence for Fig. 3. Errors in Fig. 3 followed a Uniform probability distribution function. Each frame was coded as a single slice, unless otherwise stated. Further detail of the error resilience methods compared is available in [31]. Notice that we have tested the methods separately for clarity, whereas a combination of methods in an error resilience strategy depending on channel conditions is possible, though there is a cost in extra overhead.

At around zero error packet loss-rate, FMO results in a somewhat lower video quality than omitting error resilience, because of its overhead, resulting in a lower coding efficiency. Separating into three independently-coded slices ('Slices' in Fig. 3), rather than one slice, is seen to be a little more effective at lower loss rates than in other loss regimes, as the risk of packet error is lower for shorter packet lengths.

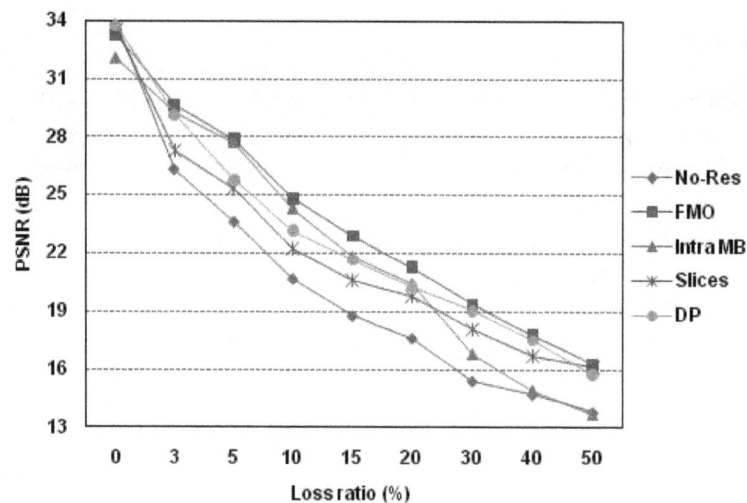

Fig. 3. Comparison between several H.264/AVC error resiliency methods and no resilience (No-Res) communication with Uniform bit errors, for Foreman QCIF sequence.

Data-partitioning ('DP' in Fig. 3 – separating configuration data and motion vectors, intra-coded, and inter-coded data into respectively A, B, and C NALUs) allows a frame to be effectively reconstructed even if the inter-coded macroblocks are missing, provided motion vectors in partition A are protected (which was simulated by assuming strong FEC protection of A NALUs). This technique is most useful at low loss rates. Insertion of Intra-coded Macroblocks ('Intra MB' in Fig. 3) allows temporal error propagation to be gradually arrested if some of a previous reference frame's macroblocks have been lost. In the tests, each row of macroblocks in rotation on a frame-by-frame basis was intra-coded. Intra-coded macroblocks in the tests was most helpful at medium loss rates. In the error conditions simulated, at medium to higher loss rates checkerboard FMO was the most effective method, though delivered video quality can no longer be considered fair for Foreman at a 10% loss rate and higher. However, users normally accept quality at 25 to 30 dB if it is in a mobile application. As the vertical axis in Fig. 3 is effectively logarithmic, a gain of less than one dB will still make a difference at the boundary between a sequence being viewable or not viewable, though at PSNRs below 25 dB that gain is of little importance. Therefore, the main conclusion from Fig. 3 is that FMO is superior to other built-in error resilience methods of H.264/AVC except at very low error rates.

We also compared a selection of alternative FMO patterns, though for reasons of brevity and because it is not central to the main theme of this paper, the results are not plotted herein. The H.264/AVC Checkerboard pattern was superior to other built-in FMO patterns at higher loss rates, when coding QCIF sequences with two slices. At lower loss rates, checkerboard FMO is somewhat weaker but the quality is anyway good with whatever FMO pattern. FMO patterns tested were: a selection of foreground in one slice group and the remainder in another; row interleaving; raster scan ordering with two groups; and selection of columns or part columns. Therefore, the Checkerboard pattern should normally be selected for FMO, especially as it can be used for up to eight slice groups in H.264/AVC and is not confined to a two-slice scheme.

6.2. Packet loss experiments

Table 2 gives the packet size features of the protection schemes in the experiments. Packet sizes grow with the inclusion of intra-coded macroblocks and, because IDR-frames are split into at least two packets,

Table 2
Protection schemes according to packet payload size

Scheme	Size range (B)	Characteristic
VRC with IDR frames	70 B to 885 B	10% over 500 B
VRC with Intra-refresh	183 B to 966 B	50% over 450 B
Redundant frames multiple path	260 B to 780 B	10% over 500 B
FMO single path	39 B to 493 B	55% below 125 B

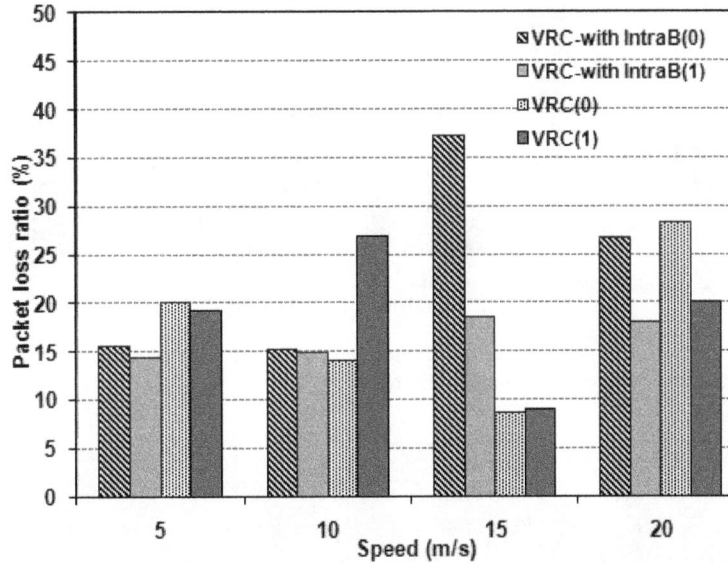

Fig. 4. Packet loss ratio for VRC for description (0) and description (1) versus vehicle speed with Table 1's settings, with IDR-frames and with intra-refresh macroblocks (marked IntraB).

larger packet sizes arise from VRC with intra-refresh than for VRC with IDR-frames. The smaller packet sizes for FMO are the result of splitting every frame into two slices, with a slice per NALU packet.

We firstly make some general remarks about the relationship between packet loss, vehicle speed and resulting video quality. As shown in Table 1, the number of emergency vehicle destinations is taken to be six, with one other video source vehicle, and the default total number of nodes is 100. At a given speed, if vehicles are on average in proximity to each other for sufficient time for packet transfer then one can expect less packet loss to occur. However, a vehicle must also avoid travelling at too slow a speed, as then that vehicle may not have sufficient time to approach another vehicle to affect a further transfer. In general, our experiments showed packet loss rates to be difficult to predict and as Section 6.3 demonstrates, an unreliable indicator of video quality when different protection schemes are compared. There now follows experimental results using the three source-coding schemes.

6.2.1. Packet loss under VRC

Figure 4 shows that packet losses under multiple-path VRC whether with IDR-frames or with Intra-refresh are generally higher than 10%. A level below about 10% is normally required for reasonable delivered video quality *unless* error protection is provided. There is only one speed (15 m/s) at which packet loss is within bounds for the classical VRC scheme. Figure 5 shows the result of altering the number of vehicles available for multi-hop relay. There is a trend towards reduction of packet losses to an acceptable level as the network size is increased, because there are greater opportunities for packet

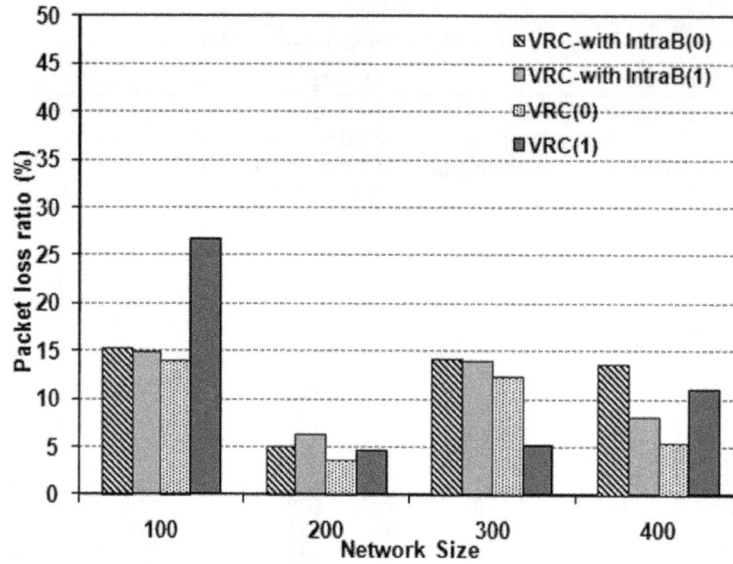

Fig. 5. Packet loss ratio for VRC for description (0) and description (1) versus network size with Table 1's settings, with IDR-frames and with intra-refresh macroblocks (marked IntraB).

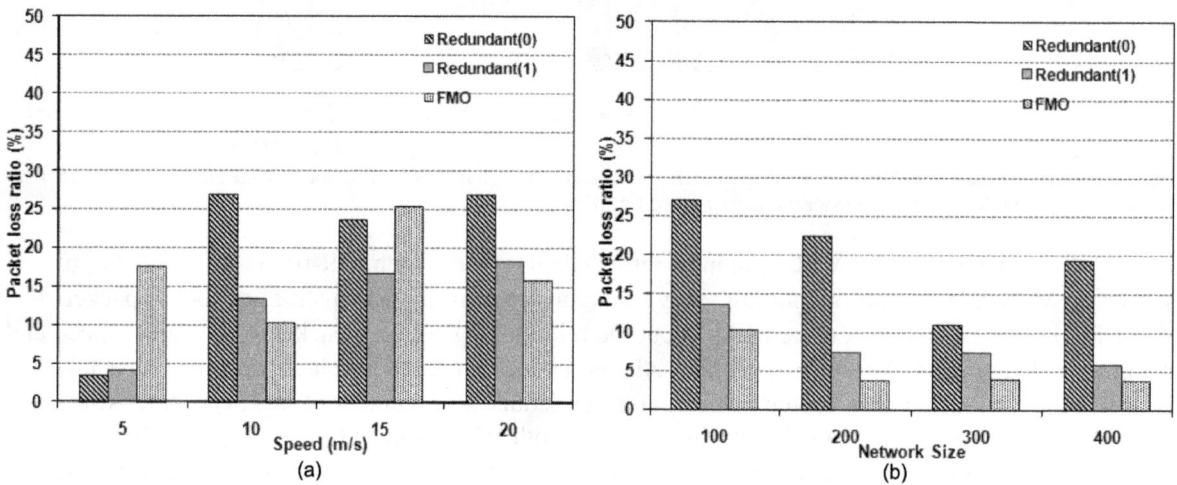

Fig. 6. Packet loss ratio for redundant frames and FMO with Table 1's settings with variation of a) vehicle speed and b) network size.

transfer to occur. In the experiment for IDR-frame VRC, each path might compensate for packet losses in the other path or description. Perhaps, because of the more regular data rate, this trend is not strongly apparent in the Intra-refresh version of VRC streaming. However, the main conclusion that can be drawn from these tests is that VRC in the form used herein is not satisfactory for video transport.

6.2.2. Packet loss under Redundant Frames or FMO

Figure 6 presents multiple-path video transfer packet loss when employing redundant frames. The packet loss levels with variation of speed or network size are generally higher than for VRC. However, packet loss rates are especially not a reliable guide to the likely video quality in this case, as, if redundant

Table 3
Packet losses by protection scheme for the selected destination node

	Packet loss ratio (%)			
	VRC (Intra-Refresh)	VRC (IDR-frames)	Redundant frames	FMO
Network size				
100	6	14	12	8
200	1	2	15	3
300	5	4	3	4
400	4	6	10	2
Speed				
5	2	3	1	15
10	6	14	12	8
15	18	6	10	25
20	8	9	15	13

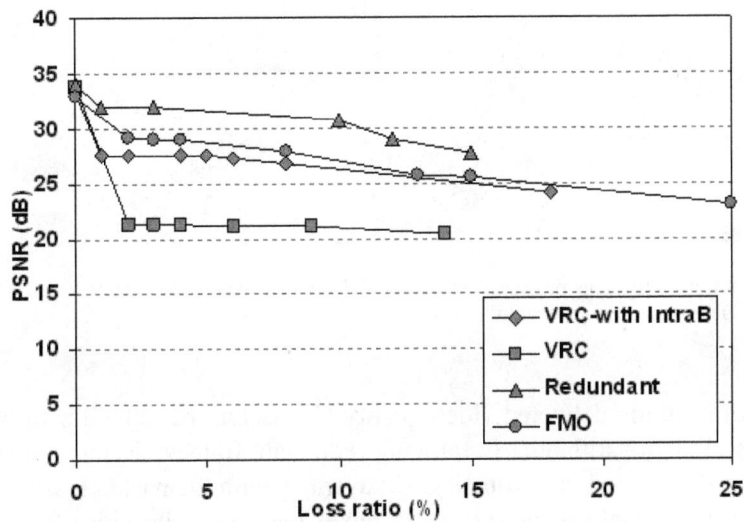

Fig. 7. Video quality by protection scheme for the selected node according to the packet loss ratios of Table 3. VRC = VRC with IDR frames, Redundant = Multiple path with redundant frames, VRC with IntraB = VRC with Intra-refresh macroblocks

frame packets are dropped, the video quality remains the same. Of course, a mixture of redundant and active frame packet loss will actually occur. Turning to single-path FMO error resilience, from Fig. 6 also, there is generally a much reduced level of packet loss for larger network sizes at the default speed, but FMO packet loss is sensitive to vehicle speed. Therefore, the main conclusions are that redundant frames lead to larger packet losses, though this may not necessarily result in worse video quality, whereas single-path FMO suffers less packet loss.

6.3. Video performance

From the six destination nodes, a 'median' node was selected in terms of packet loss, rather than one experiencing very high or very low packet loss. Table 3 reports the packet loss statistics for the chosen node. There is a decreasing trend in packet losses with network size, but no clear trends are apparent in terms of vehicle speed.

Fig. 8. Sample equivalent frames from the protection schemes. IDR = VRC with IDR frames, Rand = VRC with randomised Intra-refresh, Red = Redundant frames with multiple path

6.3.1. Video quality

Figure 7 compares resulting delivered video quality (PSNR) at the default settings. Recall that for PSNR the vertical scale is logarithmic. Employing redundant frames, despite the comparatively high packet loss ratios, results in good video quality. On average, with the packet loss ratios found from the simulation, sufficient redundant frames survive to repair the composite video frame sequence. FMO allows single stream video transfer to compete with multiple-path transfer but video quality is only fair in some settings tested. Including Intra-refresh macroblocks does improve the PSNR of VCR but not sufficiently to compete with the redundant frame scheme. Therefore, the main conclusion is that both redundant frames and FMO achieve reasonable video quality in the given scenario.

For illustrative purposes only, a sample frame in the Foreman sequence is shown in Fig. 8 for each of the schemes. In the two VRC schemes, there is considerable distortion around the face as previous frame error concealment has been affected by motion of the face between the frames. FMO results in less distortion but the cream-slated background shows some distortions.

6.3.2. Video latency and control overhead

Moreover, in broad terms from Fig. 9, average end-to-end delay (for all vehicles) is longer in duration for the VRC schemes. Generally, across all schemes packet delay times are significant and do require buffering. For other applications, e.g. a Web click and view service, there would be a perceptible start-up delay between selecting the video and starting to view the stream. For the emergency scenario, the destination emergency vehicles are unaware of the start time of the stream at the source and, therefore, would not notice the small delay. However, the time available to view the scene as a vehicle approached

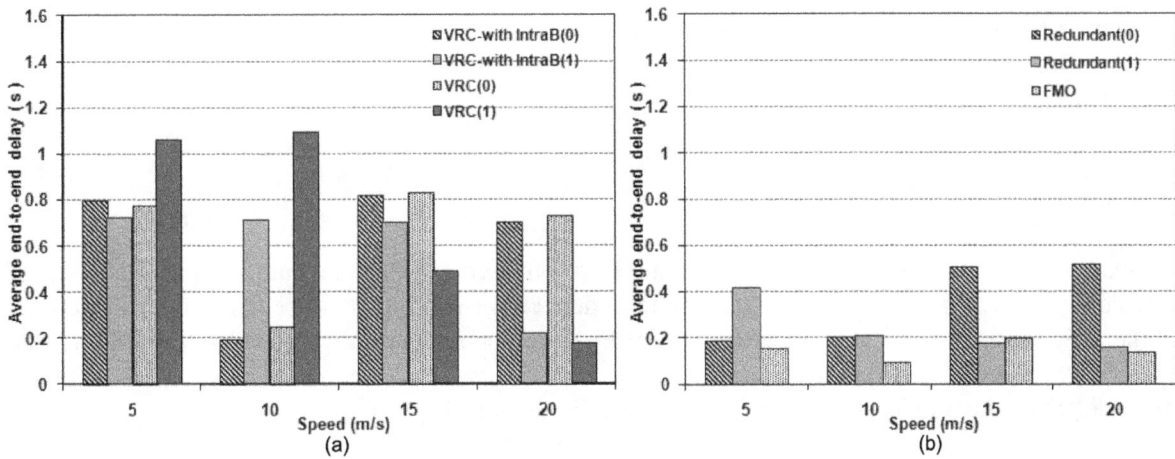

Fig. 9. End-to-end packet delay according to vehicle speed for a) VRC schemes. b) non-VRC schemes.

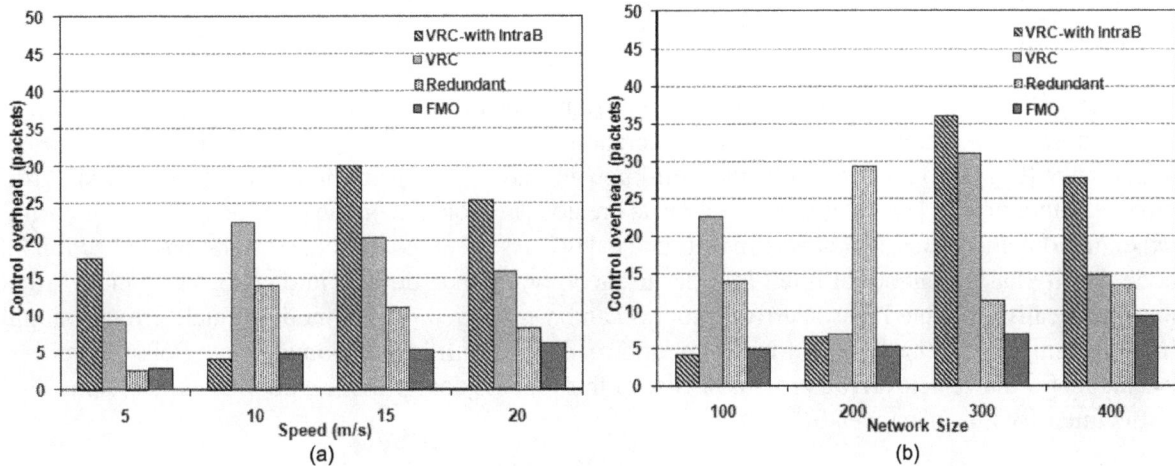

Fig. 10. Control packet overhead according to a) vehicle speed b) network size.

the emergency site would be reduced. The end-to-delay levels for VRC schemes were found to be similar to those for the FMO and redundant frames schemes when the network size was greater then 100 nodes. Therefore, Fig. 9a represents a worst case for the VRC schemes. Jitter was found to be broadly similar across the schemes and scenarios, with a range from 0.04 to 0.07 s. At 15 frame/s, this represents a need for a two-frame jitter buffer.

From Fig. 10, the number of control overhead packets was least for FMO, because only one stream was transmitted, whereas especially when speed was varied the redundant frame scheme resulted in reduced control packet overhead compared to the VRC schemes. However, the number of control packets expended to deliver a packet can be high. Moreover control packets add to the overall congestion in a network even though within a VANET the extra energy consumed in transmitting them may be discounted (as a vehicle's alternator is a convenient power source). This might be alleviated by employing a different multi-path routing protocol, e.g. [20], to route over multiple paths. However, this aspect is left to future work as the authors are aware that there are many issues involved in choice of protocol.

Therefore, combining the impact of extra latency and control overhead implies that a compromise solution is to use FMO rather than redundant frames fro transmitting video streams in this scenario. In our tests, VRC is not only weaker in terms of delivered video quality but it increases latency and overhead.

7. Tentative analytical model

This Section constructs the components of a tentative analytical model that aims to provide a way of calculating delivered video quality. The long term delivered video quality depends on the connectivity of the network, as this determines the ability of vehicles to communicate across the VANET. In turn, the speed of the vehicles is required to determine the connectivity. For example, the speed of vehicle i at time t, $\nu_i(t)$, in the Manhattan grid model [2] of the simulations is given by:

$$\nu_i(t + \Delta t) = \nu_{i+1}(t) - a/2, \text{ if } \Delta x_i(t) \leqslant \Delta x_{\min}$$
$$= \tilde{\nu}_i(t + \Delta t), \text{ otherwise} \tag{1}$$

$$\tilde{\nu}_i(t + \Delta t) = \min(\max(\nu_i(t) + \eta a \Delta t, \nu_{\min}), \nu_{\max}) \tag{2}$$

where $\Delta x_i(t)$ is the bumper-to-bumper distance from vehicle i to the vehicle in front, $\Delta x_{\min}$ is the minimum safety distance, a is the vehicle acceleration, and η is a uniformly distributed random variable in the range [0, 1]. Notice that unlike the well-known Random Waypoint model, in the Manhattan grid model the motion of a vehicle is dependent on its previous motion and on the position of the car in front. Other micro-cellular motion models can be found in [11]. As observed from the simulations in Section 6, node density has an important role. In order to incorporate node density into a stochastic model, the mathematically-tractable Poisson arrival process can be employed [18] to model vehicle arrivals along roads leading into the terrain (the Manhattan grid road layout). It is not claimed that a Poisson process is a model for the vehicle arrival process but use of this process does provide a point of comparison. The traffic intensity λ is then given by

$$\lambda = \frac{2f}{\nu_{mean}} \tag{3}$$

where ν_{mean} is the average speed of vehicles with a Poisson arrival process of intensity f in terms of vehicles/hour/road, with two lanes per road assumed in Eq. (3). In a Manhattan grid of size N × N blocks, there are 4(N-1) roads for vehicles to enter or leave the terrain. In [18], for a grid with N = 10, a road segment length of 400 m (100 m segments were used in our simulations), with a car-following driver model (dependent on the car in front with some modelling of driver behaviour [11]), it was found that intensity saturated at around $f = 800$ to 1000 vehicles/h/lane.

Assuming all vehicles are wireless-enabled, then wireless communication essentially depends on the range r. If there are some cars that are not wireless enabled then a factor ρ in the range [0,1] is introduced. The factor ρ also affects traffic congestion and in that way connectivity. However, in the simulations of this paper ρ is effectively set to unity. In steady-state, i.e. the number of vehicles leaving and entering the terrain is balanced, an upper bound on the expected number of vehicles that are *not* connected to any other vehicle, n, is:

$$E[n] = \exp(-2\lambda \rho r) \tag{4}$$

In the aforementioned setting [18], Eq. (4) was found to give a reasonable fit to simulated results for values of λ up to about 15 vehicles/km. However, this was for linear motion across the terrain, i.e. entering at one end of a road and leaving at the other end, and not to random destinations, as in this paper's results.

The range is governed by the propagation model. Thus, for the two-ray ground reflection path loss model used in our simulations, if the distance d is more than the cross-over distance [27] then the received power is:

$$P_r(d) = \frac{P_t g_t g_r h_t^2 h_r^2}{d^4 L} \tag{5}$$

where g_t and g_r are the gains of the transmitter and receiver respectively, h_t and h_r are the antenna heights of the transmitter and receiver respectively, P_t is the power at the transmitter, and L is the system loss. For distances less than the cross-over distance, the Friis free-space model [27] is applied, namely:

$$P_r(d) = \frac{P_t g_t g_r \Lambda^2}{(4\pi)^2 d^2 L} \tag{6}$$

where Λ is the wavelength. If power thresholding is applied at the receiver then this model allows a simple prediction of range. However, in the simulations packet length modelling was performed according to the form of the modulation for an AGWN channel [27]. The cross-over distance, d_c, is:

$$d_c = (4\pi h_t h_r)/\Lambda \tag{7}$$

Assuming some function, g, can be found that relates connectivity to packet loss then, from experience with the error-resilience techniques in the paper, these techniques act to additively increase the PSNR. For example, a 10% packet loss rate may achieve overall a fair video quality of 25 dB PSNR or above. However, if the packet loss rate is more that 10%, then error-resilience can restore the quality for excess (over 10%) packet loss rates up to total packet loss rate of about 20%. Thus, the effective packet loss rate is given by:

$$R_{eff} = g(E[n]) - C \tag{8}$$

where C is the reduction in packet loss rate due to error resilience (say 10%). If R_{eff} is greater than a given threshold value (say) 10% then video quality is judged to be unacceptable for all vehicles in the VANET multicast group. This simple model may be adjusted according to the number of paths employed and the strength factor of the error-resilience method. However, further development of this model is beyond the scope of the present paper.

8. Conclusions

This paper has investigated whether the operating conditions in a city are likely to permit video communication. It has found that MDC with redundant frame insertion (a new suggestion for VANETs) and single stream FMO, both source coding techniques included in the H.264/AVC codec, can support robust communication when packet loss rates are relatively high. Video communication to a group of emergency vehicles allows recognition of suspect vehicles, description of burning buildings and the like to be passed from the first vehicle on the scene to approaching vehicles. End-to-end delay remains a concern which should be addressed by reduction of hop counts by the routing protocol. As this is an outdoor scenario, location aware routing based on the global positioning system will be investigated. Further work will also consist of detailed investigation of urban wireless propagation conditions and modelling of driver behaviour.

References

[1]   J. Apostolopoulos, Reliable video communication over lossy packet networks using multiple state encoding and path diversity, *Visual Comms: Image Processing* (Jan 2001), 392–409.

[2]   F. Bai, N. Adagopan and A. Helmy, IMPORTANT: A framework to systematically analyze the Impact of Mobility on Performance of routing protocols over Ad hoc NeTworks, *IEEE INFOCOM* (April 2003), 825–835.

[3]   S. Biswas, R. Tatchiko and F. Dion, Vehicle-to-vehicle wireless communication protocols for enhancing highway traffic safety, *IEEE Communications Mag* **44**(1) (Jan 2007), 74–82.

[4]   J.J. Blum, A. Eskandarian and L.J. Hoffman, Challenges of inter-vehicle ad hoc networks, *IEEE Trans. on Intelligent Transportation Systems* **5**(4) (2004), 347–351.

[5]   P. Bucciol, E. Masala, N. Kawaguchi, K. Takeda and J.C. de Martin, Performance evaluation of H.264 video streaming over inter-vehicular 802.11 ad hoc networks, *IEEE 16th Int Symp on Personal, Indoor and Mobile Radio Comms* (Sept 2005), 1936–1940.

[6]   N. Chand, R.C. Joshi and M. Misra, Cooperative caching in mobile ad hoc networks based on data utility, *Mobile Information Systems* **3**(1) (2006), 19–37.

[7]   Y.C. Chang, M.S. Beg and T.S. Tang, Performance evaluation of MPEG-4 video error resilient tools over a mobile channel, *IEEE Trans on Consumer Electronics* **49**(1) (Feb 2003), 6–13.

[8]   S.R. Dickey, C.-L. Huang and X. Guan, Field measurements of vehicle to roadside communication performance, *IEEE Vehicular Technol Conf* (Fall 2007), 2179–2183.

[9]   J. Dunlop, D. Girma and J. Irvine, *Digital Mobile Communications and the TETRA System*, J. Wiley & Sons, Chichester, UK, 2000.

[10] M. Fiore, J. Härri, F. Filali and C. Bonnet, Vehicular mobility simulation for VANETs, *40th Annual Simulation Symposium* (Mar 2007), 301–307.

[11] M. Fiore and J. Härri, The network shape of vehicle mobility, *9th ACM Int'l Symposium on Mobile Ad Hoc Networking and Computing* (2008), 261–272.

[12] M. Gerla, R.G. Cascella, Z. Cao, B. Crispo and R. Battiti, An efficient weak secrecy scheme for network coding data dissemination in VANET, *IEEE PIMRC* (Sept 2008), 1–15.

[13] M. Ghanbari, Standard codecs: Image compression to advanced video coding, IET Press, London, UK, 2003.

[14] V.K. Goyal, Multiple description coding: Compression meets the network, *IEEE Signal Process Mag* **18**(5) (Sept 2001), 74–93.

[15] M. Guo, M.H. Ammar and E.W. Zegura, V3: A vehicle-to-vehicle live video streaming architecture, *3rd IEEE Int'l Conf. on Pervasive Computing and Comms* (Mar 2005), 171–180.

[16] A.M. Hanashi, I. Awan and M. Woodward, Performance evaluation with different mobility models for dynamic probabilistic flooding in MANETs, *Mobile Information Systems* **5**(1) (2009), 65–80.

[17] D. Jiang and L. Delgrossi, IEEE 802.11p: Towards an international standard for wireless access in vehicular environments, *IEEE Vehicular Technol Conf* (May 2008), 2036–2040.

[18] M. Kafsi, P. Papadimitratos, O. Dousse, T. Alpcan and J.-P. Hubaux, VANET connectivity analysis, *IEEE Workshop on Automotive Networking and Applications* (Dec 2008).

[19] P. Lambert, W. Deneve, Y. Dhondt and R. Vandewalle, Flexible macroblock ordering in H.264/AVC, *J of Visual Communication and Image Representation* **17** (2006), 358–375.

[20] S.-J. Lee and M. Gerla, Split multipath routing with maximally disjoint paths in ad hoc networks, *IEEE Int'l Conf. on Communications* (Jun 2001), 3201–3205.

[21] H.V. Leong and A. Si, Multi-resolution information transmission in mobile environments, *Mobile Information Systems* **1**(1) (2005), 25–40.

[22] L. Lima, M. Médard and J. Barros, Random linear network coding: A free cipher? *IEEE Int'l Symp on Info Theory* (Jun 2007).

[23] Q. Lu, L. Du, Z. Zuo and X. Xiao, Improved multi-path AODV protocols for real-time video transport over Mobile Ad Hoc Networks, *IEEE Pacific-Asia Workshop on Computational Intelligence and Industrial Application* (2008), 621–625.

[24] D. Marpe, H. Schwarz and T. Wiegand, Context-based adaptive binary arithmetic coding in the H.264/AVC video compression standard, *IEEE Trans on Circuits and Systems for Video Technol* **13**(7) (2003), 620–636.

[25] J.S. Park, L. Uichin, S.Y. Oh, M. Gerla and D. S. Lun, Emergency related video streaming in VANET using network coding, *3rd International Workshop on Vehicular Ad Hoc Networks* (2006), 102–103.

[26] I. Radulovic, Y.-K. Wang, S. Wenger, A. Hallapuro, M.H. Hannuksela and P. Frossard, Multiple description H.264 video coding with redundant pictures, *Int'l Workshop on Mobile Video* (Sept 2007), 37–42.

[27] T.S. Rappaport, *Wireless Communications*, (2nd edition), Prentice-Hall, Upper Saddle River, NJ, 20012.

[28] M. Roccetti, M. Gerla, C.E. Palazzi, S. Ferretti and G. Pau, First responders' crystal ball: How to scry the emergency from a remote vehicle, *IEEE 26th Int'l Conf on Performance of Computing and Communs* (April 2007), 556–556.

[29] S. Saponara, C. Blanch, K. Denolf and J. Bormans, The JVT advanced video coding standard: Complexity and performance analysis on a tool-by-tool basis, *Int'l Packet Video Workshop* (April 2003).

[30] R.M. Schreier and A. Rothermel, Motion adaptive intra refresh for the H.264 video coding standard, *IEEE Trans. on Consumer Electronics* **52**(1) (2006), 249–253.

[31] T. Stockhammer and W. Zia, Error-resilient coding and decoding strategies for video communication, in: *Multimedia over IP and Wireless Networks*, M. van der Schaar and P.A. Chou, eds, Academic Press, Amsterdam, 2007, pp. 59–80.

[32] D. Tian, M.M. Hannuksela, Y.-K. Wang and M. Gabbouj, Error resilience coding techniques using spare pictures, *Int'l Packet Video Workshop* (April 2003).

[33] T. Tillo, M. Grangetto and G. Olmo, Redundant slice optimal allocation for H.264 multiple description coding, *IEEE Trans Circuits and Syst for Video Technol* **18**(1) (2008), 59–70.

[34] O.K. Tonguz, W. Viriyasitavat and F. Bai, Modeling urban traffic: A Cellular automata approach, *IEEE Communications Mag* **47**(5) (May 2009), 142–150.

[35] V. Varsa, M.N. Hannuksela and Y. Wang, Non-normative error concealment algorithms, ITU-T SGI6 Doc., VCEG-N62, 2001.

[36] Y. Wang, A.R. Reibman and S. Lee, Multiple description coding for video delivery, *Proc of the IEEE* **93**(1) (2005), 57–70.

[37] W. Wei and A. Zakhor, Multipath unicast and multicast video communication over wireless ad hoc networks, *Int'l Conf on Broadband Networks* (Oct 2004), 494–505.

[38] S. Wenger and M. Horowitz, Flexible MB ordering—a new error resilience tool for IP-based video, *Int'l Workshop on Digital Communications*, Capri, Italy, (2002).

[39] S. Wenger, H264/AVC over IP, *IEEE Trans Circuits and Syst for Video Technol* **13**(7) (2003), 645–656.

[40] S. Wenger, G.D. Knorr, J. Ou and F. Kossentini, Error resilience support in H.263+, *IEEE Trans Circuits and Syst for Video Technol* **8**(7), 867–877.

[41] T. Wiegand, G.J. Sullivan, G. Bjøntegaard and A. Luthra, Overview of the H.264/AVC video coding standard, *IEEE Trans Circuits and Syst for Video Technol* **13**(7) (July 2003), 560–576.

[42] Z. Wu and J.M. Boyce, Adaptive error resilient video coding based on redundant slices of H.264/AVC, *IEEE Int'l Conf. on Multimedia and Expo* (July 2007), 2138–2141.

[43] S. Yousefi, M.S. Mousavi and M. Fathy, Vehicular ad hoc networks (VANETS): Challenges and perspectives, *6th Int'l Conf on ITS Telecommunications* (Jan 2006), 761–766.

[44] H. Yu, Streaming media encryption, in: *Multimedia Security*, B. Furht and D. Kirovski, eds, CRC Press, Boca Raton, FO, 2005, pp. 197–220.

[45] X. Zeng, R. Bagrodia and M. Gerla, GloMoSim: A library for parallel simulation of large-scale wireless networks, *12th Workshop on Parallel and Distributed Simulations* (May 1998).

Nadia N. Qadri received her PhD at the School of Computer Science and Electronics Engineering, University of Essex, UK in 2010. She received her Masters of Engineering (Communication Systems and Networks) and Bachelors of Engineering (Computer Systems), from Mehran University of Engineering and Technology, Jamshoro, Pakistan in 2004 and 2002 respectively. She has more than four years of teaching and research experience at renowned universities of Pakistan viz. Mehran University of Engineering & Technology, Fatima Jinnah Women's University and COMSATS Institute of Information Technology. Her research interests include video streaming for mobile ad hoc networks and vehicular ad hoc networks, along with P2P streaming.

Muhammad Altaf received his BSc degree from the University of Engineering and Technology, Peshawar, Pakistan in 2001 and his MSc degree in computer system engineering from the National University of Science and Technology, Rawalpindi, Pakistan in 2004. He has recently been awarded his PhD at the University of Essex, UK. His research interests are video compression and video streaming over wired and wireless networks.

Martin Fleury has a degree in Modern History (Oxford University, UK) and a Maths/Physics based degree from the Open University, Milton Keynes, UK. He obtained an MSc in Astrophysics from QMW College, University of London, UK in 1990 and an MSc from the University of South-West England, Bristol in Parallel Computing Systems in 1991. He holds a PhD in Parallel Image Processing Systems from the University of Essex, Colchester, UK. He is currently employed as a Senior

Lecturer at the University of Essex. Martin has authored over 160 articles, and other publications on the subjects of low-level image- and signal-processing algorithms (including document and image compression algorithms), performance prediction of parallel systems, software engineering, and vision systems. His current research interests are video communication over MANS, WLANs, PANs, BANs, MANETs, and VANETs.

Mohammed Ghanbari is best known for his pioneering work on two-layer video coding for ATM networks (which earned him an IEEE Fellowship in 2001), now known as SNR scalability in the standard video codecs. He has served as an Associate Editor for IEEE Trans. on Multimedia. He has registered for eleven international patents on various aspects of video networking and was the co-recipient of A.H. Reeves prize for the best paper published in the 1995 Proc. of IEE on the theme of digital coding. He is the co-author of "Principles of Performance Engineering", a book published by IET press in 1997, the author of "Video Coding: An Introduction to Standard Codecs", a book also published by IET press in 1999, which received the year 2000 best book award by the IEE, and the author of "Standard Codecs: Image Compression to Advanced Video Coding" also published by the IET press in 2003. Prof. Ghanbari has authored or co-authored about 450 journal and conference papers, many of which have had a fundamental influence in this field.

7

Modelling medium access control in IEEE 802.15.4 nonbeacon-enabled networks with probabilistic timed automata

Tatjana Kapus

Faculty of Electrical Engineering and Computer Science, University of Maribor, Smetanova ul. 17, SI-2000 Maribor, Slovenia
E-mail: tatjana.kapus@um.si

Abstract. This paper concerns the formal modelling of medium access control in nonbeacon-enabled IEEE 802.15.4 wireless personal area networks with probabilistic timed automata supported by the PRISM probabilistic model checker. In these networks, the devices contend for the medium by executing an unslotted carrier sense multiple access with collision avoidance algorithm. In the literature, a model of a network which consists of two stations sending data to two different destination stations is introduced. We have improved this model and, based on it, we propose two ways of modelling a network with an arbitrary number of sending stations, each having its own destination. We show that the same models are valid representations of a star-shaped network with an arbitrary number of stations which send data to the same destination station. We also propose how to model such a network if some of the sending stations are not within radio range of the others, i.e. if they are hidden. We present some results obtained for these models by probabilistic model checking using PRISM.

Keywords: Wireless personal area networks, medium access control, hidden stations, formal specification, probabilistic model checking

1. Introduction

The IEEE 802.15.4 standard specifies the *Medium Access Control* (MAC) sublayer and the physical layer for low-rate Wireless Personal Area Networks (WPANs) [22,23]. The well-known WPANs that rely on it are the wireless sensor networks (WSNs) following the ZigBee standard, which defines the upper protocol layers for WPANs [1]. The IEEE 802.15.4 standard allows WPANs with star and peer-to-peer operation. In the former, there is a relaying device called a coordinator, and all the other devices can only communicate through it. In the latter, there is also a coordinator, but all the devices can communicate directly if they are within radio range of one another. The MAC sublayer specification is needed because those WPAN devices that are within range of each other share a common channel. Two modes of the MAC operation are defined: beacon- and nonbeacon-enabled. In the former, all communication takes place via the coordinator, which controls any access of the devices to the common channel by sending beacon frames to them. After obtaining a beacon frame, the devices compete for the transmission channel towards the coordinator by executing a slotted Carrier Sense Multiple Access with Collision Avoidance (CSMA-CA) algorithm. No beacon frames are sent in the nonbeacon-enabled mode. All the devices within the same range, including the coordinator, compete for the common channel by executing an unslotted CSMA-CA algorithm.

Traditional methods for the performance evaluation of networks and protocols comprise discrete-event simulation with simulators such as OPNET or ns-2 and mathematical analysis using stochastic modelling. These have extensively been used for studying the performance of beacon-enabled (e.g. [13, 18,21]) and nonbeacon-enabled IEEE 802.15.4 networks (e.g. [9,12,26]). Over the last decade, formal methods, probabilistic model checking in particular, have increasingly been investigated as a means of studying the performance and reliability of probabilistic systems [6,11,16]. Probabilistic model checking is an algorithmic method implemented in software called a *model checker*. Basically, it takes a formal description of the system in the form of a finite-state model in which transitions between states can fire with certain probabilities and, in contrast to the simulators, automatically checks the complete state space for determining the validity of a required performance or reliability property expressed in a kind of logic. Depending on the kinds of model and logic, it can for example be verified as to whether an event will happen with a given probability or within a given time-limit, or even what the probability or the time-limit is for that event.

Fruth was the first to use probabilistic model checking for the performance analysis of IEEE 802.15.4 MAC protocols [7,8]. The PRISM model checker was used [24], and both the beacon- and nonbeacon-enabled networks were analysed. In this paper, we are interested only in the nonbeacon-enabled networks. In [7,8], their MAC operation is basically modelled by using probabilistic timed automata. However, as probabilistic timed automata were not supported by PRISM before 2011, the models are in fact written by using Markov decision processes, in which there is no time. The time is represented by additional variables. In this paper, we explain how to write the models directly by using probabilistic timed automata which are supported by new versions of PRISM [14]. In [7,8], only a network consisting of two stations sending to two different destination stations within the same radio range is modelled, and we found that the model is slightly inaccurate. We first created an improved model of such a network. Based on it, we propose two ways of modelling a network with n sending and n receiving stations, for an arbitrary n. We show that the same models can be used to represent a star-shaped network with n stations sending to the same destination. We also show how these models can be adapted for those cases when some sending stations are hidden, i.e. not within radio range of the others. Using the proposed models, we carried out some experiments with the PRISM model checker regarding the probability of successfully sending data and the effect of the duration of the clear channel assessment, which is part of the MAC protocol.

This paper is organised as follows. Section 2 briefly introduces probabilistic timed automata. Section 3 contains a description of the MAC operation in the case where a station wants to send a data frame within a nonbeacon-enabled network. In Section 4, we explain the improved probabilistic timed automata model of the MAC operation for two sending and two receiving stations. In Section 5, we present the models for n pairs of communicating stations. Section 6 explains why the suggested models are also valid representations of the star topologies with one receiving station and how to model a star-shaped network with hidden stations. In Section 7, we present the results of model checking using PRISM. Section 8 concludes the paper with a discussion and suggestions for future work.

2. Probabilistic timed automata

We present the probabilistic timed automata rather informally (cf. [14]) and as supported by PRISM. For formal definitions please see e.g. [15]. A Probabilistic Timed Automaton (PTA) has a finite set of *states*, including an initial one. It is, in fact, a finite-state automaton enriched with non-negative real-valued variables, called *clocks*, and with discrete probabilistic choice. Additionally, it can have finite-range *data variables*. Initially, the values for all the clocks are zero. The values of the clocks increase

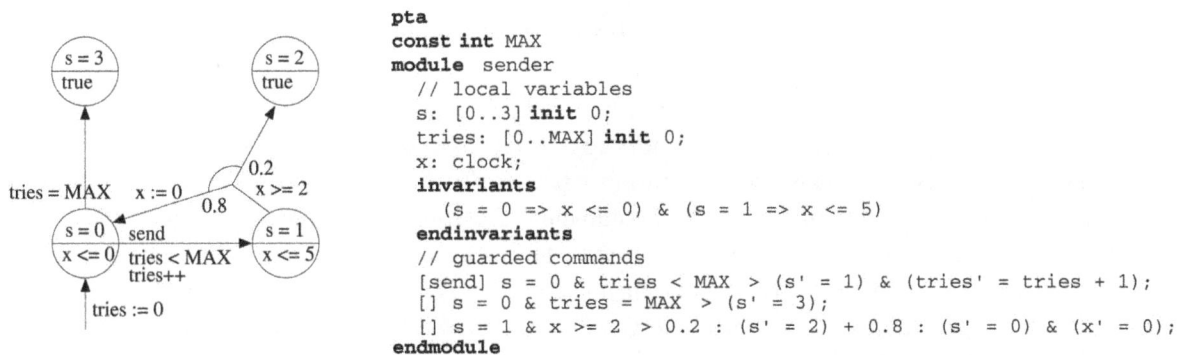

```
pta
const int MAX
module sender
    // local variables
    s: [0..3] init 0;
    tries: [0..MAX] init 0;
    x: clock;
    invariants
        (s = 0 => x <= 0) & (s = 1 => x <= 5)
    endinvariants
    // guarded commands
    [send] s = 0 & tries < MAX > (s' = 1) & (tries' = tries + 1);
    [] s = 0 & tries = MAX > (s' = 3);
    [] s = 1 & x >= 2 > 0.2 : (s' = 2) + 0.8 : (s' = 0) & (x' = 0);
endmodule
```

Fig. 1. A graphical representation of a PTA and its corresponding PRISM code.

simultaneously over time. The execution of *transitions* between the states is controlled by two kinds of predicates. *Invariants* are predicates over the clock variables and are associated with the states. The PTA may stay in a state only as long as the invariant associated with it is true. *Guards* are predicates over the clock variables and the data variables, and are associated with the transitions. Only non-negative integer values may be used in invariants and guards. A transition may occur only if its guard is true. A transition can reset some of the clocks (to integer values), update the data variables, and lead the automaton to a new state. The effect of the transition is chosen probabilistically. The possible choices are specified by a discrete probability distribution. If multiple transitions can be taken from a state, one of them is chosen nondeterministically. Instead of a transition occurrence, the values of the clocks may increase implicitly. Formally, this is called a delay transition. Whether a delay transition will occur as well as the amount of the delay are chosen nondeterministically, subject to invariant satisfaction.

In Fig. 1, on the left, there is a graphical representation of a PTA, which represents the sending of a message over an unreliable channel. It has four states. In the lower half of the nodes, the invariants of the states are indicated. *tries* is a data variable, which counts the number of transmission attempts, and x is a clock. New sending attempts are made until *tries* is equal to a constant *MAX*. The state $s = 0$ must be left immediately because its invariant does not allow x to be greater than 0. The transition labelled with *action send* represents the beginning of a new transmission attempt. It increments the counter and leads to state $s = 1$. The invariant associated with this state and the guard of the transition leading from it mean that the transmission can take between 2 and 5 time units. This transition has two possible effects. With a probability of 0.2, the transmission succeeds, otherwise, the transmission fails, the clock is reset to 0, and the PTA returns to $s = 0$.

In PRISM, PTAs and other supported probabilistic models are specified by using a uniform textual language. The PRISM code for the presented PTA is given on the right of Fig. 1. A PTA is specified as a module. It should be noticed that in PRISM, the states are represented by local variables. The initial value of a variable is the lowest value within the declared range if it is not set by using `init`. As usual, primed variables denote the values in the next state. Transitions are described by guarded commands and can be labelled with actions. A system can be specified as a collection of modules. By default, they represent a parallel composition of PTAs with synchronisation on common actions, which means that all the PTAs are executed concurrently and that the transitions of different PTAs with the same label must be executed simultaneously. Each module may read local variables (i.e. use them in its guards) of the others, but can only write to its own. The values of all the clocks for all the modules increase simultaneously over time.

3. Operation of medium access control in nonbeacon-enabled mode

When a station wants to send a data frame in the nonbeacon-enabled mode, it starts to execute unslotted CSMA-CA algorithm as follows, in order to get access to the channel [22]. It first waits for a random number of backoff periods, chosen uniformly between 0 and $2^{BE} - 1$. Initially, BE, the *backoff exponent*, is equal to the value of the standard parameter *macMinBE* (in the sequel denoted *BE_MIN*), the default value of which is 3, but can range from 0 to 3. The *backoff period*, in the sequel denoted *BO_PERIOD*, is equal to the time needed to transmit 20 symbols with a chosen standard bit rate. After the waiting time, the station performs the *clear channel assessment* (CCA) for the duration of 8 symbols (denoted as *CCA* in the sequel).

If the channel is found to be busy at any time during the CCA period, it is indicated as busy by the CCA at the end of this period, and otherwise as free. If it is indicated as busy, the station increments BE by one if it is smaller than *macMaxBE* (denoted by *BE_MAX*; its default value being 5) and leaves it unchanged otherwise, and again randomly backs off and carries out the CCA. This backoff procedure is repeated until either the CCA succeeds, i.e. indicates that the channel is free, or the number of backoffs reaches *macMaxCSMABackoffs* (denoted by *NB_MAX*; it can range from 0 to 5, its default value being 4). If the backoff procedure is executed *NB_MAX* times without success, the station ends the execution of CSMA-CA by declaring a *channel access failure*.

If the CCA succeeds, the CSMA-CA is taken to be finished successfully, the station switches from the *receive* mode to the *transmit* mode (*RX-to-TX*) and transmits the frame. As usual in the literature, we shall assume that any RX-to-TX or TX-to-RX turnaround lasts exactly 12 symbols, which is the maximal possible value allowed by the standard (*aTurnaroundTime*, denoted as *TURNAROUND* in this paper). Optionally, the sending station can require an acknowledgement within the data frame. If the destination station receives such a frame, it performs a RX-to-TX turnaround and immediately after this transmits the acknowledgement. After the data frame transmission, the sending station performs a TX-to-RX turnaround and waits for *macAckWaitDuration* symbols, denoted *ACK_TIMEOUT* in the sequel and equal to the sum of the acknowledgement length in symbols, *TURNAROUND* and *BO_PERIOD*. If the acknowledgement does not come within this time, the sending station sets BE to *BE_MIN* and repeats the CSMA-CA algorithm as if sending the data frame for the first time. If the acknowledgement does not come even after repeating the algorithm *aMaxFrameRetries* times (denoted as *MAX_RETRIES* in the sequel; 3 by default, but can range from 0 to 7), the sending station declares a *collision failure* and stops trying. If the acknowledgement is received, the sending station declares *success*. If the sending station does not use the acknowledgement option, it declares *success* immediately after the data frame transmission.

After the transmission of an acknowledgement, the destination station performs a TX-to-RX turnaround. If a data frame starts to arrive at the destination during this time, the destination ignores it [26].

The length of the acknowledgement frames is 11 octets. The data frames can be from 15 to 133 octets in length. The rest of this paper supposes that one octet corresponds either to 8 symbols (this is the case if the standard bit rate of 20 or 40 kbps is used for transmission) or 2 symbols (this is the case if the bit rate is 250 kbps).

Unless otherwise stated, in the sequel we consider formal models for the MAC using acknowledgements. The models for the MAC without acknowledgements can easily be obtained from them.

4. Improved formal model for two pairs of stations

In this section, we introduce a PTA model of an IEEE 802.15.4 network consisting of sending stations s_i, $i = 1,2$, and receiving stations r_i, $i = 1, 2$, all being within radio range of each other. As in [7,8], we suppose that the air is an ideal medium except for the possibility of collisions, that station s_i sends one message to station r_i by using the unslotted CSMA-CA and the acknowledgements as described in Section 3, and that both sending stations have a message to send at the same time. Throughout this paper, we assume that whenever two stations actually transmit a frame at the same time, a collision occurs, and that propagation delays are negligible.

In [7,8], the model of this network is a parallel composition of a PTA representing the common communication channel between the stations and of two PTAs representing the pair s_1, r_1, and, respectively the pair s_2, r_2. In fact, the PTAs are given only in graphical form. In the PRISM language, they are represented as Markov decision processes. Our aim is to write the PTA model of the network in PRISM. The PTA model given in [7] (and repeated in [8]) contains come features unsupported by the PRISM language. It contains urgent states and transitions. An *urgent state* is one in which time must not advance and must, therefore, be left immediately. It can easily be represented in PRISM by introducing a clock variable which is set to 0 on all its ingoing transitions and by requiring in its invariant that the clock be less or equal to 0. For example, state $s = 0$ in Fig. 1 is urgent. An *urgent transition* forbids the execution of delay transitions in its starting state if its guard is true, i.e. the transition must be taken as soon as it is enabled [15]. Suppose, for instance, that in Fig. 1 we would want the transition from state $s = 1$ to be urgently executed at $x = 2$. Then, it would be insufficient to only write $x = 2$ instead of $x \geqslant 2$ in its guard. The transition should additionally be labelled as urgent, but PRISM does not support this. A solution would be to impose the invariant $x \leqslant 2$ on the starting state of this transition.

Since PRISM neither supports the description of systems in graphical form nor the simulation of PTAs, we used UPPAAL as an auxiliary tool for the preparation of PRISM PTA descriptions. UPPAAL is a tool for the modelling, simulation, and verification of timed automata [2]. These are basically like PTAs, except that they do not support probabilistic choice in transitions. Since in our case, the only effects with probabilities less than 1 are those for choosing the *backoff* values, we can model them in UPPAAL by nondeterministic choice (cf. [17]). Figure 2 shows the timed automaton, drawn in UPPAAL, which corresponds to our PTA model of a pair of stations s_i, r_i, written in PRISM. It is very similar to the PTA drawn in [7,8], except that inaccuracies are eliminated and that urgent states and transitions are represented as explained above. UPPAAL, unlike PRISM, allows writing of state names. The latter and the invariants are written in bold at the nodes. Although UPPAAL contains a special label to indicate a state as urgent, we modelled urgent states with help of invariants. UPPAAL descriptions also contain declarations, but they are not shown in this paper. For instance, the constants ($DATlen$, BO_PERIOD, ...) and the variables (*be*, *backoff*, ...) from Fig. 2 are declared there. Variable *be* is initialised to BE_MIN. UPPAAL uses the C-style notation for equality and assignment.

The transitions in Fig. 2 between the initial state *SET_BACKOFF* and *BACKOFForNOT* represent the setting of variable *backoff* to a random number as prescribed by the CSMA-CA algorithm. The states containing letter C are so-called committed states, which are known in UPPAAL, but not in PTAs in PRISM. For example, if *be* is equal to 2 in state *SET_BACKOFF*, then the automaton passes to the committed state in which the outgoing transition nondeterministically chooses an integer i between including 0 and 3, and assigns it to *backoff*. The committed state guarantees that this transition is executed immediately after the transition to this state and with no other transition (of another automaton in the parallel composition) inbetween. It should be noticed that some of the transitions in the automaton are

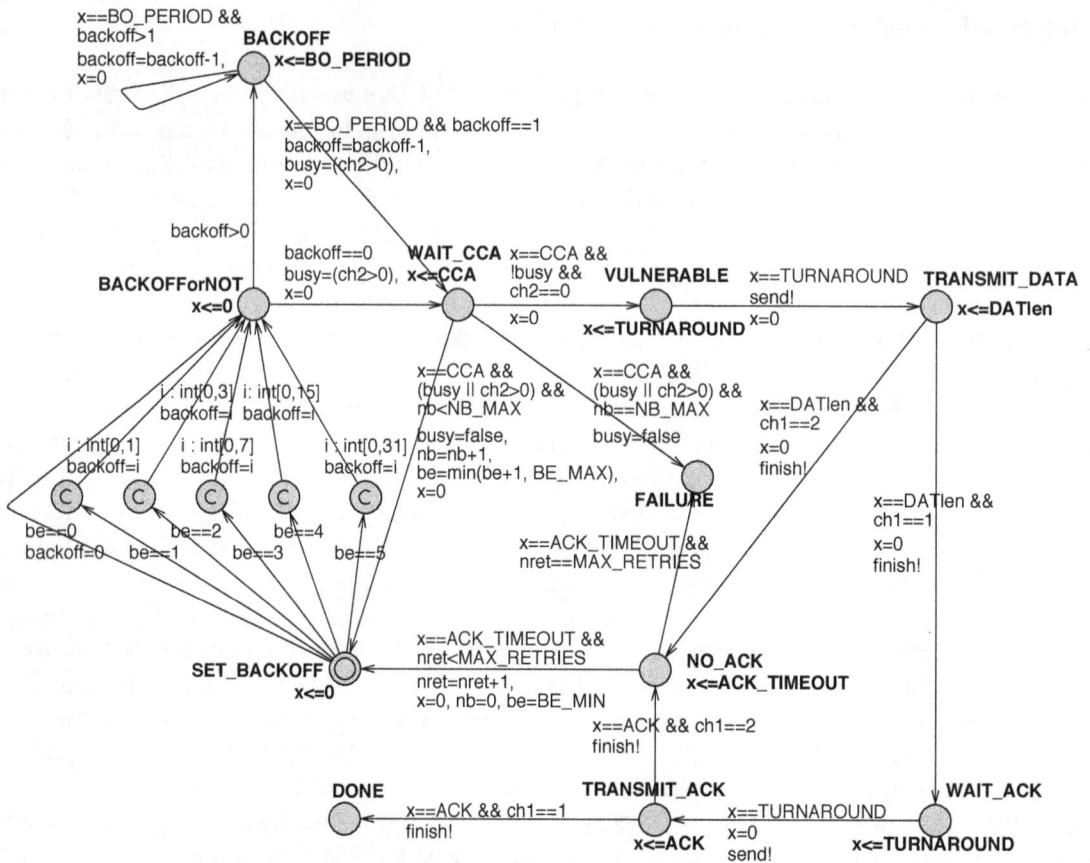

Fig. 2. Timed automaton model of a sending/receiving station pair in case of two sending stations.

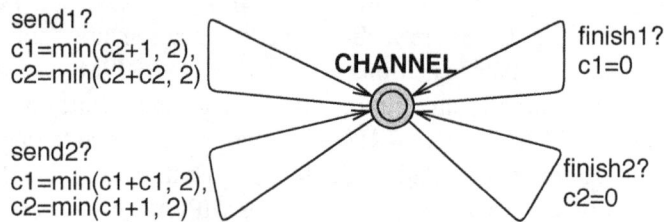

Fig. 3. Model of medium for two sending stations.

labelled with *send*! and, respectively, *finish*!. These labels have a role similar to the *send* mentioned in Section 2, except that in UPPAAL, every label ends with ! ('output') or ? ('input'). In contrast to PRISM, in a parallel composition of timed automata in UPPAAL, there is binary synchronisation of transitions with complementary labels: a transition with a label ending with ! is executed simultaneously with a transition labelled with the same label, but ending with ?.

The automaton shown is a so-called template. Let us call it *Station*. $ch1$, $ch2$, $send$, and $finish$ are its parameters. The model of the pair of stations s_1, r_1 (respectively, s_2, r_2) is obtained by setting $ch1$ to $c1$ (respectively, $c2$), $ch2$ to $c2$ ($c1$), $send$ to $send1$ ($send2$), and $finish$ to $finish1$ ($finish2$). The network model is the parallel composition of these two models, let us call them *Station1* and *Station2*, and of the timed automaton representing the communication channel. The latter is shown in Fig. 3. Let us call it *Medium*. Except for the uses of ! and ? indications, it is the same as the graphical representation of the PTA in [7], with the exception that in the latter its only state should not be indicated as urgent. $c1$ and $c2$ are (in UPPAAL global and in PRISM local) variables initialised to 0.

The idea is that the values of $c1$ and $c2$ can be either 0, 1, or 2. $c1 = 0$ (and analogous for $c2$) indicates that nothing is being sent by *Station1* (i.e. between s_1 and r_1). $c1 = 1$ (note that in the text, we use the PRISM notation for equality and similar) means that a (data or acknowledgement) frame is being sent by *Station1*, but not by *Station2*, and that the frame has not been garbled by a collision. $c1 = 2$ means that a frame is being sent by *Station1* and that it has been garbled by a collision with a frame sent by *Station2*.

The transitions labelled with $send1!$ and, respectively, $finish1!$ starting in states *VULNERABLE* and, respectively, *TRANSMIT_DATA* represent the start and, respectively, the end of sending the data frame by s_1 as described in Section 3, whereas the similarly labelled transitions from state *WAIT_ACK* and, respectively, *TRANSMIT_ACK* represent the start and, respectively, the end of sending the acknowledgement frame by r_1 (and at the same time receiving it by s_1). The analogous holds for sending and receiving in *Station2*. Every such transition is executed simultaneously with the transition of *Medium* with the complementary label, which properly sets variables $c1$ and $c2$.

The representation of the pair s_1, r_1 (and analogously for $i = 2$) with one automaton can be justified as follows. Once the last bit of a data frame sent by s_1 arrives (successfully, i.e. with no collision during the frame sending) at r_1, it shall inevitably send an acknowledgement after switching from RX to TX within a *TURNAROUND* interval because it is impossible that it ignores this frame. As told in Section 3, it would ignore it if it was in the course of turning from the TX mode to RX after sending an acknowledgement. It is, however, impossible for a new data frame to arrive during that time because for every data frame sent, the sending station waits longer than the time needed to completely send the acknowledgment, and after this waiting does not send the same data frame or a new one for at least *CCA + TURNAROUND* time units. Please note, that in this paper we model only the sending of one fresh data frame per station but nevertheless take into account the possibility of sending more frames when reasoning about the validity of the models. Consequently, all the models for sending one data frame in this paper can be readily used to build models for sending more than one data frame by adding appropriate transitions leading from the success/failure states back to the beginning of the CSMA-CA algorithm. Different delays before the latter could also easily be introduced in different stations.

It suffices to represent the start (respectively, the end) of sending of a data frame from s_i to r_i and the start (respectively, the end) of sending of the acknowledgement in the other direction with the same labels in the (P)TAs representing the stations and with the same transitions in the (P)TA representing the medium for the following reason. s_i waits a sufficient time before (re)sending a data frame so that the acknowledgement for the previous one from this station does not occupy the medium anymore, or shortly, because it is impossible for a data frame from s_i and an acknowledgement from r_i to be transmitted simultaneously.

There are four inaccuracies in the PTA model of the pair of stations given graphically in [7,8]. The model in Fig. 2 does not contain them anymore. Three of them are also no longer present in the PRISM code for Markov decision processes in [8]. One of the latter is that the length of the data frame to be sent is chosen anew each time the sending is retried because of the absence of an acknowledgement. This is not in accordance with the standard. We use a constant value $DATlen$. The length of the fresh data frame in the model could, of course, be chosen nondeterministically and remembered for retransmission as in the code in [8]. Please note, that for convenience throughout this paper it is assumed that all the stations send data frames of equal length.

Another inaccuracy in [7] is the following. The length of each data frame is chosen nondeterministically between $DATA\_MIN$ and $DATA\_MAX$ units, and assigned to a variable $data$. The state which is analogous to state $TRANSMIT\_DATA$ in Fig. 2 has the invariant $x \leqslant data$, which is analogous to the invariant in Fig. 2. On the transitions from this state, in the model for the station pair i, ci ($ch1$ in Fig. 2) is checked to see if the data frame sent has collided ($ci = 2$) or not ($ci = 1$). However, the transition which leads to state NO_ACK, meaning that a collision has occurred, has the time guard $x \geqslant DATA\_MIN$ in [7]. This, together with the invariant, means that the station can stop sending the data frame as soon as the $DATA\_MIN$ units are sent, even if the frame is longer than that. This is not in accordance with the standard. In the CSMA-CA algorithm, the station sends the complete frame even if a collision has occurred, because it detects collisions from the absence of acknowledgements. In [7], the time guard should be $x = data$ (or equivalently, $x \geqslant data$, as in the code in [8]). In our model, the analogous guard $x = DATlen$ is applied.

One inaccuracy in [7] is intentional: the data frame is resent until an acknowledgement arrives instead of only $MAX\_RETRIES$ times as in the standard and optionally in the PRISM code in [8].

The following inaccuracy, however, is unsolved in [8]. In the PTA graphs in [7,8], testing whether the channel is clear or not is carried out only at the end of the CCA period. As mentioned in Section 3, it should be assessed from the beginning to the end of the CCA period and declared as clear at the end of the period iff it was clear during the whole period. This inaccuracy has already been detected and corrected in the model of nonbeacon-enabled MAC written in ns-2. In [25], it is written that in that model the channel is tested at the end of the first symbol of the CCA period, but its status is reported after the last symbol. It is unclear whether the channel is also tested at the end of the last symbol as in the previous version. According to [5], in the OMNeT++ model the channel is tested at the beginning and at the end of the CCA period and reported busy iff it is busy at least at one of these moments. We came to the same solution independently for the PTA model. In Fig. 2, the channel is tested on both transitions to state $WAIT\_CCA$, as well as in this state when CCA time units pass. This is equivalent to testing for the whole CCA period because the minimal lengths of the data and acknowledgement frames are greater than the CCA period duration. Consequently, if a CCA period overlaps with a data or acknowledgement frame sending period, at least the start or the end of the CCA period lies within the latter.

It is easy to obtain the description of PTAs in PRISM from the timed automata in UPPAAL by following the pattern shown in Fig. 1. In PRISM, there are no templates, but a given module can be used to define another one which differs from it only in the module name and the names of some variables, constants, and actions by calling the first one and specifying the renaming of the original names. Another difference from UPPAAL is in the fact that those variables and constants meant to be local to a module must be named differently than variables and constants in other modules. Naturally, the PRISM module representing *Station1* gets all of them, as well as actions *send* and *finish* indexed with 1. The module representing *Station2* can be obtained by replacing them with the names indexed with 2:

```
module station2=station1[x1=x2, s1=s2, c1=c2, c2=c1, backoff1=backoff2,
                         be1=be2, nb1=nb2, nret1=nret2, busy1=busy2,
                         send1=send2, finish1=finish2]
endmodule
```

There, $s1$ and $s2$ are the variables used to represent the states of *Station1* and, respectively, *Station2* by integer values in a similar way as the states are represented in Fig. 1.

The essential difference between *Station*s in UPPAAL and the station modules in PRISM is, of course, that the latter represent *probabilistic* timed automata. Therefore, the nondeterministic choice of the *backoff* value in the initial state is modelled by specifying a uniform probabilistic distribution. Suppose that $s1 = 2$ denotes state *SET_BACKOFF* and $s1 = 3$ state *BACKOFForNOT* from Fig. 2 for *Station1*. The setting of *backoff1* in the PTA in PRISM is specified by requiring equal probabilities for all the possible values of *backoff1* for a given *be1* as follows (notice that we do not show all the updates for space reasons) [8,24]:

```
[] s1=2 & be1 = 0 -> (s1'=3) & (backoff1' = 0);
[] s1=2 & be1 = 1 ->
    1/2 : (s1'=3) & (backoff1' = 0) + 1/2 : (s1'=3) & (backoff1' = 1);
[] s1=2 & be1 = 2 ->
    1/4 : (s1'=3) & (backoff1' = 0) + 1/4 : (s1'=3) & (backoff1' = 1)
  + 1/4 : (s1'=3) & (backoff1' = 2) + 1/4 : (s1'=3) & (backoff1' = 3);
[] s1=2 & be1 = 3 ->
    1/8 : (s1'=3) & (backoff1' = 0) + ... + 1/8 : (s1'=3) & (backoff1' = 7);
[] s1=2 & be1 = 4 ->
    1/16 : (s1'=3) & (backoff1' = 0) + ... + 1/16 : (s1'=3) & (backoff1' = 15);
[] s1=2 & be1 = 5 ->
    1/32 : (s1'=3) & (backoff1' = 0) + ... + 1/32 : (s1'=3) & (backoff1' = 31);
```

Here we also give the PRISM guarded commands representing the transitions of PTA *Station1* (cf. Fig. 2) from states *BACKOFForNOT* and *BACKOFF* ($s1 = 4$) that start the CCA period, as well as the transitions from *WAIT_CCA* ($s1 = 5$) because these (and the analogous ones for any other *Station*) will mainly be the ones changed in the sequel:

```
// start of CCA
[] s1=3 & backoff1=0 -> (s1'=5) & (busy1'=(c2>0)) & (x1'=0);
[] s1=4 & x1=BO_PERIOD & backoff1=1 ->
    (s1'=5) & (backoff1'=backoff1-1) & (busy1'=(c2>0)) & (x1'=0);

// end of CCA
[] s1=5 & x1=CCA & (busy1 | c2>0) & nb1<NB_MAX ->
    (s1'=2) & (busy1'=false) & (nb1'=nb1+1) & (be1'=min(be1+1,BE_MAX)) & (x1'=0);
[] s1=5 & x1=CCA & (busy1 | c2>0) & nb1=NB_MAX -> (s1'=12) & (busy1'=false);
[] s1=5 & x1=CCA & !busy1 & c2=0 -> (s1'=6) & (x1'=0);
```

Since the PTA representing the medium has only one state (one node in UPPAAL), no variable is needed to represent it within the PRISM language. Its transitions can either be specified by using the expressions from Fig. 3 (with guards set to true) or as in Fig. 4 [8,24]. It can easily be seen that the effect of the expressions used in Figs 3 and 4 is equivalent if the medium is used in the parallel composition with the stations.

5. Models for two or more station pairs

In this section, we first generalise the model for two pairs of stations s_i, r_i with the common channel to the case of n pairs, $n \geqslant 2$. The model for two pairs, in particular the model of the medium in Fig. 4, suggests that the PTA model of the network is the parallel composition of n PTAs similar to *Station* in Fig. 2, and of a PTA representing the medium with variable ci for messages associated with the pair s_i, r_i, $i = 1, \ldots, n$. The idea is that each ci still ranges over 0, 1, and 2, and that the meaning of these values is analogous to what it was before. It follows that for each i, the transition labelled with $send_i$

```
module medium

    // medium status
    c1 : [0..2];
    c2 : [0..2];
    // ci corresponds to messages associated with station i
    // 0 nothing being sent
    // 1 being sent correctly
    // 2 being sent garbled

    // begin sending message and nothing else currently being sent
    [send1] c1=0 & c2=0 -> (c1'=1);
    [send2] c2=0 & c1=0 -> (c2'=1);
    // begin sending message and something is already being sent
    // in this case both messages become garbled
    [send1] c1=0 & c2>0 -> (c1'=2) & (c2'=2);
    [send2] c2=0 & c1>0 -> (c1'=2) & (c2'=2);
    // finish sending message
    [finish1] c1>0 -> (c1'=0);
    [finish2] c2>0 -> (c2'=0);

endmodule
```

Fig. 4. A possible description of medium for two sending stations in PRISM.

should set ci to 1 and all the other c variables should remain equal to 0 if no station is sending. If at least one message is being sent (garbled or not), this transition should leave the c variables of the stations that are not sending equal to 0 and set all the others to 2. For each i, the transition labelled with $finishi$ should set ci to 0.

Figure 5 shows the first proposal for the PTA model of the medium for $n = 3$. The template for a station pair differs from the one in Fig. 2 in that in the states *BACKOFForNOT*, *BACKOFF*, and *WAIT_CCA*, *ch3* has to be checked besides *ch2* to see if the medium is busy (i.e. $ch2 > 0 \mid ch3 > 0$) or not (i.e. $ch2 = 0 \quad \& \quad ch = 0$). Clearly, for an arbitrary n, all the ch variables except $ch1$ have to be checked in the model. Speaking in terms of PRISM, in module *Station1* the commands for these states for an arbitrary n become as follows:

```
// start of CCA
[] s1=3 & backoff1=0 -> (s1'=5) & (busy1'=(c2>0 | ... | cn>0)) & (x1'=0);
[] s1=4 & x1=BO_PERIOD & backoff1=1 ->
    (s1'=5) & (backoff1'=backoff1-1) & (busy1'=(c2>0 | ... | cn>0)) & (x1'=0);

// end of CCA
[] s1=5 & x1=CCA & (busy1 | c2>0| ... | cn>0) & nb1<NB_MAX ->
    (s1'=2) & (busy1'=false) & (nb1'=nb1+1) & (be1'=min(be1+1,BE_MAX)) & (x1'=0);
[] s1=5 & x1=CCA & (busy1 | c2>0| ... | cn>0) & nb1=NB_MAX ->
    (s1'=12) & (busy1'=false);
[] s1=5 & x1=CCA & !busy1 & c2=0 & ... & cn=0 -> (s1'=6) & (x1'=0);
```

Given the new module *Station1* in PRISM, the module representing the station pair i, $1 < i \leqslant n$, can be obtained by renaming as before, except that now, assuming that we list $c_1, \ldots, c_n$ as the parameters of *Station*$_1$ in this order (from now on we will write the indices subscripted for convenience), c_1 must be renamed to c_i, whereas the other c variables of *Station*$_1$ must be renamed disjointly to those c variables other than c_i. Generally, the ordering does not matter, but could if the kind of symmetry present in this protocol could be exploited by the PRISM model checker. We suggest one of the fol-

```
module medium

    // medium status
    c1 : [0..2];
    c2 : [0..2];
    c3 : [0..2];
    // ci corresponds to messages associated with station i
    // 0 nothing being sent
    // 1 being sent correctly
    // 2 being sent garbled

    // begin sending message and nothing else currently being sent
    [send1] c1=0 & c2=0 & c3=0 -> (c1'=1);
    [send2] c2=0 & c1=0 & c3=0 -> (c2'=1);
    [send3] c3=0 & c1=0 & c2=0 -> (c3'=1);
    // begin sending message and something is already being sent
    // in this case all messages sent become garbled
    [send1] c1=0 & c2>0 & c3>0 -> (c1'=2) & (c2'=2) & (c3'=2);
    [send1] c1=0 & c2>0 & c3=0 -> (c1'=2) & (c2'=2);
    [send1] c1=0 & c2=0 & c3>0 -> (c1'=2) & (c3'=2);
    [send2] c2=0 & c1>0 & c3>0 -> (c2'=2) & (c1'=2) & (c3'=2);
    [send2] c2=0 & c1>0 & c3=0 -> (c2'=2) & (c1'=2);
    [send2] c2=0 & c1=0 & c3>0 -> (c2'=2) & (c3'=2);
    [send3] c3=0 & c1>0 & c2>0 -> (c3'=2) & (c1'=2) & (c2'=2);
    [send3] c3=0 & c1>0 & c2=0 -> (c3'=2) & (c1'=2);
    [send3] c3=0 & c1=0 & c2>0 -> (c3'=2) & (c2'=2);
    // finish sending message
    [finish1] c1>0 -> (c1'=0);
    [finish2] c2>0 -> (c2'=0);
    [finish3] c3>0 -> (c3'=0);

endmodule
```

Fig. 5. A possible description of medium for three sending stations in PRISM.

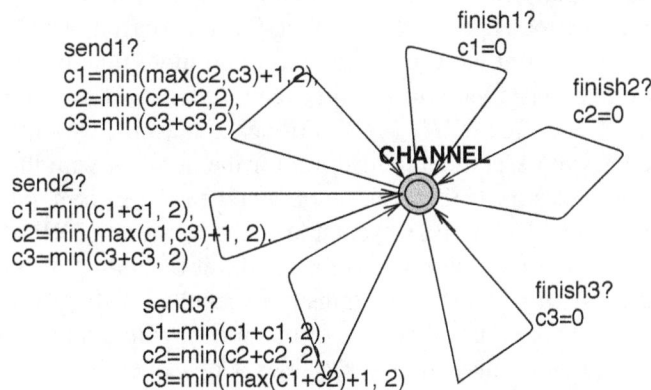

Fig. 6. Simpler description of medium for three sending stations.

lowing (the renaming of c_1 included): $Station_i = Station_1(c_i, c_1, \ldots, c_{i-1}, c_{i+1}, \ldots, c_n)$ or $Station_i = Station_1(c_i, c_{i+1}, \ldots, c_n, c_1, \ldots, c_{i-1})$.

We could, of course, derive the description of the medium for an arbitrary n from the one in Fig. 5. We do, however, leave this to the reader and strive for a short description similar to the one in Fig. 3.

From Fig. 5, it can be seen that to properly set c_i in action $send_i$, the largest from all the other c variables has to be taken and incremented by 1, but the result must not be larger than 2. The PTA in Fig. 6 is an equivalent of the PTA in Fig. 5 if used together with the PTAs representing the stations adapted as just explained. For the network with n sending stations, all the commands of the medium can have guards equal to true and the following effect for $i = 1, \ldots, n$:

- the effect of $send_i$:
  * $c_i' = \min\left(\max\left(c_1, \ldots, c_{i-1}, c_{i+1}, \ldots, c_n\right) + 1, 2\right)$,
  * $c_j' = \min(c_j + c_j, 2)$, for $j = 1, \ldots, n, j \neq i$;
- the effect of $finish_i$: $c_i' = 0$.

Now, we propose a simpler model for the network consisting of n station pairs. In [10], a beacon-enabled network without acknowledgements is modelled by using timed stochastic automata. An integer variable *sending* initialised to 0 is used. Every station increments it by 1 when it starts to send a data frame and decrements it by 1 when it finishes. This variable suffices for the purpose of CCA – the channel is busy iff it is greater than 0. It, however, does not suffice for the detection in all situations as to whether a frame has collided. The detection of collisions is needed because we have to know whether the acknowledgment will be received successfully or not. In the current models, this is checked in the outgoing transitions of states *TRANSMIT_DATA* and *TRANSMIT_ACK* by checking c_i for *Station$_i$* (see Fig. 2). Clearly, if *sending* is greater than 1, it indicates a collision, but it cannot be used instead of c_i because if *sending* is equal to 1, there are two possibilities. One is that exactly one station is sending and that the frame is not garbled because it started to send when the channel was clear and no other station has started to send thereafter. Another one is that only one station is sending, but the frame is garbled. This could happen as follows. Suppose that only two stations are sending, meaning that the frames being sent have collided. When one of them finishes sending, the variable *sending* is decremented to 1.

We therefore introduce boolean variable *colind*, which is set by the model of the medium in transitions labelled with $send_i$ to indicate whether the sending of a new (data or acknowledgement) frame by *Station$_i$* causes a collision (in this case, it is set to true). A similar idea is used in [3]. This suffices in order for the model to properly represent the protocol. However, it is our wish that *colind* be true exactly when there are garbled messages in the channel. For this reason, we also set it to true in every transition labelled with $finish_i$ if at the time of its execution more than one station is sending, i.e. if *sending* > 1, and to false otherwise. Variable *colind* is checked instead of c_i in the outgoing transitions of states *TRANSMIT_DATA* and *TRANSMIT_ACK*. For space reasons, we only show the template for *Station*s (Fig. 7) and the PTA representing the medium for the network with three station pairs (Fig. 8) drawn in UPPAAL. The modules in PRISM can be obtained easily. In principle, the transitions of the medium can be the same as in UPPAAL. However, at least the version of PRISM we have used requires a limiting of the possible values of variables in them. We did it by adding the conditions on the values of *sending* in their guards (*sending* > 0 in the transitions labelled with $finish_i$ and *sending* $< n$ in the transitions labelled with $send_i$). The advantage of this model is that the common channel for any n is represented by only two variables instead of n variables. Consequently, the model of the medium for n has transitions labelled with $send_i$, $i = 1, \ldots, n$, all with the same effect, and likewise for $finish_i$. All the n modules representing the station pairs are instances following the template in Fig. 7. They all check the variables *sending* and *colind* of the medium in the same way, and differ only in the indices of the other variables and the action labels.

In fact, UPPAAL allows even a simpler model. The same label, for example *send!* (respectively, *send?*) can be used in the automata representing the stations (respectively, the medium) instead of $send_i$! (respectively, $send_i$?) for $i = 1, \ldots, n$, and likewise for $finish_i$. Consequently, the medium in UPPAAL

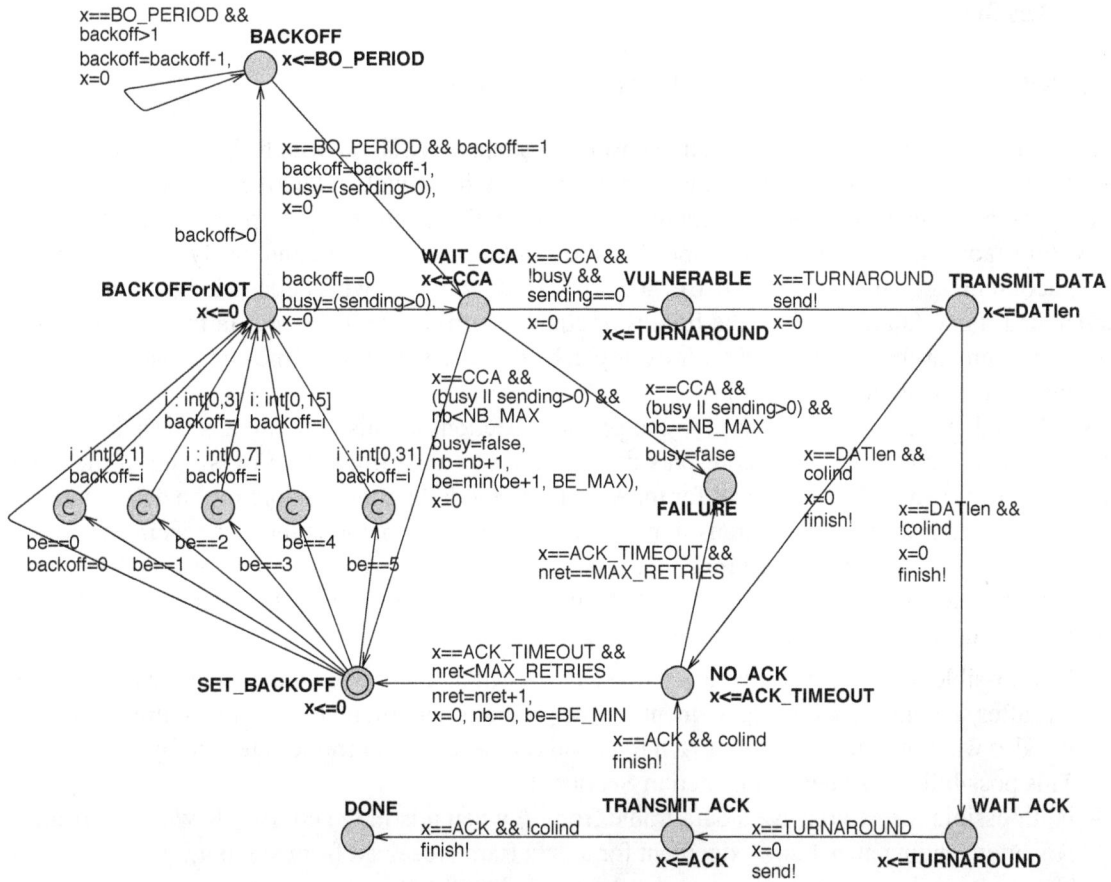

Fig. 7. Model of a station pair using variable *sending*.

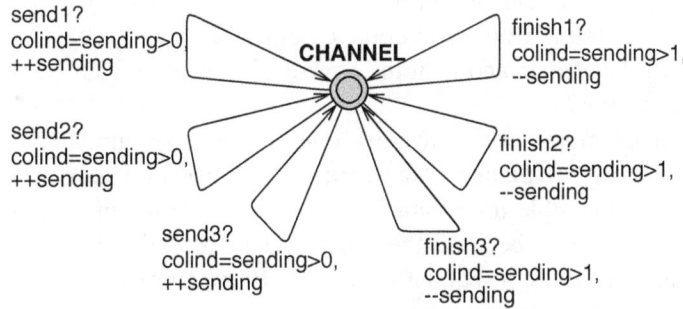

Fig. 8. Medium for three sending stations using variable *sending*.

can be represented with only two transitions. This simplification is possible because of the binary synchronisation in the parallel composition in UPPAAL. In PRISM, by default, all the transitions with the same label must synchronise. It is possible to achieve other kinds of synchronisations with special operators, but we could not find a solution that would allow only labels *send* and *finish* in the model in PRISM.

6. Models for star topology

6.1. Models for star topology networks with no hidden stations

The IEEE 802.15.4 standard is often used in star-shaped WSNs in which there is a central node (a coordinator called 'receiving station' and denoted by r in the sequel) collecting sensor data sent directly to it by nodes ('sending stations', denoted as s_i) located around it [4]. Suppose that all the stations are within radio range of each other and that the sending stations simultaneously start to execute the nonbeacon-enabled version of the IEEE 802.15.4 MAC protocol with acknowledgements in order to send a data frame (all the stations the frames of equal length) to the receiving station, which only sends acknowledgements back. This could, for example, be the case in a WSN if the coordinator sent a request for sensor data to all the stations [4].

We claim that the models proposed in the previous sections are also the models of such networks. In all of them, automaton *Station$_i$* represents the sending of a data frame to r by s_i and of the acknowledgement from r to s_i. The argument for the validity of modelling the sending of the data frame and the acknowledgement with the same transitions in the *Medium* automaton is the same as in Section 4. The question remains whether it is true that once s_i successfully (i.e. without a collision) sends a data frame to r, the latter necessarily sends an acknowledgement to it as in the model. In order for the answer to be positive, we must eliminate the following three possibilities:

– Is it possible for r to receive a data frame from s_i when it is in the course of switching from TX to RX after sending an acknowledgement for the previous data frame from s_i? (As already mentioned, we also want our modelling approach to be valid if several data frames are sent by s_i.)
  This possibility has been eliminated in Section 4.

– Is it possible for r to receive a data frame from s_i when it is in the course of switching from TX to RX after sending an acknowledgement for a data frame received from some s_j, $j \neq i$?
  The answer is no because even the minimal length of the data frames is larger than the *TURNAROUND* interval. Consequently, it is impossible for the last bit of the data frame from s_i to be received by r inside the *TURNAROUND* interval without part of that frame overlapping and thus colliding with the acknowledgement sent by r until the start of that interval.

– Is it possible for a data frame from s_i to come complete and not garbled to r when the latter has just received another data frame and is in the course of switching from RX to TX in order to send an acknowledgement for it?
  It is impossible for a data frame from s_i to come immediately after another data frame from the same station because s_i does not send a new data frame before the acknowledgement for the previous one is finished. It is also impossible for a data frame from s_i to come immediately after a data frame from some station s_j, $j \neq i$, because the minimal length of the data frames is larger then the *TURNAROUND* interval, and consequently, the data frame from s_i ending in that interval would necessarily collide with the data frame from s_j.

6.2. Models for star topology networks with hidden stations

In contrast to Section 6.1, we now suppose that not all the sending stations in a star-shaped network are within the range of each other and propose accordant adaptions of the models. Suppose a network as sketched in Fig. 9 consisting of the receiving station r, which receives data frames from and sends acknowledgements to the sending stations s_i, $i = 1, \ldots, n$. They are all within the range of r and

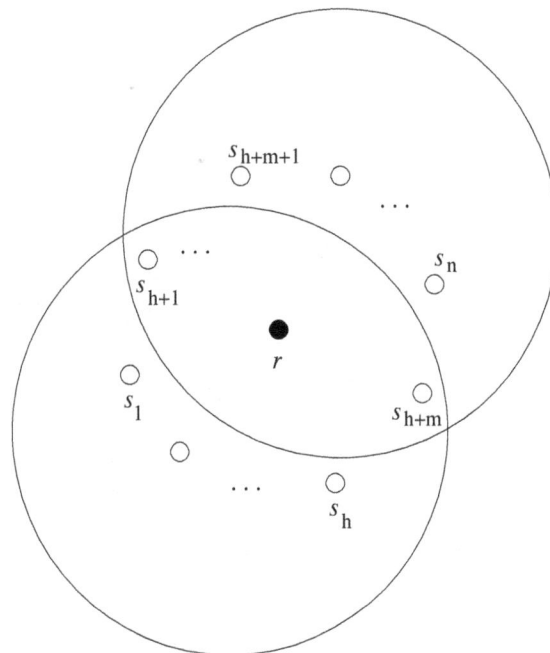

Fig. 9. A star-shaped network with hidden stations.

vice versa. There are three different kinds of sending stations. Stations s_i with the lowest indices, $i = 1, \ldots, h$, are within the range of each other, but outside the range of stations s_i with the highest indices, $i = h + m + 1, \ldots, n$. Likewise, the latter are within the range of each other, but outside the range of the former. Stations s_i, $i = h + 1, \ldots, h + m$, are in a common range with all the others, like r. It follows that the stations with the lowest indices are *hidden* from the stations with the highest ones and vice versa.

The problem with the hidden stations with the lowest indices is in that when they perform CCA, they do not "hear" whether anything is being sent by the stations with the highest ones, and vice versa. A hidden station, therefore, might send a data frame after the *TURNAROUND* time from the end of CCA even if a hidden station from the other range is sending. This might increase the probability of collisions because once a data frame from a hidden station is sent, it propagates into the other range. As already mentioned, we neglect the propagation delay, which means that such a data frame comes into that range immediately after the start of sending.

We first propose a way of modelling the star-shaped network with the hidden stations by adapting the model for n stations presented in Section 5 in which the medium is represented by variables c_i, $i = 1, \ldots, n$.

Now, the model in PRISM is a collection of a PTA (i.e. module) representing the medium as well as of two kinds of PTAs, the ones representing the hidden stations and the ones representing the sending stations that are within the range of all the others.

The PTA in PRISM representing a hidden station s_i is the same as the PTA representing station s_i in the network with no hidden stations (and similar to the UPPAAL template shown in Fig. 2), except for the following. In the former model, in states *BACKOFForNOT*, *BACKOFF* and *WAIT_CCA*, the variables c_j of all the other *Station*s are tested. However, c_j represents the sending of either a data frame by station s_j or an acknowledgement by r. Now, s_i is only reached by acknowledgements sent by r and by data frames sent by the stations that are within the same range as s_i. We, therefore, introduce a new variable

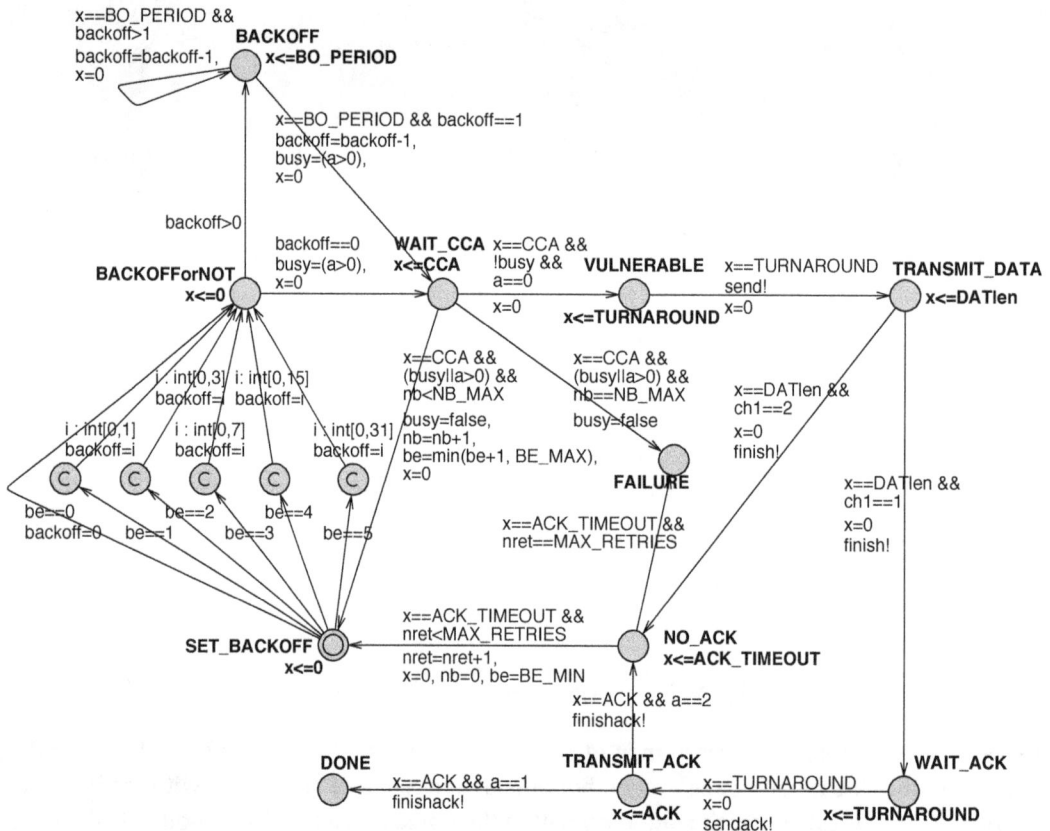

Fig. 10. Model of a hidden station using variable a.

a which is equal to 0 if no acknowledgement is being sent, 1 if being sent but not garbled, and 2 if garbled. Consequently, in states *BACKOFForNOT*, *BACKOFF* and *WAIT_CCA* of the PTA representing the hidden station s_i, the following variables are tested:

- if $1 \leqslant i \leqslant h$: a and c_j for all j, $1 \leqslant j \leqslant h + m$, $j \neq i$,
- if $h + m + 1 \leqslant i \leqslant n$: a and c_j for all j, $h + 1 \leqslant j \leqslant n$, $j \neq i$.

Furthermore, in the PTA representing any hidden station s_i, i.e. for $1 \leqslant i \leqslant h$ and $h + m + 1 \leqslant i \leqslant n$:

- the transition leading from state *WAIT_ACK* is labelled with action $sendack_i$,
- in the guards of the transitions from state *TRANSMIT_ACK*, a is tested instead of c_i to see whether the acknowledgement sent has collided or not, and the transitions are labelled with action $finishack_i$.

Figure 10 shows the template of the timed automaton model of a sending station which is hidden from all the other stations in the network except the destination one.

The PTAs representing stations s_i which are within the range of all the other stations, i.e. for $h + 1 \leqslant i \leqslant h + m$, are the same as those representing the hidden ones, except that in the guards of the transitions from states *BACKOFForNOT*, *BACKOFF* and *WAIT_CCA*, a and c_j for all j, $1 \leqslant j \leqslant n$, $j \neq i$, are tested.

If we do not want to write all the (P)TAs representing the hidden stations in PRISM or, respectively, in UPPAAL from scratch, we need a template for the hidden stations with the lower indices and another

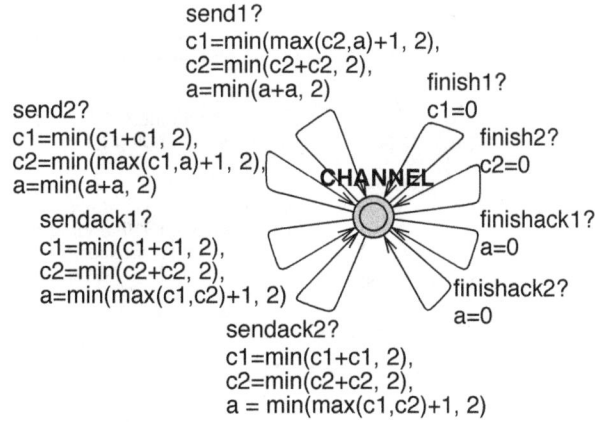

Fig. 11. Model of medium using a for a star network with two sending stations.

one for those with the higher ones if the number of the former is not the same as the number of the latter because in this case the automata differ in the number of the c variables they must check.

The PTA representing the medium differs from the PTA described in Section 5 in that all the transitions take variable a into account besides variables c_i, $i = 1, \ldots, n$, and that transitions labelled with $sendack_i$ and $finishack_i$ for $i = 1, \ldots, n$ are added. Figure 11 shows the model of the medium for the case $n = 2$.

All the commands of the medium can have the guards equal to true and have the following effect for $i = 1, \ldots, n$:

- the effect of $send_i$:
    * $c'_i = \min(\max(c_1, \ldots, c_{i-1}, c_{i+1}, \ldots, c_n, a) + 1, 2)$,
    * $c'_j = \min(c_j + c_j, 2)$, for $j = 1, \ldots, n, j \neq i$,
    * $a' = \min(a + a, 2)$;
- the effect of $finish_i$: $c'_i = 0$;
- the effect of $sendack_i$:
    * $c'_j = \min(c_j + c_j, 2)$, for $j = 1, \ldots, n$,
    * $a' = \min(\max(c_1, \ldots, c_n) + 1, 2)$;
- The effect of $finishack_i$: $a' = 0$.

It can be seen that the transitions labelled with $sendack_i$ and, respectively, $finishack_i$ have the same effect for $i = 1, \ldots, n$. Again, because of binary synchronisation, in the model in UPPAAL built by following this approach it suffices to replace them with one transition labelled with $sendack$ and, respectively, $finishack$, and use the same labels in all the PTAs representing the stations.

Since the solution with variable a requires a lot of variables, we next propose one which is an adaption of the solution for n stations from Section 5 with the variables $sending$ and $colind$. The idea is to have the latter indicate whether the currently sent frames are garbled as before, and two variables similar to $sending$, both initialised to 0. Given the network shown in Fig. 9, let variable $sending1$ denote the number of frames being sent at a time by stations r and s_i, $i = 1, \ldots, h + m$, i.e. within the common range denoted by the lower circle in that figure (let us call it range 1). Likewise, let variable $sending2$ denote the number of frames being sent at a time by stations r and s_i, $i = h + 1, \ldots, n$, i.e. within the common range denoted by the upper circle in the figure (let us call it range 2). When a data frame is sent by a station that is only within range 1 (respectively, range 2), only $sending1$ (respectively,

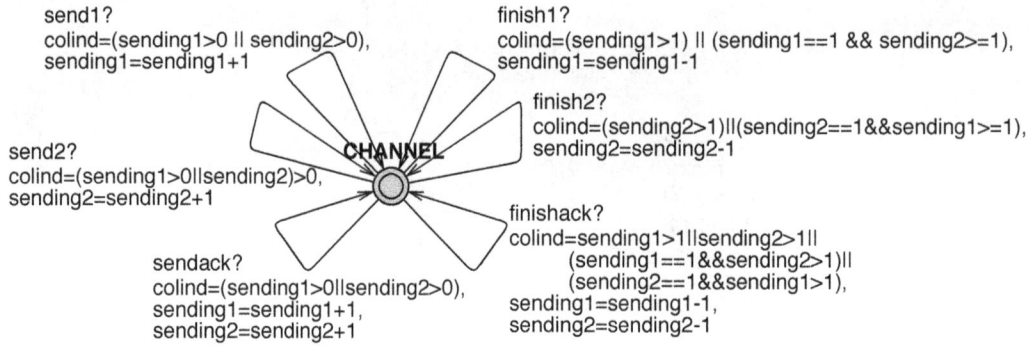

send1?
colind=(sending1>0 ll sending2>0),
sending1=sending1+1

finish1?
colind=(sending1>1) ll (sending1==1 && sending2>=1),
sending1=sending1-1

finish2?
colind=(sending2>1)ll(sending2==1&&sending1>=1),
sending2=sending2-1

CHANNEL

send2?
colind=(sending1>0llsending2)>0,
sending2=sending2+1

finishack?
colind=sending1>1llsending2>1ll
(sending1==1&&sending2>1)ll
(sending2==1&&sending1>1),
sending1=sending1-1,
sending2=sending2-1

sendack?
colind=(sending1>0llsending2>0),
sending1=sending1+1,
sending2=sending2+1

Fig. 12. UPPAAL model of medium using *sending* variables for star network with two mutually hidden stations.

$sending2$)) is incremented by one. *colind* is set to true if something is already being sent in range 1 or range 2 because of the assumption of zero propagation delay. When a data frame or, respectively, an acknowledgement frame is sent by a station which is within both ranges, the variables $sending1$ and $sending2$ are incremented by one and *colind* is set as in the former case. As in the case with no hidden stations, once *colind* is true, it must stay so until all the stations involved in the collision finished sending. During the CCA period, those stations that are only within one range only check the *sending* variable representing the latter. Those stations that are within both ranges check both *sending* variables before sending a data frame to see if the channel is clear.

Again, the model in PRISM consists of n *Station$_i$* modules representing PTAs of the same form as the TA in Fig. 2, except for the following, and of a medium module.

For $i = 1, \ldots, n$, in the guards of the transitions from states *TRANSMIT_DATA* and *TRANSMIT_ACK* of the module for s_i, *colind* is checked as in the solution with variable *sending* for the network with no hidden stations (Fig. 7). The transition from state *WAIT_ACK* is labelled with action *sendack$_i$* instead of *send$_i$*, and the transitions from state *TRANSMIT_ACK* with the action *finishack$_i$* instead of *finish$_i$*.

For $i = 1, \ldots, h$ (respectively, for $i = h + m + 1, \ldots, n$), $sending1$ (respectively, $sending2$) is checked in states *BACKOFForNOT*, *BACKOFF* and *WAIT_CCA* of the PTA representing s_i instead of *sending* as in the model with no hidden stations. For $i = h + 1, \ldots, h + m$, $sending1$ and $sending2$ are checked – if at least one of them is greater than 0, the channel is busy.

The model of the medium differs from the one for the network with no hidden stations (see e.g. Fig. 8) in that it has actions *sendack$_i$* and *finishack$_i$* besides *send$_i$* and *finish$_i$* for $i = 1, \ldots, n$. All the guards can be true, but for the reason already mentioned in Section 5, we have added conditions on $sending_1$ and/or $sending_2$ in the guards to prevent the complaints of the version of PRISM we used.

The transitions labelled with *sendack$_i$* and *finishack$_i$* have the same effect for any i, $1 \leqslant i \leqslant n$:

– the effect of *sendack$_i$*:
  * $colind' = (sending1 > 0 \mid sending2 > 0)$,
  * $ending1' = sending1 + 1$,
  * $sending2' = sending2 + 1$;

– the effect of *finishack$_i$*:
  * $colind' = sending1 > 1 \mid sending2 > 1 \mid (sending1 = 1 \;\&\; sending2 > 1) \mid (sending2 = 1 \;\&\; sending1 > 1)$,
  * $sending1' = sending1 - 1$,
  * $sending2' = sending2 - 1$.

The transitions labelled with $send_i$ or $finish_i$ for $i = 1, \ldots, h$ have the following effect:

- the effect of $send_i$:
  * $colind' = (sending1 > 0 \mid sending2 > 0)$,
  * $sending1' = sending1 + 1$;
- the effect of $finish_i$:
  * $colind' = sending1 > 1 \mid (sending1 = 1 \ \& \ sending2 \geqslant 1)$,
  * $sending1' = sending1 - 1$.

The effect of the transitions labelled with $send_i$ and, respectively, $finish_i$ for $i = h + m + 1, \ldots, n$ is obtained from the one for $i = 1, \ldots, h$ by substituting $sending2$ for $sending1$ and vice versa in the above expressions.

The transitions labelled with $send_i$ and, respectively, $finish_i$ for $i = h + 1, \ldots, h + m$ have the same effect as the transitions labelled with $sendack_i$ and, respectively, $finishack_i$ for $i = h + 1, \ldots, h + m$.

Figure 12 shows such a model of the medium for a star topology network with two sending stations hidden from each other, i.e. for $n = 2$, $h = 1$, and $m = 0$ in Fig. 9, but note that there is only one $sendack$ and, respectively, *finishack* transition. For the reason already mentioned, this is only possible in UPPAAL.

7. Some results of probabilistic model checking

In UPPAAL, the properties to be verified by model checking are expressed in its query language based on Computation Tree Logic (CTL). In PRISM, for PTAs they are expressed in Probabilistic CTL (PCTL). For a quick overview of the property specification, please consult [2,24]. Within the framework of this paper, the experiments were carried out by using the default values of *BE_MAX*, *NB_MAX*, and *MAX_RETRIES*, and the values from 0 to 3 for *BE_MIN*. The values for the other constants (note that they are all related to time) were first set to the standard values mentioned in Section 3, but we soon realised that we had to scale them down by using timescale abstraction in order for the PRISM model checker not to run out of memory for larger values of *BE_MIN* and $DATlen$ at least for networks with two sending stations. We performed the model checking with PRISM 4.0.3 by using the MTBDD engine and the PTA model checking method with digital clocks [15] on a computer with 2 GB of memory. We focused on the networks with bit rate 250 kbps. This means the minimal (respectively, maximal) length of data frames 30 (respectively, 266) symbols (i.e. time units, speaking in terms of time in (P)TA models), the length of acknowledgement frames 22 symbols, and the length of *ACK_TIMEOUT* 54 symbols. An exact timescale abstraction consists in choosing a new unit of time such that it is a common divisor of all constants in clock constraints in the model and dividing all these constants by it [7]. Such an abstraction preserves the properties we want to verify. We wanted to carry out model checking for the minimal and maximal lengths of data frames. We, therefore, chose the new time unit to be 2, giving the new *BO_PERIOD* equal to 10, *CCA* equal to 4, *TURNAROUND* to 6, $DATlen$ to 15 for the shortest frames and 133 for the longest, *ACK* to 11, and *ACK_TIMEOUT* to 27. Since the new time unit is the time needed to transmit two symbols, i.e. 8 bits, it is equal to 32 μs for the bit rate 250 kbps.

This scaling enabled us to verify networks with no hidden stations, as well as with hidden ones, for *BE_MIN* and $DATlen$ up to their maximal values (with a few exceptions for the latter) for $n = 2$ and $n = 3$. We carried out experiments for $DATlen$ equal to 15, 45, 75, 105, and 133. However, for $n = 3$, the verification was feasible only for the MAC without acknowledgements.

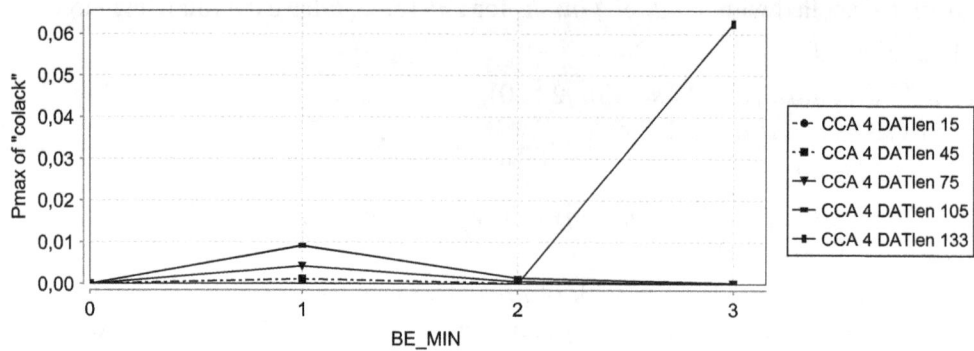

Fig. 13. Probability of collisions involving an acknowledgement frame for standard *CCA* in network with two sending stations and acknowledgements.

We usually first checked the prepared network model for validity with UPPAAL and/or PRISM, typically by verifying whether certain combinations of states or values of variables are possible. For instance, for the models with two stations with acknowledgements, we checked whether it is possible that two acknowledgements are transmitted simultaneously. In UPPAAL, we checked, for example, that formula `E<>(Station1.TRANSMIT_ACK and Station2.TRANSMIT_ACK)` (E meaning 'there exists a run' and <> meaning 'eventually') was false. In PRISM, the same was expressed with formula `P>0[F((s1=9)&(s2=9))]` (P>0 meaning 'non-zero probability' and F meaning 'eventually').

For all the networks considered, PRISM reported the same number of global states and transitions for the model using the c (and possibly a) variables and the one using variable(s) *sending*. As this is a good indication that these models are equivalent (and valid), we used them interchangeably, and shall, therefore, not indicate which results were obtained using which of them.

7.1. Model checking of networks with no hidden stations

We first experimented with the models of IEEE 802.15.4 MAC with acknowledgements for two stations. In [12], it is claimed that the length of the CCA period being greater than the turnaround time eliminates the possibility for a data frame to collide with an acknowledgement frame from the coordinator. By model checking, we found that this claim cannot be confirmed in the models given in [7,8] because of inaccurate modelling of the clear channel assessment. In our models with no hidden stations, the claim was found to be true. We first checked the probability of collisions involving a data frame and an acknowledgement for the standard value of *CCA*, i.e. 8 symbols or 4 new time units. The results are shown in Fig. 13. This probability is greater than zero for all the combinations of parameter values considered except those with *BE_MIN* = 0 and those with the maximal data frame length, for which it is greater than zero only at *BE_MIN* = 3. For *CCA* equal to 16 symbols (i.e. 8 new units of time) and to 14 symbols (i.e. 7 units), the result is zero in all the cases.

Further experiments were performed for *CCA* equal to 8 and 16 symbols. We were interested in the probability that both (i.e. all) sending stations successfully send the data frame (let us denote it *Pd*), expressed, for example, as `Pmin=?[F "done"]` with "done" defined to mean that all the sending stations are in state *DONE* (please, note that for all our PTA models, `Pmin` is equal to `Pmax` because they do not contain nondeterminism [15]). For all the models for which we present the verification results in the rest of this paper, *Pd* for *BE_MIN* = 0 is 0 (cf. [8]).

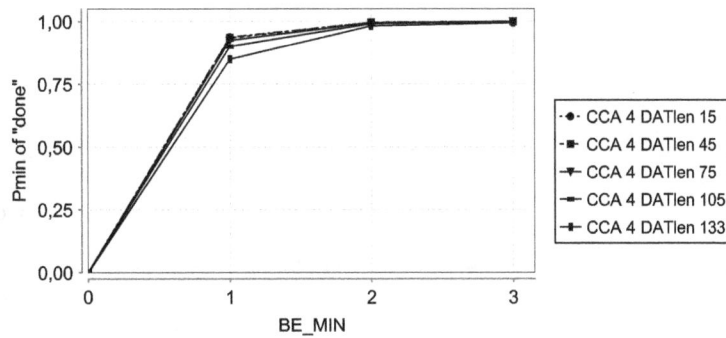

Fig. 14. Probability of "done" for standard *CCA* in network with two sending stations and acknowledgements.

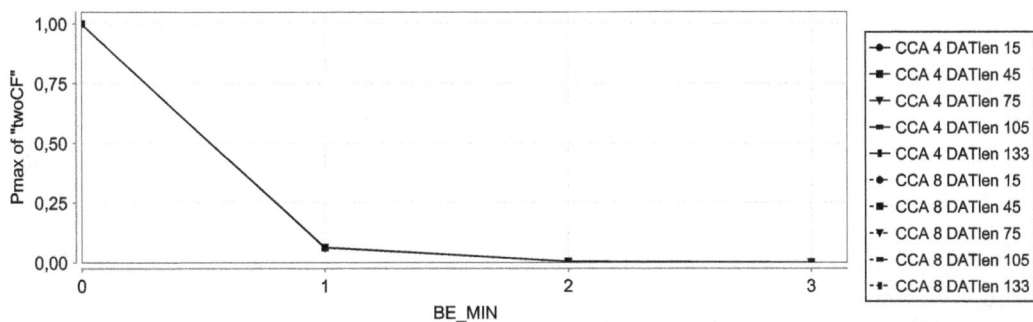

Fig. 15. Probability of a collision failure in network with two sending stations and acknowledgements.

For the network with two sending stations and $CCA = 4$, Pd at $BE\_MIN = 1$ is from around 90 (respectively, 85) percent for $DATlen = 105$ (respectively, 133) to 93 percent for $DATlen = 15$ (Fig. 14). At BE_MIN equal to 2 and 3 it is more than 99 percent for all these frame lengths, the largest being at $BE\_MIN = 3$. Although $CCA = 8$ prevents collisions of data with acknowledgements, it improves Pd for all the considered data frame lengths by at most a few tenths of a percent, except at $BE\_MIN = 1$ and $DATlen = 133$, where Pd increases by about 1.6 percent. This might seem surprising because for the longest data frames and the standard CCA, the probability of collisions involving an acknowledgement at $BE\_MIN = 1$ is zero, whereas it is more than 6 percent at $BE\_MIN = 3$ (Fig. 13).

Consequently, in both (P)TAs representing the stations, we introduced two states instead of the *FAIL-URE* state, one representing the channel access failure and pointed to by the transition that tests the condition $nb = \text{NB\_MAX}$, and another one representing the collision failure and being the target state of the transition that tests the condition $nret = MAX\_RETRIES$. By using UPPAAL and PRISM, we verified that for all the combinations of parameter values, for CCA equal to 4 and 8, the executions can terminate in one of the following three ways: either both station s_1 and s_2 are successful (denoted "done"), both stations declare a collision failure (denoted "twoCF" in the sequel), or one of the stations declares success and another one a channel access failure (denoted "oneCAFoneD"). Figures 15 and 16 show the probability of the latter two cases. Interestingly, $CCA = 8$ causes a little decrease (by at most about a tenth of a percent) in the probability of a collision failure at all the values of BE_MIN greater than 0 and $DATlen$ except at $BE\_MIN = 1$ and $DATlen = 133$, where that probability does not change. For the latter combination, the probability of a channel access failure drops by about 1.6 percent,

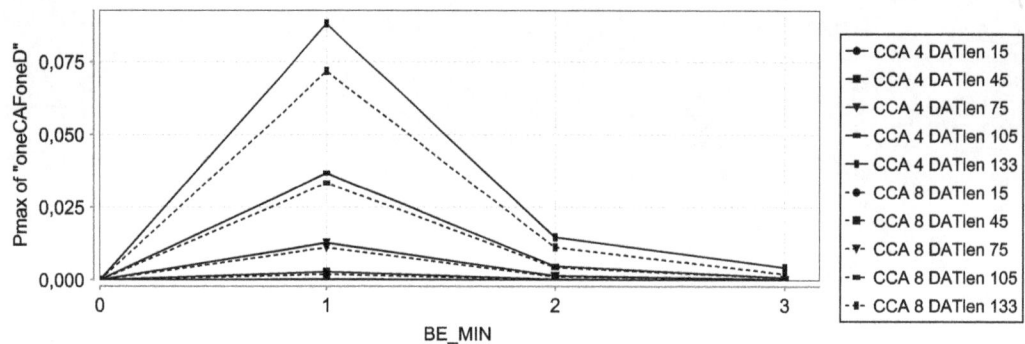

Fig. 16. Probability of a channel access failure in network with two sending stations and acknowledgements.

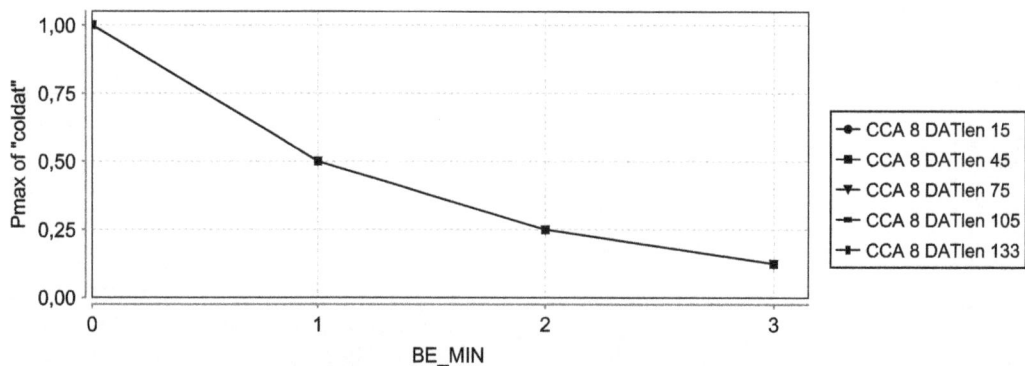

Fig. 17. Probability of collisions of data frames for *CCA* equal to 16 symbols in network with two sending stations and acknowledgements.

which explains the increase in Pd. By looking at (the results used to draw) Figs 15 and 16, it can be seen that generally, the increase of *CCA* to 8 mostly affects the probability of a channel access failure. It should also be mentioned that for *CCA* = 8, at each value of *BE_MIN* the probabilities in Fig. 15 are the same for all the data frame lengths considered. The same holds for the probability that two data frames collide: it is equal to 1 at *BE_MIN* = 0 and drops by a half for every subsequent *BE_MIN* (Fig. 17).

CCA = 8 might prolong the expected time needed to successfully finish the transmission from both stations. Unfortunately, PRISM allows for verification of the expected time needed to reach a certain global state only if the probability of reaching it is 1. So, we could only query about the maximal expected finishing time in the models with limits *NB_MAX* and *MAX_RETRIES* removed as suggested in [7]. To compute the expected time, the reward structure

```
rewards "time"
        true : 1;
endrewards
```

was added to the models, which causes the so-called reward in each state to increase by 1 for each 1 unit of time elapsed [14].

The maximal expected finishing times returned by PRISM upon the query R{"time"}max=?[F "done"] for *CCA* equal to 8 (i.e. 256 μs) are from 7 to 14 units (i.e. from 224 to 448 μs) longer than

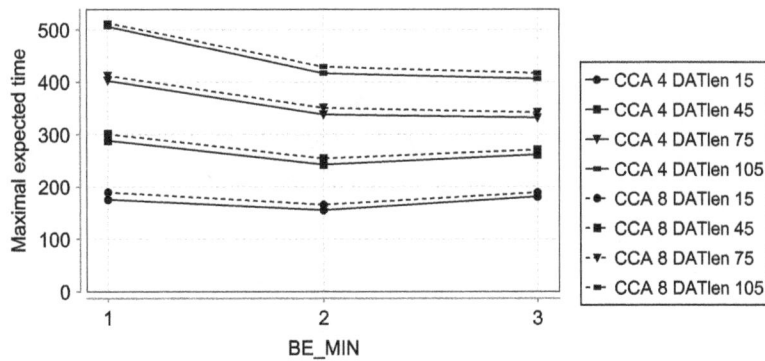

Fig. 18. Maximal expected time to reach "done" for *CCA* equal to 8 and 16 symbols in network with two sending stations and acknowledgements.

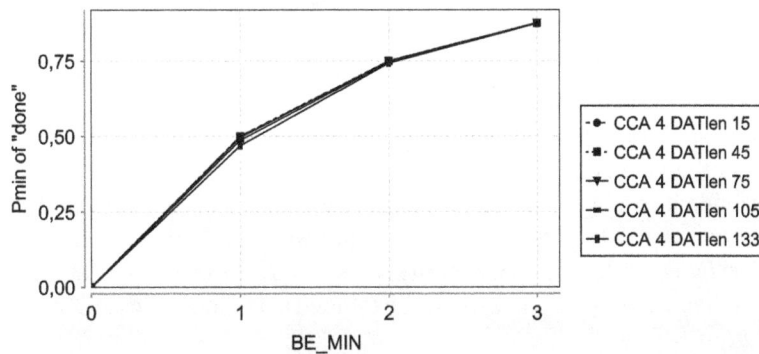

Fig. 19. Probability of "done" for standard *CCA* in network with two sending stations without acknowledgements.

for the standard *CCA* (Fig. 18). In both cases, the shortest time for $DATlen$ equal to 15 and 45 was returned at $BE\_MIN = 2$. For $DATlen$ equal to 75 and 105 (due to the memory limitation, we could not obtain it for 133), it is the best at $BE\_MIN = 3$. The absolute maximal expected times for $DATlen = 15$ (i.e. 480 μs of transmission) are less than 200 units (i.e. 6400 μs) and increase by around 70 (2240 μs) to 110 units (3520 μs, i.e. 11 backoff periods) for every additional 30 units (i.e. representing 960 μs of transmission) in $DATlen$.

From the models proposed in this paper, we constructed the models of the MAC without acknowledgements by removing the states *WAIT_ACK*, *TRANSMIT_ACK*, and *NO_ACK*, and their outgoing transitions. The transitions leading from the state *TRANSMIT_DATA* were redirected to state *FAILURE* (for the case of collision) and, respectively, to state *DONE* (for the case of no collision). Please note, that we wanted to verify the probability that the data frames successfully reach the destination and not the probability of success as defined for the protocol without acknowledgements (see e.g. Section 3).

In the model without acknowledgements with $n = 2$, for both values of *CCA*, Pd for $DATlen = 15$ is 0.5 at $BE\_MIN = 1$, 0.75 at $BE\_MIN = 2$, and 0.875 at $BE\_MIN = 3$ (Fig. 19 shows the probabilities for $CCA = 4$). For $DATlen$ from 45 to 105 and also 133, Pd is a few tenths of a percent smaller. Pd is a little better in the case *CCA* is 8 than when it is 4, but of course, not because $CCA = 8$ would eliminate the collisions with acknowledgements.

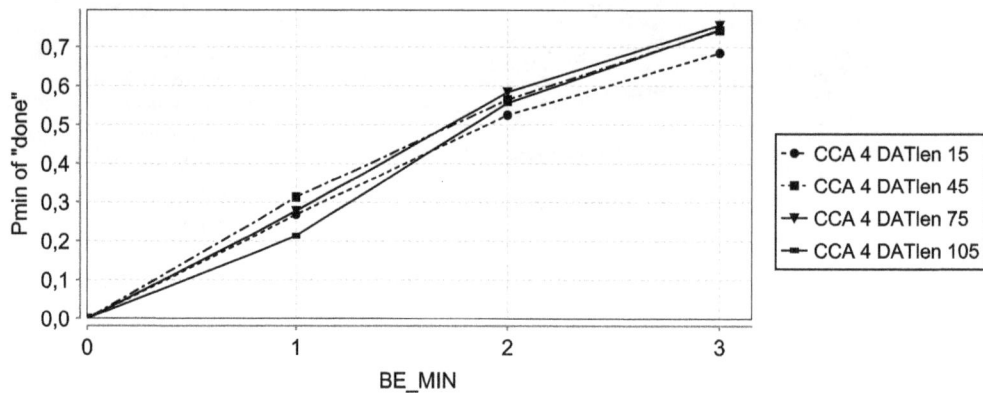

Fig. 20. Probability of "done" for standard *CCA* in network with three sending stations without acknowledgements.

Pd is also a little better for $CCA = 8$ than for $CCA = 4$ if $n = 3$. Again, only the values for $CCA = 4$ are shown in Fig. 20 for clarity. It is from around 0.15 to 0.3 at $BE\_MIN = 1$, between 0.5 and 0.6 at the latter being 2, and around 0.7 at 3 (we could not verify this for $DATlen = 133$ – the state space contained more than 14 million global states; therefore, the results for $DATlen = 133$ are not shown in the figure). In the networks for $n = 2$ with acknowledgements (Fig. 14) and without them (Fig. 19), the following holds for all the values of BE_MIN considered: the greater the value of $DATlen$, the smaller the value of Pd. As evident from Fig. 20, this is not the case for $n = 3$. As expected, at $BE\_MIN = 1$, Pd is the worst for $DATlen = 133$ (it is more than 15 percent smaller than the best one, i.e. for $DATlen = 45$), whereas at BE_MIN equal to 2 and 3, it is the worst for $DATlen = 15$ (at 3 it is around 6 percent smaller than for the other values of $DATlen$, 133 excluded). The best Pd at 2 and 3 (133 excluded for the latter) is at $DATlen = 75$.

7.2. Model checking of networks with hidden stations

Next, we were interested in Pd for the MAC with acknowledgements in a star network with two sending stations hidden from each other, i.e. in a network of the form shown in Fig. 9 with $n = 2$, $h = 1$, and $m = 0$.

However, we first checked whether *CCA* greater than the turnaround time also prevents collisions of data with acknowledgements in the case of hidden stations. We found that generally, the answer is negative. The answer of PRISM as well as UPPAAL for $CCA = 8$ (i.e. 16 symbols) and $CCA = 7$ (i.e. 14 symbols) was, for example, negative for $BE\_MIN = 1$ and $DATlen = 15$ within the mentioned network. Following the negative answer to the query A[]not(Station1.TRANSMIT_DATA and Station2.TRANSMIT_ACK) (A[] meaning 'on all runs always'), UPPAAL drew a diagnostic trace which confirmed our expectations on how the collision could occur. It can happen that the channel is free and that one station starts to send a data frame. Eventually, the other station finishes the backoff and executes the CCA, but assesses the channel as idle because *it does not hear the transmission*. Before the end of the RX-to-TX turnaround after the CCA at this station, the first station successfully ends the transmission of the data frame. Eventually, the turnaround at the other station is finished and it starts to send a data frame. Soon, the RX-to-TX turnaround at the destination is finished and it starts to send an acknowledgement for the data frame of the first station. It collides with the data frame being sent. Another possibility, also confirmed with UPPAAL, is that the destination starts to send the acknowledgement before the turnaround at the other station is finished. Eventually, the latter starts to send the data

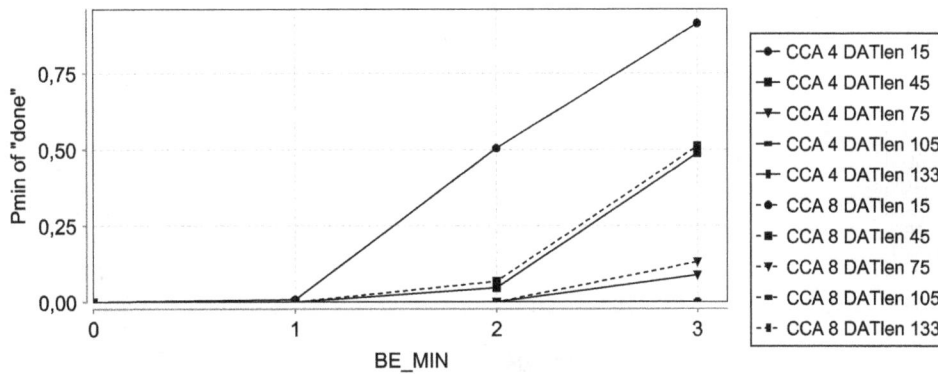

Fig. 21. Probability of "done" in star network with two mutually hidden stations and acknowledgements.

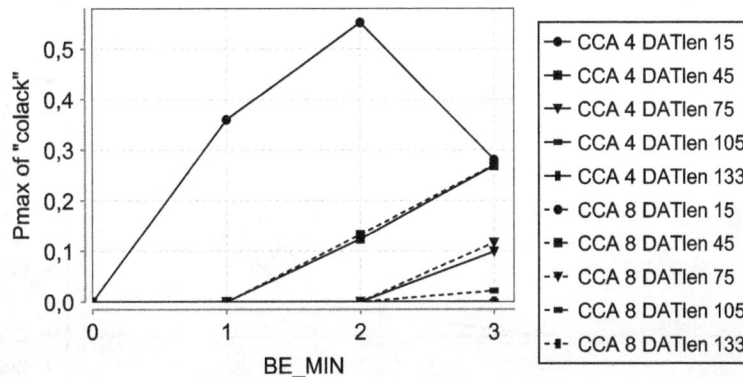

Fig. 22. Probability of collisions involving an acknowledgement frame in star network with two mutually hidden stations and acknowledgements.

frame, which collides with the acknowledgement. It follows that the assumption made in [26], that CCA of 16 symbols also ensures an acknowledgement is never involved in a collision in the case of hidden stations, is wrong. In spite of these findings, we conducted further experiments for networks with hidden stations for $CCA = 4$ and $CCA = 8$.

Figure 21 shows that the only non-zero (or close to zero) values of Pd in the network with two hidden stations with acknowledgements were obtained for BE_MIN equal to 2 and 3 and $DATlen$ equal to 15, 45, and 75. For the latter, Pd is zero at $BE\_MIN = 2$ and around 9 (respectively, 13) percent for $CCA = 4$ (respectively, $CCA = 8$) at $BE\_MIN = 3$. Pd for $DATlen = 15$ is the same for both values of CCA: around 50 percent at $BE\_MIN = 2$ and 91 percent at $BE\_MIN = 3$. For $DATlen = 45$, Pd is around 5 percent at $BE\_MIN = 2$ and around 50 percent at $BE\_MIN = 3$, and is circa 2 percent better for $CCA = 8$ than for $CCA = 4$. We can see that $CCA = 8$ improves Pd by around 4 percent for $DATlen = 75$ and $BE\_MIN = 3$.

Figure 22 shows the probability of collisions involving an acknowledgement frame in the network with two hidden stations for both values of CCA. For all the combinations of BE_MIN and $DATlen$, it either does not change (e.g. in the case of the minimal data frame length) or *increases* for $CCA = 8$. We can see that the changes of Pd in Fig. 21 largely correspond to the changes of the probability in Fig. 22.

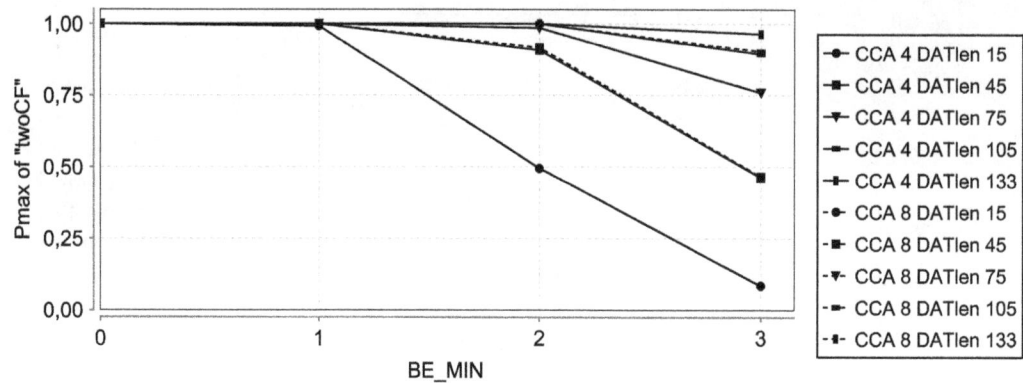

Fig. 23. Probability of a collision failure in both stations in star network with two mutually hidden stations and acknowledgements.

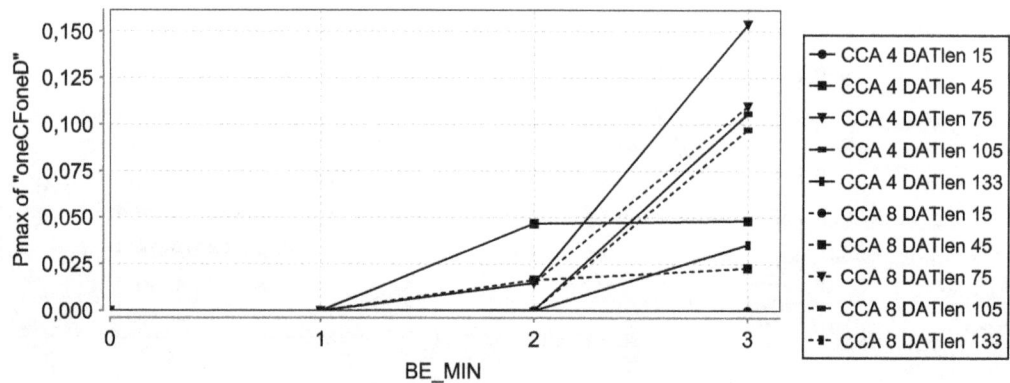

Fig. 24. Probability of a collision failure in only one station in star network with two mutually hidden stations and acknowledgements.

The former improves for $DATlen = 45$ at BE_MIN equal to 2 and 3 as well as for $DATlen = 75$ and $BE\_MIN = 3$ although the latter increases for these combinations.

In order to find an explanation for this, we introduced two states instead of the *FAILURE* state in the models of both hidden stations in the same way as described in Section 7.1. We verified that for all the combinations of parameter values, for *CCA* equal to 4 and 8, the executions of the network can terminate in one of the following three ways: either both station s_1 and s_2 are successful, both stations declare a collision failure, or one of the stations declares success and another one a collision failure (denoted "oneCFoneD"). Figures 23 and 24 show the probability of the latter two cases. Interestingly, there is no chance of a channel access failure, but in contrast to the network with both stations within the same range, it is possible that only one station finishes by a collision failure and another one successfully sends a data frame. By negating this possibility in an UPPAAL query, we obtained a diagnostic trace revealing how this can happen: One station, for example s_1, randomly chooses a smaller *backoff* value than the other, s_2. Station s_1 starts to transmit the data frame first. Eventually, s_2 finishes backoff and since it does not hear s_1, it also starts to transmit the data frame. Eventually, s_1 transmits the complete frame, but does not receive an acknowledgement because its data frame has collided with the data frame

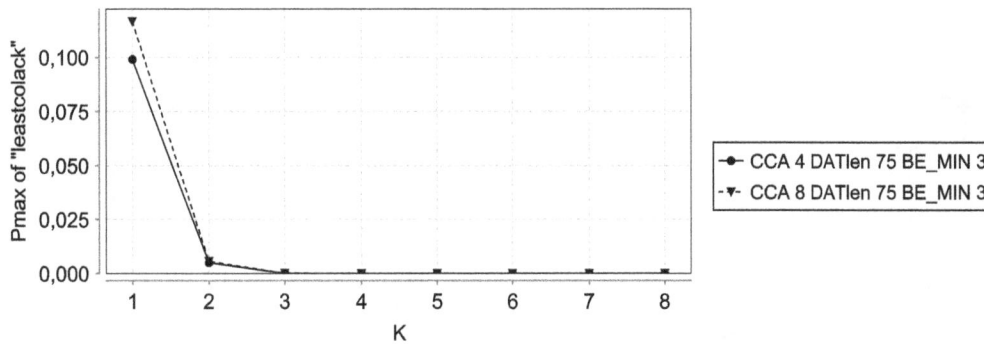

Fig. 25. Probability of at least K collisions involving an acknowledgement in case of two mutually hidden stations, *BE_MIN* equal to 3, data frame length 150 symbols, and *CCA* 8 or 16 symbols.

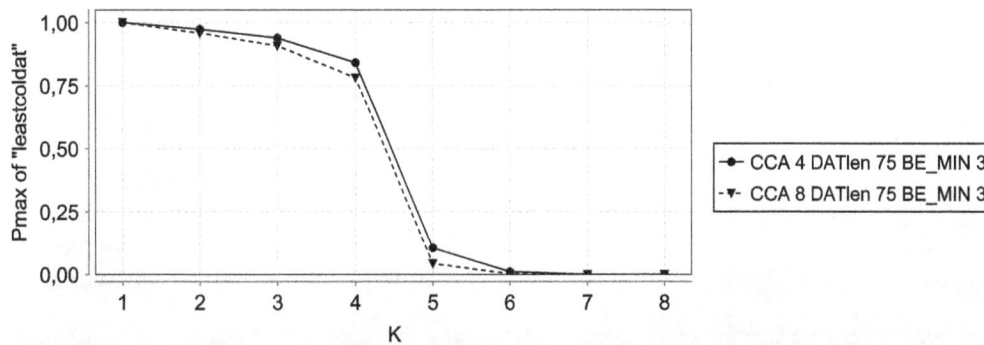

Fig. 26. Probability of at least K collisions not involving an acknowledgement in case of two mutually hidden stations, *BE_MIN* equal to 3, data frame length 150 symbols, and *CCA* 8 or 16 symbols.

of s_2. Hence, s_1 repeats the CSMA-CA algorithm. Before it sends the data frame again, s_2 finishes the transmission and repeats the algorithm for the same reason. A similar situation as at the first attempt of both stations can happen for additional three times (if *MAX_RETRIES* = 3). However, the values of *backoff* may be chosen in such a way that when s_1 finishes the data frame transmission for the fourth time, s_2 finishes the data frame transmission for the third time. An acknowledgement is sent neither to s_1 nor to s_2. Station s_1 eventually declares a collision failure, whereas s_2 can successfully accomplish the last allowed attempt.

Altogether, we can see that the improvement of Pd for $CCA = 8$ in Fig. 21 is the consequence of a smaller probability of collision failures in the network (please note, that some of the probabilities in Fig. 23 increase, but only by less than a percent). This probability is the consequence of a smaller probability of a larger number of collisions. The fact of the latter in spite of the increased probability of collisions involving acknowledgements can be explained as follows: It is important to notice that Fig. 22 shows the probability of at least one collision involving an acknowledgement. By recording the number of such collisions in the model, up to some limit K (cf. [8]), we found that for all the cases where Pd improves, the increase in the probability of at least K collisions of this kind is negligible for K greater than 1 (see e.g. Fig. 25 for the case of $DATlen = 75$ and *BE_MIN* = 3). In a similar way, we found that the probability of at least K collisions involving only data drops for the larger values of K (between 3

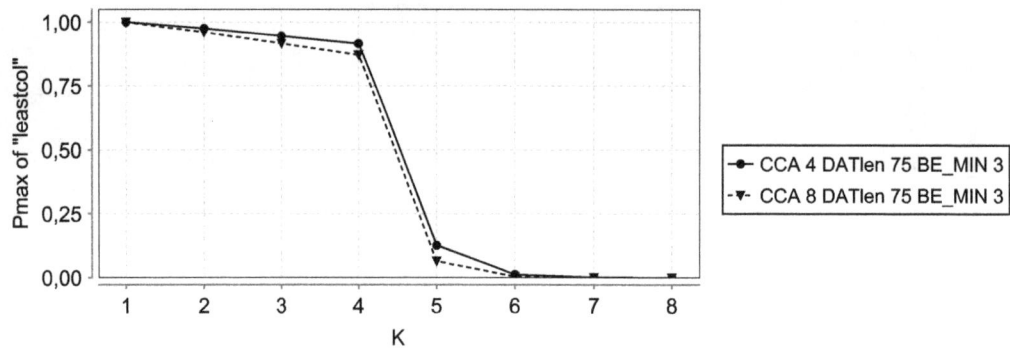

Fig. 27. Probability of at least K collisions in case of two mutually hidden stations, *BE_MIN* equal to 3, data frame length 150 symbols, and *CCA* 8 or 16 symbols.

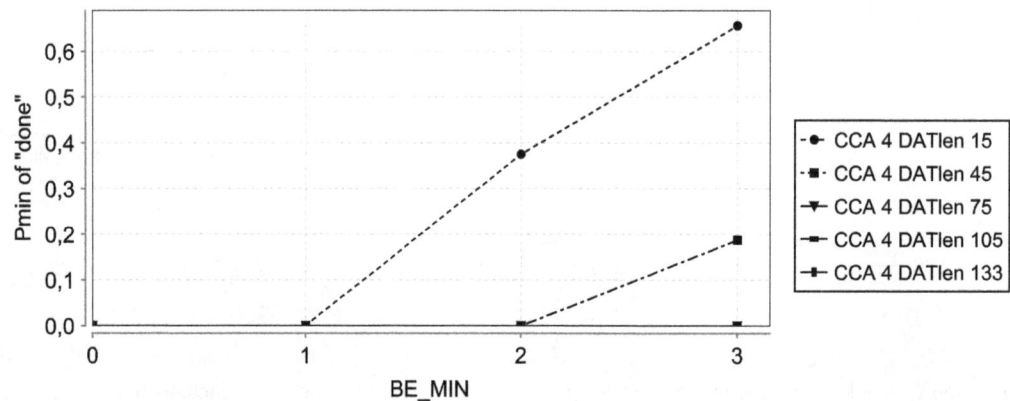

Fig. 28. Probability of "done" for standard *CCA* in star network with two hidden stations without acknowledgements.

and 5 or so) by an amount close to the increase of Pd (see e.g. Fig. 26). Consequently, the probability of at least four collisions of either kind (four being the minimal number that causes a collision failure if *MAX_RETRIES* is equal to 3) decreases by a similar amount (see e.g. Fig. 27).

As can be seen from Fig. 28, Pd within the network consisting of two mutually hidden stations not using the acknowledgement option is non-zero only for $DATlen = 15$ in the case of $BE\_MIN = 2$ and $BE\_MIN = 3$, as well as for $DATlen = 45$ at $BE\_MIN = 3$. It is the same for *CCA* equal to 4 and 8.

Finally, we conducted experiments for two networks with 3 sending stations, some of them being hidden, and no acknowledgements. One network contained, besides the destination node, two mutually hidden stations like the last one considered, and an additional one between the two, being within the range of both like the destination. This corresponds to the case with $n = 3$, $h = 1$, and $m = 1$ in Fig. 9. From Fig. 29 we can see that in contrast to the network with $n = 3$ and non-hidden stations (Fig. 20), it holds for any *BE_MIN* greater than zero (with one little exception at 1 and 2), that the smaller the $DATlen$, the greater the Pd. In all the cases, Pd is some tenths of a percent better at $CCA = 8$.

The other network was of the form shown in Fig. 9 with $n = 3$, $h = 1$, and $m = 0$. This means a network with the destination and one sending station hidden from a pair of sending stations, the latter being able to hear one another. As shown in Fig. 30, Pd within this network is non-zero only at *BE_MIN*

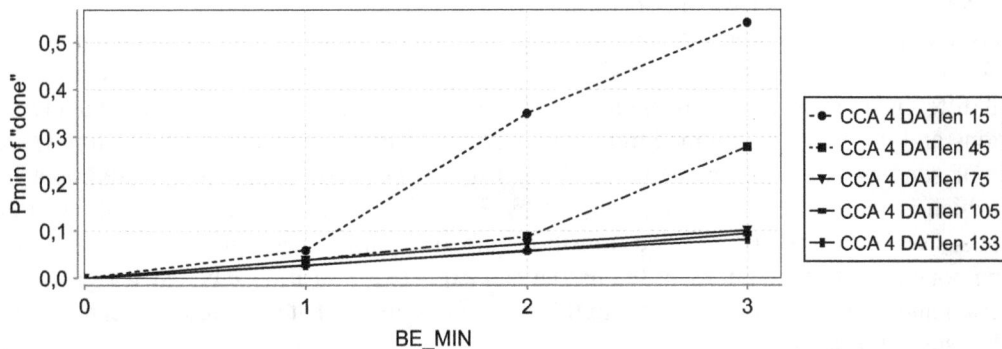

Fig. 29. Probability of "done" for standard *CCA* in star network with a sending station between two mutually hidden stations and without acknowledgements.

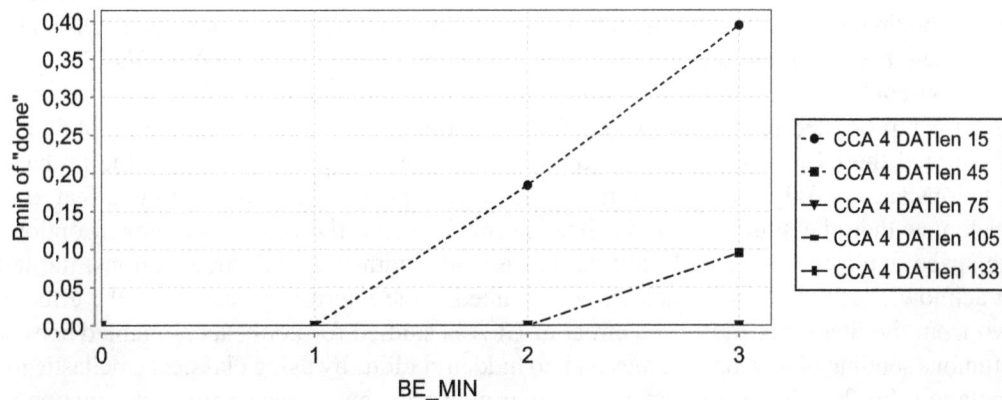

Fig. 30. Probability of "done" for standard *CCA* in star network with a sending station hidden from a pair of stations and without acknowledgements.

equal to 2 and 3 for $DATlen = 15$ and at $BE\_MIN = 3$ for $DATlen = 45$ as within the network with two mutually hidden stations without acknowledgements (Fig. 28). It is, however, around 10 to 15 percent smaller than in the latter. For $CCA = 8$, it is some tenths of a percent greater than for $CCA = 4$.

By comparing Figs 29 and 30 we can see that the probability that the data frames of all the three stations come through within the network with one sending station within the range of two mutually hidden stations is better than in the network with two stations on one side of the coordinator and the other one hidden from them on the other for all the checked values of $DATlen$. It is around 5 (at $BE\_MIN = 1$) to 15 percent (at the larger values of BE_MIN and $DATlen$) better. The latter observation as well as the comparison of Fig. 29 with Fig. 28 suggest that the presence of a sending station within the range of two mutually hidden (groups of) sending stations helps that Pd is not zero for the longer data frames.

8. Discussion and conclusions

Based on the model given graphically in [7,8], in this paper we first prepared an accurate PTA model of a nonbeacon-enabled IEEE 802.15.4 network consisting of two sending/receiving station pairs in

PRISM. We subsequently derived the models for networks with n station pairs. We proposed two kinds of models, one with a c variable for every sending station and one with common variables *sending* and *colind* for all the stations. The latter is similar to the UPPAAL model recently presented in [3]. The essential difference between them is in that the UPPAAL model consists only of automata representing the stations and no model of the medium. This is achieved in such a way that the action representing the start of sending is equally named (e.g. *busy!*) in all the stations and is declared as a broadcast action, which is allowed by UPPAAL, but not by PRISM. A transition labelled with such an action synchronises with all currently enabled transitions labelled with the complementary action (*busy?*) in other components. All the stations in the model can execute a transition labelled with *busy?* over the CCA period and during the transmission. In this way, any stations that are interested can hear if another station has started to send.

We further argued that the presented models for n sending/receiving station pairs are also valid models of star topology networks with n sending stations and a coordinator if the latter only sends acknowledgements. Finally, we showed that by relatively small variations, models of typical star-shaped networks with mutually hidden stations can be obtained from them. We could not find formal models of such networks in the literature. As for the star topology networks without hidden stations, MODEST [10] and SPIN [20] have been applied to model the MAC in beacon-enabled mode. A MODEST specification for nonbeacon-enabled mode without acknowledgements can be found at [19]. In all these models, the coordinator is represented separately, and the CCA is represented accurately in neither of them.

We illustrated the usages of the proposed models by presenting verification of the probability that all the stations successfully send a data frame. We were especially interested in the effect of the CCA duration longer than the standard one in those networks using the acknowledgement option. Due to the state-space explosion, we could only verify networks with two and three stations, the latter only without acknowledgements. Nevertheless, we obtained some interesting results on the effect of *CCA*, unknown from the literature. In [9], the effect of *CCA* is studied for nonbeacon-enabled networks with the continuous sending of new data frames and no hidden stations by using classical stochastic modelling and simulation. In [26], the performance of similar networks, but consisting of a destination node and mutually hidden stations located within two different ranges around it, is studied by using simulation. The value of 16 symbols is used for *CCA* because of the conviction that it entirely prevents collisions of acknowledgements.

The main contribution of this paper, besides the proposed modelling approaches, is the evidence obtained by model checking that a longer CCA period does not ensure the absence of collisions involving an acknowledgement frame within those networks with hidden stations, but may nevertheless improve the probability of successful sending. A 'side result' of this paper is the models in UPPAAL. Its capability of interactive simulation and diagnostic trace generation, together with the graphical user interface, proved to be a useful supplement to PRISM.

The modelling approach for networks with hidden stations using the c variables can readily be applied to networks containing sending stations within more than two different ranges around the destination. It only needs to be seen which c variables have to be checked within the CCA period in the models of the individual stations with respect to their ranges. The approach using the *sending* variables could be generalised to more than two different ranges by introducing as many *sending* variables as there are ranges.

In the future, we will further investigate the effect of *CCA* in nonbeacon-enabled networks containing hidden stations by using probabilistic model checking as well as classical network simulation tools. Clearly, even with better computer equipment it cannot currently be expected that with the model checking of PTAs the networks containing tenths of nodes could be handled. Analytical performance results are usually validated by comparing them with the results of simulation tools. We, however, intend to check to what degree the analytical results in [4] for nonbeacon-enabled networks in which all the stations send one data frame simultanously to the coordinator, as in the presented models, could be confirmed by model checking using PRISM.

References

[1] P. Baronti, P. Pillai, V.W.C. Chook, S. Chessa, A. Gotta and Y. Fun Hu, Wireless sensor networks: A survey on the state of the art and the 802.15.4 and ZigBee standards, *Computer Communications* **30** (2007), 1655–1695.

[2] G. Behrmann, A. David and K.G. Larsen, A tutorial on UPPAAL 4.0 (Updated November 28, 2006), http://www.uppaal.org/.

[3] P. Bulychev, A. David, K.G. Larsen, A. Legay and M. Mikučionis, Computing Nash equilibrium in wireless ad hoc networks: a simulation-based approach, in: *Proc. 2nd International Workshop on Interactions, Games and Protocols (IWIGP2012)*, J. Reich and B. Finkbeiner, eds, EPTCS 87, 2012, pp. 1–14.

[4] C. Buratti and R. Verdone, Performance analysis of IEEE non beacon-enabled mode, *IEEE Transactions on Vehicular Technology* **58** (2009), 3480–3493.

[5] F. Chen and F. Dressler, A simulation model of IEEE 802.15.4 in OMNeT++, in: *6. Fachgespräch Sensornetzwerke der GI/ITG Fachgruppe "Kommunikation und Verteilte Systeme"*, Technischer Bericht der RWTH Aachen AIB 2007-11, Distributed Systems Group, RWTH Aachen University, 2007, pp. 35–38.

[6] L. de Alfaro, *Formal verification of probabilistic systems*, Ph.D. Dissertation, University of Stanford, 1997.

[7] M. Fruth, Probabilistic model checking of contention resolution in the IEEE 802.15.4 low-rate wireless personal area network protocol, in: *Proc. 2nd International Symposium on Leveraging Applications of Formal Methods, Verification and Validation (ISoLA 2006)*, 2006, pp. 290–297.

[8] M. Fruth, *Formal methods for the analysis of wireless network protocols*, Ph.D. Dissertation, University of Oxford, 2011.

[9] M. Goyal, W. Xie and H. Hosseini, IEEE 802.15.4 modifications and their impact, *Mobile Information Systems* **7** (2011), 69–92.

[10] C. Groß, H. Hermanns and R. Pulungan, Does clock precision influence ZigBee's energy consumptions? in: *Proc. 11th International Conference on Principles of Distributed Systems (OPODIS 2007)*, E. Tovar, P. Tsigas and H. Fouchal, eds, LNCS 4878, Springer-Verlag, 2007, pp. 174–188.

[11] J.-P. Katoen, Perspectives in probabilistic verification, in: *Proc. 2nd IFIP/IEEE International Symposium on Theoretical Aspects of Software Engineering*, IEEE CS Press, 2008, pp. 3–10.

[12] T.O. Kim, J.S. Park, H.J. Chong, K.J. Kim and B.D. Choi, Performance analysis of IEEE 802.15.4 non-beacon mode with the unslotted CSMA/CA, *IEEE Communications Letters* **12** (2008), 238–240.

[13] A. Koubaa, M. Alves and E. Tovar, A comprehensive simulation study of slotted CSMA/CA for IEEE 802.15.4 wireless sensor networks, in: *Proc. 2006 IEEE International Workshop on Factory Communication Systems*, 2006, pp. 183–192.

[14] M. Kwiatkowska, G. Norman and D. Parker, PRISM 4.0: verification of probabilistic real-time systems, in: *Proc. 23rd International Conference on Computer Aided Verification (CAV'11)*, LNCS 6806, Springer-Verlag, 2011, pp. 585–591.

[15] M. Kwiatkowska, G. Norman, D. Parker and J. Sproston, Performance analysis of probabilistic timed automata using digital clocks, *Formal Methods in System Design* **29** (2006), 33–78.

[16] M. Kwiatkowska, G. Norman, D. Parker and J. Sproston, Verification of Real-time Probabilistic Systems, in: *Modeling and Verification of Real-Time Systems: Formalisms and Software Tools*, S. Merz and N. Navet, eds, John Wiley and Sons, 2008, pp. 249–288.

[17] M. Kwiatkowska, G. Norman and J. Sproston, Probabilistic model checking of the IEEE 802.11 wireless local area network protocol, in: *Proc. 2nd Joint International Workshop on Process Algebra and Probabilistic Methods and Performance Modeling in Verification (PAPMPROBMIV 2002)*, LNCS 2399, Springer-Verlag, 2002, pp. 169–187.

[18] J. Mišić, V.B. Mišić and S. Shafi, Performance of IEEE 802.15.4 beacon enabled PAN with uplink transmissions in non-saturation modes – access delay for finite buffers, in: *Proc. First International Conference on Broadband Networks*, 2004, pp. 416–425.

[19] MoDeST Tutorial, http://depend.cs.uni-sb.de/modesttutorial/index.html.

[20] Z.L. Németh, Model checking of the slotted CSMA/CA MAC protocol of the IEEE 802.15.4 standard, in: *Proc. 2010 Mini-Conference on Applied Theoretical Computer Science (MATCOS-10)*, A. Brodnik and G. Galambos, eds, University of Primorska Press, 2011, pp. 121–126.

[21] T. Park, T. Kim, J.Y. Choi, S. Choi and W. Kwon, Throughput and energy consumption analysis of IEEE 802.15.4 slotted CSMA/CA, *Electronics Letters* **41** (2005), 1017–1019.

[22] Part 15.4: Wireless Medium Access Control (MAC) and Physical Layer (PHY) Specifications for Low-Rate Wireless Personal Area Networks (WPANs), IEEE Std 802.15.4-2006, IEEE Computer Society, 2006.

[23]   Part 15.4: Low-Rate Wireless Personal Area Networks (LR-WPANs), IEEE Std 802.15.4-2011, IEEE Computer Society, 2011.

[24]   PRISM model checker homepage, http://www.prismmodelchecker.org/.

[25]   I. Ramachandran, Changes made to the IEEE 802.15.4 NS-2 implementation, February 7, 2006.

[26]   D. Rohm, M. Goyal, H. Hosseini, A. Divjak and Y. Bashir, A simulation based analysis of the impact of IEEE 802.15.4 MAC parameters on the performance under different traffic loads, *Mobile Information Systems* **5** (2009), 81–99.

Tatjana Kapus received her M.Sc. and Ph.D. degrees from the Faculty of Electrical Engineering and Computer Science, University of Maribor, Slovenia, in 1991 and 1994, respectively. She is currently a professor there. She teaches courses on communications networks, formal methods, and programming. Her primary research interest lies in formal methods for the specification and verification of reactive systems, such as communications protocols.

8

A scheduling method to reduce waiting time for close-range broadcasting

Yusuke Gotoh[a,b,*], Tomoki Yoshihisa[c], Hideo Taniguchi[a], Masanori Kanazawa[d],
Wenny Rahayu[b] and Yi-Ping Phoebe Chen[b]
[a]*Okayama University, Okayama, Japan*
[b]*La Trobe University, Melbourne, VIC, Australia*
[c]*Osaka University, Osaka, Japan*
[d]*The Kyoto College of Graduate, Studies for Informatics, Kyoto, Japan*

Abstract. Due to the recent popularization of digital broadcasting systems, close-range broadcasting using continuous media data, i.e. audio and video, has attracted great attention. For example, in a drama, after a user watches interesting content such as a highlight scene, he/she will watch the main program continuously. In close-range broadcasting, the necessary bandwidth for continuously playing the two types of data increases. Conventional methods reduce the necessary bandwidth by producing an effective broadcast schedule for continuous media data. However, these methods do not consider the broadcast schedule for two types of continuous media data. When the server schedules two types of continuous media data, waiting time that occurs from finishing the highlight scene to starting the main scene, may increase. In this paper, we propose a scheduling method to reduce the waiting time for close-range broadcasting. In our proposed method, by dividing two types of data and producing an effective broadcast schedule considering the available bandwidth, we can reduce the waiting time.

Keywords: Close-range broadcasting, scheduling, waiting time, continuous media data

1. Introduction

Due to the recent popularization of digital broadcasting systems, close-range broadcasting using continuous media data, i.e. audio and video, has attracted great attention. Examples follow:

- In news programs, after watching a news headline, the user watches the main news continuously.
- In dramas, the user first watches a highlight scene. If the user has an interest in this drama, he/she watches the main program.

In our paper, we consider the case of delivering data for close-range broadcasting. In broadcasting systems, clients generally have to wait until their desired data is broadcasted. Therefore, there is many research on reducing the waiting time. We have proposed several scheduling methods to reduce the waiting time without interruption. These methods reduce the waiting time before playing data. However, we did not consider the case of close-range broadcasting. In close-range broadcasting, the server needs to deliver two types of continuous media data: one is the highlight scene and the other is the main program. When the client watches the highlight scene and the main program sequentially, in

---

*Corresponding author: Yusuke Gotoh, Okayama University, Okayama, Japan. E-mail: gotoh@cs.okayama-u.ac.jp.

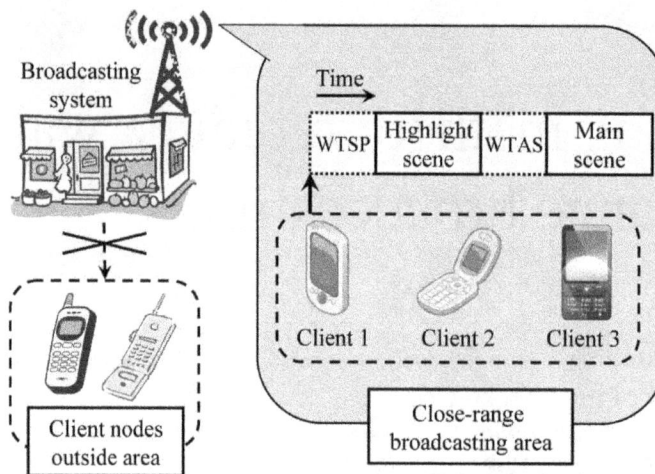

Fig. 1. Assumed structure of close-range broadcasting.

addition to the waiting time before starting the highlight scene, waiting time occurs between finishing the highlight scene and starting the main program. By increasing the available bandwidth for delivering it, waiting time before starting the highlight scene can be reduced. On the other hand, since the available bandwidth for delivering the main program becomes relatively short, the waiting time between finishing the highlight scene and starting the main program increases significantly and the total waiting time may increase.

In this paper, we propose a scheduling method to reduce the waiting time in close-range broadcasting. Our scheduling method considers the waiting time between finishing the highlight scene and starting the main program. Also, the total waiting time is reduced by calculating the available bandwidth of each channel considering the playing time of the highlight scene and the main program and making the broadcast schedule.

The remainder of this paper is organized as follows. We explain the research interest in close-range broadcasting in Section 2. Related works are introduced in Section 3. Our proposed method is explained in Section 4 and evaluated in Section 5. Finally, we conclude the paper in Section 6.

2. Close-range broadcasting

2.1. Basic idea

In this section, we explain close-range broadcasting. The assumed structure of close-range broadcasting is shown in Fig. 1. In close-range broadcasting, interference between channels can be reduced by controlling the output of radio waves and limiting the range for receiving data. The server can deliver continuous media data such as audio and video in a limited area. Users can improve convenience by receiving information according to the watching area using their nodes. Examples of such situations follow:

- In sport events, users can acquire the information of their favorite team and the situation of the game in real time.
- In the city, users can get the coupon at their favorite shops and watch the highlight scene of their interested movie.

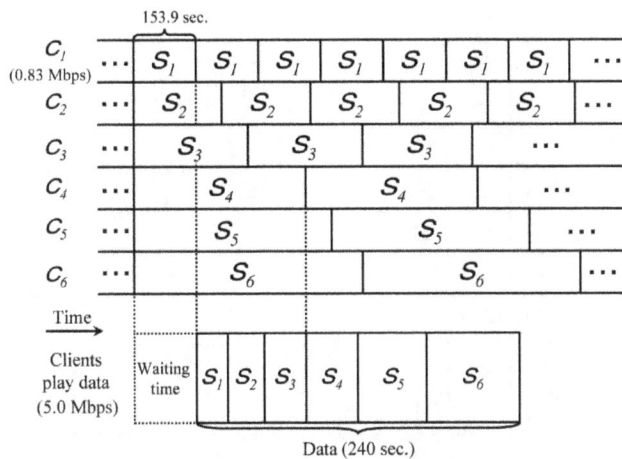

Fig. 2. Example of division-based broadcasting under BE-AHB method.

The server can deliver a program and advertisements effectively to many clients in a limited area, and clients can watch the program using their devices. However, since users move frequently, they may interrupt the program by moving out of the broadcast area. Therefore, to maintain a user's interest and encourage him/her to stay as long as possible in the area, the server needs to deliver two types of continuous media data: the highlight scene and the main program. Users first watch the highlight scene in the close-range broadcasting area. If the users are interested in the highlight scene, they would watch the main program continuously.

In close-range broadcasting, there are two types of waiting time. We call the waiting time that occurs from selecting the highlight scene to starting to play "waiting time after selection" (WTAS). The server can set an upper limit for WTAS. Also, we call the waiting time that occurs from requiring the program to starting to play "a waiting time for starting the program" (WTSP). WTAS occurs when requiring a user's selection. On the other hand, WTSP occurs when starting to play the data.

If the waiting time occurs after starting to watch the main program, continuity of the program is interrupted and users feel annoyed. Conventional methods do not distinguish WTAS and WTSP. In this paper, we reduce the waiting time effectively by making the broadcast schedule considering WTAS. The duration of the acceptable waiting time depends on the users, but we suppose that acceptable waiting time is given considering audience rating and so on.

2.2. Mechanism for waiting time

In this subsection, we explain the mechanism for waiting time generation. When the server broadcasts continuous media data repetitively, clients have to wait until the first portion of the data is broadcasted. For example, when the server broadcasts MPEG2-encoded music clip data whose receiving time is 60 min, waiting time is 60 min. To reduce the waiting time, many methods employ the division-based broadcasting technique, which reduces the waiting time by dividing the data into several segments and broadcasting precedent segments frequently. In division-based broadcasting, these methods suppose data delivery such as radio wave broadcasting and Internet multicast.

In division-based broadcasting, since the data are divided into several segments, segments must be scheduled without interrupting the clients' continuous play. In the conventional methods, clients can play data without interruption considering available bandwidth and several channels.

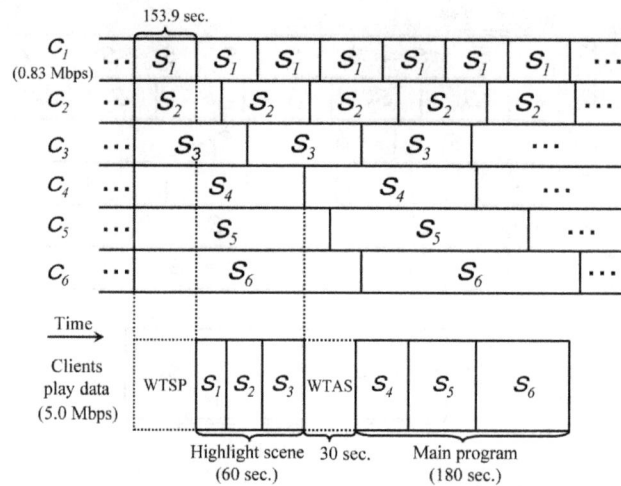

Fig. 3. Example of close-range broadcasting under BE-AHB method.

An example of division-based broadcasting is shown in Fig. 2. The example uses the Bandwidth Equivalent-Asynchronous Harmonic Broadcasting (BE-AHB) method [15] to explain the problem of waiting time. This example divides data into six segments relatively. The playing time is 240 sec. The data is divided into six segments, S_i $(i = 1, \cdots, 6)$. When the available bandwidth for clients is 5.0 Mbps, the bandwidth of each channel is $5.0/6 = 0.83$ Mbps. The consumption rate is 5.0 Mbps. Under the BE-AHB method, when the total playing time is $60 + 180 = 240$ sec, the playing time of S_1 is 25.5 sec, S_2 is 29.8 sec, $\cdots$, and S_6 is 55.3 sec. The server repetitively broadcasts S_i by broadcast channels C_i. Clients can play each segment after receiving it. While playing the data, clients receive the broadcast data and store it in their buffers. In this case, clients can play the data continuously until it has ended, even if they start playing it immediately after completely receiving S_1. When clients finish playing S_1, they have finished receiving S_2 and can play S_2 continuously. Also, when they are finished playing S_3, they have finished receiving S_4 and can play S_4 continuously. In this case, waiting time is $25.5 \times 5.0/0.83 = 153.9$ sec, which is the same as the time needed to receive only S_1. In the simple method, since the server broadcasts data without dividing it, waiting time is $240 \times 5.0/5.0 = 240$ sec. Therefore, $(240 - 153.9)/240 \times 100 = 35.9\%$, which is shorter than the simple method.

Next, in close-range broadcasting, the broadcast schedule under the BE-AHB method is shown in Fig. 3. We set WTAS to be 30 sec. In Fig. 3, the server broadcasts partitioned data repetitively under the BE-AHB method using both the highlight scene and main program. In close-range broadcasting, WTAS occurs while playing the program. On the other hand, in the BE-AHB method, since clients play segments without interruption, the BE-AHB method does not make the broadcast schedule considering WTAS and cannot reduce waiting time effectively.

3. Related works

In Japan, several services using close-range broadcasting have been proposed. Area One segment-Broadcasting (Area One-Seg) [2] is a broadcasting service that delivers data by applying one-segment technology in a limited area within a one-kilometer radius. One-segment technology is used for terrestrial digital broadcasting in Japan. By delivering content suitable for the place and time, the server can deliver information effectively in a gathering place such as a station or a stadium. Spot one-segment broadcasting [3] is a service in which the server can broadcast data to specific users. In spot one-segment broadcasting, everybody can manage the broadcast service without a broadcast license. However, electric waves in one-segment broadcasting are weaker than in Area One-Seg, and the range of transmission is limited to about a one-meter radius. In these broadcasting systems, the broadcast scheduling has not been proposed to reduce waiting time.

Several methods to reduce waiting time have been proposed in continuous media data broadcasting [4, 6–9]. These methods reduce waiting time by dividing the data into several segments and producing an efficient broadcast schedule. In Heterogeneous Receiver-Oriented Broadcasting (HeRO) [12], the data are divided into different sizes. Let J be the data size for the first segment. The data sizes for the segments are $J, 2J, 2^2 J, \ldots, 2^{K-1} J$. However, since the data size of the Kth channel becomes half of the data, clients may experience waiting time and interruptions.

In BroadCatch [13], the server divides the data into 2^{K-1} segments of equal size and broadcasts them periodically using K channels. The bandwidth for each channel is the same as the data bit rate. By adjusting K based on the available bandwidth for clients, waiting time is effectively reduced. However, since the available bandwidth is proportional to the number of channels, when an upper limit exists in the server's bandwidth, the server might not be able to acquire enough channels to broadcast the data.

In Harmonic Broadcasting (HB) [14], the data is separated into $S_1, \cdots, S_N$ of equal size. The server sets $C_1, \cdots, C_N$ channels considering the available bandwidth and schedules $S_1, \cdots, S_N$. The client can play $S_1, \cdots, S_N$ without interruption using $C_1, \cdots, C_N$. For example, when the server broadcasts continuous media data whose playing time is 60 min and whose consumption rate is 5.0 Mbps using 24 Mbps, which is identical to that under digital broadcasting systems, we need 67 channels.

In Asynchronous Harmonic Broadcasting (AHB) [15], waiting time is reduced more than the HB method by scheduling the playing unit time, such as Group of Pictures (GOP) in MPEG2 or the frame in MP3. Since the server divides the data of each playing unit time, the number of channels is the same as the number of playing units. For example, when the server broadcasts a piece of continuous media data whose playing time is 60 min and whose consumption rate is 5.0 Mbps using 24 Mbps, which is identical to that under digital broadcasting systems, the number of channels is 6,000. The BE-AHB method is an extended version of the AHB method, where continuous media data is separated by a fixed data size. The server can reduce the necessary number of channels, which is more realistic than the original AHB method.

Kua et al. proposed an algorithm for answering spatial nearest neighbor search queries by leveraging results from neighboring nodes within a mobile environment [16]. This algorithm allows a mobile node to locally verify whether candidate objects received from the neighbors are indeed part of its own nearest neighbor data set.

We previously proposed scheduling methods to reduce waiting time for division-based broadcasting [17,18]. These methods make the schedule using single continuous media data. In addition, our assumed structure is close-range broadcasting.

4. Proposed method

4.1. Basic idea

We propose a scheduling method in close-range broadcasting called the "Close-Range Harmonic Broadcasting (CR-HB)" method. We previously proposed the BE-AHB method [15]. The BE-AHB

Table 1
Formulation symbols

Sign	Explanation
r	Consumption rate
N	Number of segments
S_1	Data of highlight scene
S_2	Data of main program
s_{i-j}	j^{th} segment in S_i, $i = 1, 2 \parallel j = 1, \cdots, N$
C_i	Available bandwidth to broadcast S_i
c_{i-j}	Available bandwidth to broadcast s_{i-j}
T_i	Playing time of S_i
t_{i-j}	Playing time of s_{i-j}
B	Total available bandwidth of server
p_{i-j}	Ratio of playing time
δ	Maximum WTAS

method reduces waiting time based on the idea explained in Section 2.2. The main difference between the CR-HB and the BE-AHB methods is considering WTAS, which occurs between finishing the highlight scene and starting the main program.

4.2. Assumed environment

Our assumed system environment is listed below.

– Clients play the highlight scene and the main program sequentially.
– The waiting time occurs when starting to play the highlight scene and the main program.
– Bandwidth is stable while broadcasting the data.
– Clients can start playing a segment after they have completely received it.
– The server broadcasts segments repetitively using multiple channels.
– Once clients start playing the data, they can play it without interruption.
– Clients have adequate buffer to store the received data.

4.3. Modeling to reduce waiting time

We developed an expression to reduce waiting time for continuous media data in close-range broadcasting. The formulation values are summarized in Table 1.

In continuous media data broadcasting, we need to schedule segments based on a receiving time of s_{1-1} and interruption time between finishing the highlight scene and starting the main program. Actually, many network structures exist for division-based broadcasting systems. However, since the number of patterns is excessive, evaluating the performance of our proposed method for all of them is not realistic. Therefore, in this paper, we use the network configuration shown in Fig. 1. Although the practical programs do not always match these patterns, there are enough to show the effectiveness of our proposed method.

In our proposed method, the server reduces waiting time by calculating the available bandwidth of each broadcast channel to minimize the average waiting time. By including WTAS in the receiving time of the data, the server can lengthen the receiving time of the main program and reduce the necessary bandwidth for delivering the highlight scene. By considering WTAS, calculating the available bandwidth of each channel becomes slightly more complex. However, the available bandwidth of each channel can be calculated by a simple simulation.

4.4. Scheduling process

In this section, we explain the CR-HB method. In the CR-HB method, the total available bandwidth of the server is divided into several channels. Clients can reduce waiting time effectively by scheduling segments using several channels.

The scheduling process continues as follows. Notation is defined in Table 1. We explained the basic idea of our scheduling process in Subsection 4.3.

1. According to N, S_i separates into $s_{i-1}, \cdots, s_{i-N}$, for which the playing time is $t_{i-1}, \cdots, t_{i-N}$.
2. $s_{1-1}, \cdots, s_{1-N}$ must be scheduled so that clients that can concurrently receive data from N channels, can play the data continuously. Since the playing time of s_{1-1} is $\frac{C_1}{Nr}$, s_{1-2} must play until finishing to receive s_{1-1}, which is $(1 + \frac{C_1}{Nr})$. Next, clients can wait starting of s_{2-1} until finishing time of WTAS. Also, p_{2-j} is calculated considering p_{2-1}. Therefore, when p_{1-1} is 1.0, p_{i-j} can obtain the following equation.

$$p_{i-j} = \begin{cases} (1 + \frac{C_1}{Nr})^{j-1} \ (i = 1, \ 2 \leqslant j \leqslant N). \\[2mm] \frac{B-C_1}{Nr}\{(1 + \frac{Nr}{C_1}) \times t_{1-N} + \delta\}. \ (i = 2, \ j = 1). \\[2mm] (1 + \frac{B-C_1}{Nr})^{j-1} \times t_{2-1} \ (i = 2, \ 2 \leqslant j \leqslant N), \end{cases}$$

3. t_{i-j} is calculated as follows:

$$t_{i-j} = T_i \times \frac{p_{i-j}}{\sum_{j=1}^{N} p_{i-j}}.$$

4. Decide C_i based on t_{i-j} calculated in step 3. As shown in Subsection 4.5, we can calculate C_i in the simple computational simulation.

4.5. A simple example to reduce waiting time

In the CR-HB method, when the available bandwidth of highlight scene is C_1, the broadcast is scheduled so that clients that can receive data from C_1 can play the data without interruption even if they start playing the data just after receiving the first segment S_1. The waiting time for clients that receive data using larger bandwidth than C_1 cannot be improved. Let C_1^* be the value of optimal C_1. Then, we have to find the value of C_1^* that makes the average waiting time minimum.

When C_1 is the same as the available bandwidth given by the CR-HB method, it becomes optimum. This is because the broadcast schedule set up in order to receive the next segment just before the client finishes playing the current segment. Otherwise, the broadcast schedule is not optimal. If we call the state that there is no buffer in the meantime as a starvation, we have to make the broadcast schedule so that there is no starvation as much as possible. The deadline means the time to finish playing the current segment.

Here, the procedure for calculating C_1^* continues as follows:

1. Produce the broadcast schedule for all C_1.
2. Calculate the waiting time for all produced broadcast schedules.
3. Find C_1^* that gives the minimum average waiting time.

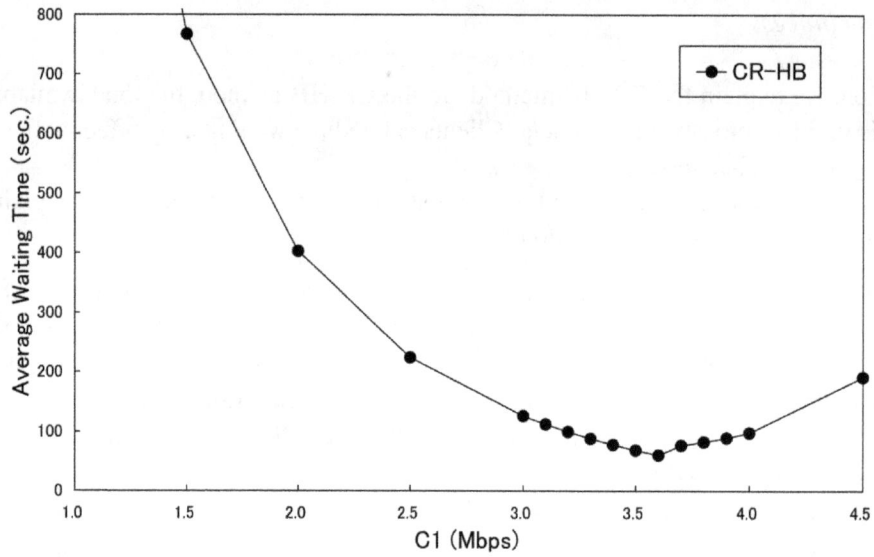

Fig. 4. Average waiting time and C_1.

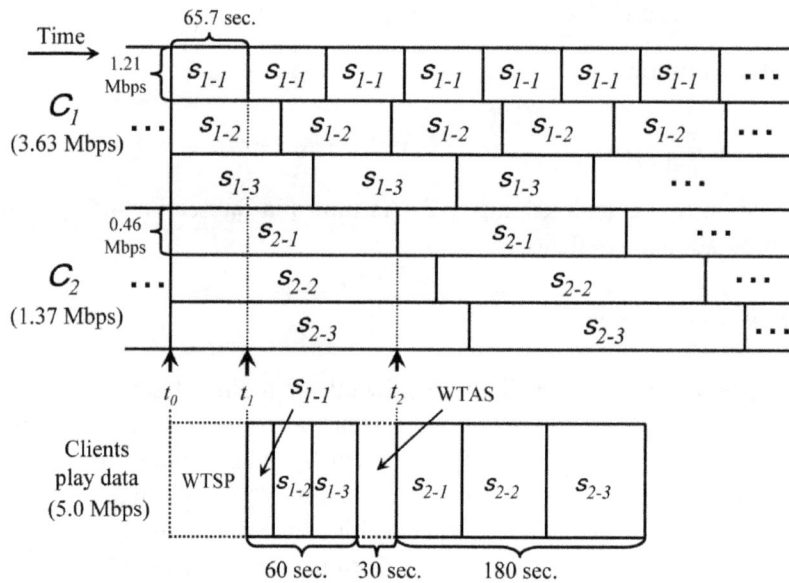

Fig. 5. Example of broadcast schedule under CR-HB method.

This is a naive, but steady approach. We show the average waiting time under different C_1 in Fig. 4. The horizontal axis is the bandwidth of C_1, and the vertical axis is the average waiting time. Total available bandwidth of server is 5.0 Mbps, the number of segments is 3, and the consumption rate is 5.0 Mbps. The playing time of the highlight scene is 60 sec, the main program is 180 sec, and WTAS is 30 sec. In Fig. 4, we can see that the average waiting time is minimum when $C_1^* = 3.63$ Mbps. In this way, we can find C_1^*. In our evaluation, we use the value of C_1^* that makes the average waiting time minimum.

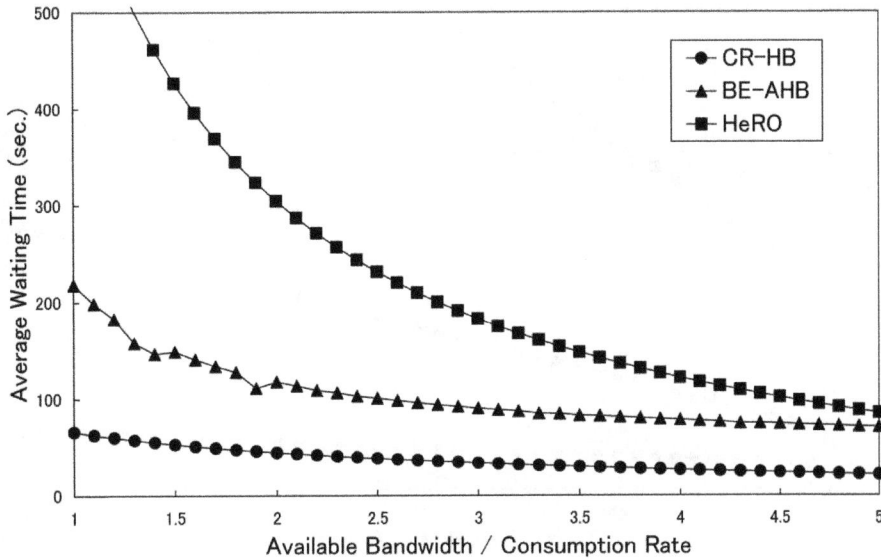

Fig. 6. The average waiting time and the available bandwidth.

4.6. Implementation

The server broadcasts segments following the procedure were explained in Subsection 4.4. Continuous media data are partitioned into several segments by the broadcast schedule in the CR-HB method, where the bandwidth of each subchannel is adjusted based on the available bandwidth of the server. In addition, the server broadcasts data repeatedly considering the broadcast schedule.

When the clients require continuous media data from the server, they start to receive them from several broadcast channels. Clients start playing the highlight scene after completely receiving s_{1-1}. They receive the data while playing it and store it in their buffer. Clients play s_{1-2} continuously, which is stored in their buffer after finishing the playing of s_{1-1}. Clients can play continuous media data without interruption until completely finished receiving it.

For example, a situation that delivers data under the CR-HB method is shown in Fig. 5. The client starts receiving data at t_0 and starts playing it after receiving s_{1-1} at t_1. In this case, since the bandwidth of c_{1-1} is 1.21 Mbps, the receiving time of s_{1-1} is 65.7 sec. Next, when WTAS is 30 sec, since the client finishes receiving s_{2-1} at t_2, it can play s_{2-1} without interruption after finishing WTAS. In Fig. 5, the waiting time under the CR-HB method is reduced $(153.9 - 66.2)/153.9 \times 100 = 57.0\%$ compared to the BE-AHB method.

5. Evaluation

5.1. Basic idea

In this section, we evaluate the performance of the CR-HB method with a computational simulation. Actually, there are many network structures for continuous media data broadcasting. However, since the number of patterns is excessive, evaluating the performance of our proposed method for all these patterns was not realistic. Therefore, we used the network configuration shown in Fig. 1 and compared the CR-HB method, the BE-AHB method [15], and the HeRO [12].

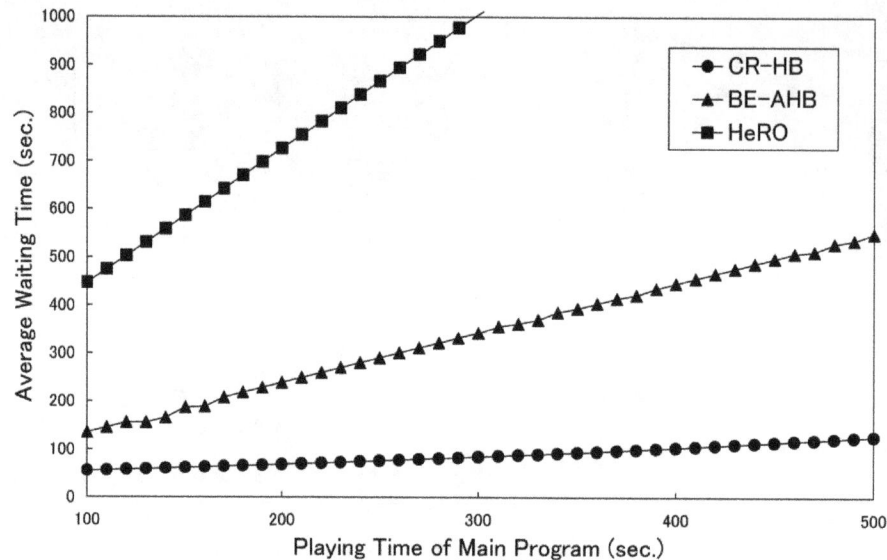

Fig. 7. Average waiting time and playing time of main program.

5.2. Effect of bandwidth

Users prefer shorter waiting times. Since the waiting time accepted by users varies, we show that it is reduced with the CR-HB method compared with conventional methods.

The result is shown in Fig. 6. The horizontal axis is the available bandwidth divided by the consumption rate. The vertical axis is the average waiting time. The playing time of the highlight scene is 60 sec, that of the main program is 180 sec, and WTAS is 30 sec. Consumption rate is 5.0 Mbps.

In Fig. 6, the average waiting time under the CR-HB method is shorter than the conventional BE-AHB method and HeRO. The CR-HB method schedules segments so that the available bandwidth of c_{1-1} that broadcasts s_{1-1} becomes large. Also, we can effectively reduce the waiting time by scheduling segments considering WTSP. For example, when the available bandwidth is 15 Mbps, the waiting time is 33.7 sec under the CR-HB method, 90.5 sec under the BE-AHB method, and 182.9 sec under the HeRO. The average waiting time under the CR-HB method is reduced 62.8% compared to the BE-AHB method, and 81.6% compared to the HeRO.

5.3. Influence of playing time of main program

Waiting time can be reduced by decreasing the playing time of the highlight scene. However, interruption time may be caused by increasing the playing time of the main program. To evaluate the influence of the playing time of the main program, we calculated the waiting time under different playing times of the main program. The result is shown in Fig. 7. The horizontal axis is the playing time of the main program, and the vertical axis is the average waiting time. The playing time of the highlight scene is 60 sec and WTAS is 30 sec. Consumption rate is 5.0 Mbps.

In Fig. 7, when the playing time of the main program increases, the effect of the average waiting time in the CR-HB method is relatively small. The CR-HB method makes the broadcast schedule considering the playing time of the highlight scene and WTAS. On the other hand, the conventional BE-AHB method and HeRO make the broadcast schedule considering both playing time of the highlight scene and the

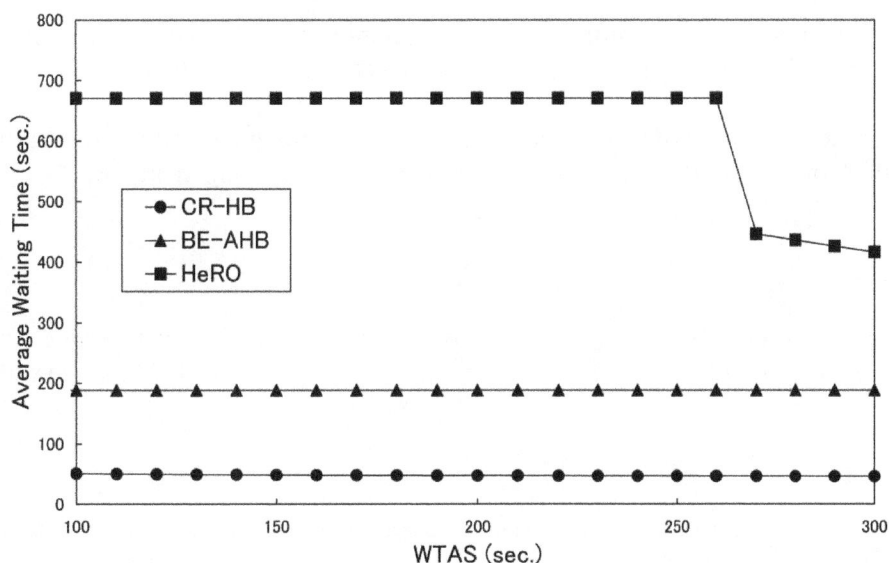

Fig. 8. Average waiting time and waiting time for starting main program.

main program. Therefore, the effect of increasing the playing time of the main program becomes large and waiting time increases.

5.4. Influence of WTAS

When we manage a close-range broadcasting service in a public area such as a shopping center or an event hall, there are a lot of users watching content while walking. In these situations, we need to shorten the upper limit of WTAS. On the other hand, when users watch sport in a stadium or standing in a long line, WTAS can be lengthened. To evaluate the influence of WTAS, we calculated the waiting time under different WTAS. The result is shown in Fig. 8. The horizontal axis is WTAS, and the vertical axis is the average waiting time. The available bandwidth is 15 Mbps. The playing time of the main program is 180 sec, and consumption rate is 5.0 Mbps.

In Fig. 8, waiting time under the CR-HB method is shorter than the conventional BE-AHB method and HeRO. In the CR-HB method, waiting time is reduced by making the broadcast schedule considering WTAS. In the conventional BE-AHB method and HeRO, waiting time is not changed significantly when WTAS is between 100 sec and 260 sec. These methods make the broadcast schedule considering only waiting time in the receiving time of s_{1-1}, and waiting time is not changed by increasing WTAS. However, when WTAS is more than 270 sec, waiting time under the HeRO is reduced. In the HeRO, when the available bandwidth for the server has an upper limit, interruption time occurs while playing data.

6. Conclusion

In this paper, we proposed a scheduling method to reduce waiting time in close-range broadcasting. In our proposed method, waiting time while playing data is effectively reduced by making the schedule considering the waiting time after the main program. In our evaluations, we showed the available bandwidth for which the average waiting time has a minimum value by the computational simulation. Also, we confirmed that the average waiting time has reduced more than the BE-AHB method and the HeRO.

A future direction of this study will involve creating a scheduling method in broadcasting selective contents in which the user watches several content stream while selecting them [18].

Acknowledgment

This research was supported in part by JSPS Postdoctoral Fellowships for Research Abroad. Also, this work was partially supported by JSPS Grant-in-Aid for Young Scientists (A) numbered 23680007.

References

[1]   WHITE PAPER Information and Communications in Japan (2011). http://www.soumu.go.jp/johotsusintokei/white paper/eng/WP2011/2011-index.html.

[2]   Digital TV Broadcasting in Japan, http://www.soumu.go.jp/main_sosiki/joho_tsusin/eng/presentation/pdf/070613_1.pdf.

[3]   N. Suzuki and H. Tsujimura, The World's First One-Segment Transmission System Using Weak Radio Waves, Fujitsu, vol.60, no.4, 2009, pp. 341–346.

[4]   B. Jinsuk and F.P. Jehan, A Tree-Based Reliable Multicast Scheme Exploiting the Temporal Locality of Transmission Errors, Proc. IEEE Int. Performance, Computing, and Communications Conference (IPCCC 2005), 2005, pp. 275–282.

[5]   L.-S. Juhn and L.M. Tseng, Fast data broadcasting and receiving scheme for popular video service, *IEEE Trans Broadcasting* **44**(1) (1998), 100–105.

[6]   J.-F. Paris, S.W. Carter and D.D.E. Long, A hybrid broadcasting protocol for video on demand, Proc. Multimedia Computing and Networking Conference (MMCN '99), 1999, pp. 317–326.

[7]   J.-F. Paris, D.D.E. Long and P.E. Mantey, Zero-delay broadcasting protocols for video-on-demand, Proc. ACM Int. Multimedia Conf. (Multimedia '99), 1999, pp. 189–197.

[8]   J.-F. Paris, An Interactive Broadcasting Protocol for Video-on-Demand, Proc. IEEE Int. Performance, Computing, and Communications Conference (IPCCC '01), 2001, pp. 347–353.

[9]   S. Viswanathan, and T. Imilelinski, Pyramid broadcasting for video on demand service, Proc. SPIE Multimedia Computing and Networking Conf. (MMCN '95), 1995, pp. 66–77.

[10] L. Shi, P. Sessini, A. Mahanti, Z. Li and D.L. Eager, Scalable Streaming for Heterogeneous Clients, Proc. ACM Multimedia, 2006, pp. 22–27.

[11] Y. Zhao, D.L. Eager and M.K. Vernon, Scalable On-Demand Streaming of Non-Linear Media, Proc. of IEEE INFOCOM, vol.3, 2004, pp. 1522–1533.

[12] K.A. Hua, O. Bagouet and D. Oger, Periodic Broadcast Protocol for Heterogeneous Receivers, Proc. of MMCN, 2003, pp. 220-231.

[13] M. Tantaoui, K. Hua and T. Do, BroadCatch: A Periodic Broadcast Technique for Heterogeneous Video-on-Demand, *IEEE Trans Broadcasting* **50**(3) (2004), 289–301.

[14] R. Janakiraman and M. Waldvogel, Fuzzycast: Efficient Video-on-Demand over Multicast, Proc. IEEE INFOCOM, 2002, 920–929.

[15] T. Yoshihisa, M. Tsukamoto and S. Nishio, A Broadcasting Scheme for Continuous Media Data with Restictions in Data Division, Proc. IPSJ International Conference on Mobile Compuring and Ubiquitous Networking (ICMU'05), 2005, pp. 90–95.

[16] W.-S. Kua and R. Zimmermannb, Nearest neighbor queries with peer-to-peer data sharing in mobile environments, *Pervasive and Mobile Computing* **4** (2008), 775–788.

[17] Y. Gotoh, K. Suzuki, T. Yoshihisa and M. Kanazawa, A Scheduling Method to Reduce Waiting Time for P2P Streaming Systems, *Journal of Mobile Multimedia* **5**(3) (2009), 255–270.

[18] Y. Gotoh, T. Yoshihisa, M. Kanazawa and Y. Takahashi, A Broadcasting Scheme for Selective Contents Considering Available Bandwidth, *IEEE Trans Broadcasting* **55**(2) (2009), 460–467.

9

Implementation of CAVENET and its usage for performance evaluation of AODV, OLSR and DYMO protocols in vehicular networks

Evjola Spaho[a,*], Leonard Barolli[b], Gjergji Mino[a], Fatos Xhafa[c], Vladi Kolici[d] and Rozeta Miho[d]

[a]*Graduate School of Engineering, Fukuoka Institute of Technology (FIT), 3-30-1 Wajiro-Higashi, Higashi-Ku, Fukuoka 811-0295, Japan*

[b]*Department of Information and Communication Engineering, Fukuoka Institute of Technology (FIT), 3-30-1 Wajiro-Higashi, Higashi-Ku, Fukuoka 811-0295, Japan*

[c]*Department of Languages and Informatics Systems, Polytechnic University of Catalonia, Jordi Girona 1-3, 08034 Barcelona, Spain*

[d]*Department of Electronic and Telecommunication, Polytechnic University of Tirana, Mother Teresa Square, Nr.4, Tirana, Albania*

Abstract. Vehicle Ad-hoc Network (VANET) is a kind of Mobile Ad-hoc Network (MANET) that establishes wireless connection between cars. In VANETs and MANETs, the topology of the network changes very often, therefore implementation of efficient routing protocols is very important problem. In MANETs, the Random Waypoint (RW) model is used as a simulation model for generating node mobility pattern. On the other hand, in VANETs, the mobility patterns of nodes is restricted along the roads, and is affected by the movement of neighbour nodes. In this paper, we present a simulation system for VANET called CAVENET (Cellular Automaton based VEhicular NETwork). In CAVENET, the mobility patterns of nodes are generated by an 1-dimensional cellular automata. We improved CAVENET and implemented some routing protocols. We investigated the performance of the implemented routing protocols by CAVENET. The simulation results have shown that DYMO protocol has better performance than AODV and OLSR protocols.

1. Introduction

During recent years, there has been an unprecedented growth in wireless networks. This can be attributed to high demand for wireless multimedia services such as data, voice, video, and the development of new wireless standards. There are lots of other driving factors that have led to the rapid and continuous change of the wireless networks worldwide. Mobility is a major driver for mobile networks because mobile users continue to demand access remotely anywhere and anytime. The ever growing need for mobile Internet access, interactive services, training, and entertainment; the need for a single standard for seamless roaming; interoperability across networks; and upward integration of earlier wireless network technologies are also driving factors for new developments in wireless networks. Other driving factors

---

*Corresponding author. E-mail: evjolaspaho@hotmail.com.

are improvements in RF performance that are attributable to improved antennas, reduction in sources of interference, and the ability to support multiple frequency bands. In recent years, wireless networks are continuing to attract attention for their potential use in several fields such as ad-hoc networks, sensor networks, mesh networks, and vehicular networks [5,9,13,14,17,18,25–27].

Vehicular communication is seen as a key technology for improving road safety and comfort through Intelligent Transportation Systems (ITS). There are many possible application of wireless technologies for vehicular environment [10].

Vehicular Ad Hoc Networks (VANETs) are an instance of ad-hoc networks, which are general-purpose distributed wireless networks interconnected without the need of any centralized infrastructure. VANETs are expected to be massively deployed in upcoming vehicles, because their use can improve the safety of driving and makes new forms of inter-vehicle communications possible as well. Given a mobility model of vehicles, usually a simulator is used to test networking protocols. In this regard, we present a lightweight simulator which can be used to understand the properties of the mobility models of vehicular traffic and their impact on the performance of VANETs. We call this simulator Cellular Automaton based VEhicular NETwork (CAVENET), because its mobility model is built upon a 1-dimensional Cellular Automaton (CA).

The CAVENET separates the problem of mobility model from that of the protocol evaluation, which is performed by means of a network simulator. The properties of the mobility model, e.g. the average transient time towards the stationary state, can be analysed independently of the protocol simulation. Eventually, the movement patterns generated by the mobility model can be mapped into a trace file format suitable for the network simulator.

The Random Waypoint (RW) model has been the earliest mobility model for ad-hoc networks. Basically, in RW every node picks up a random destination and a random velocity at certain points called waypoints. This model has been extended in a number of ways in order to take into account more realistic movements. The simulation of such models has shown the problem of velocity decay, which posed some doubts about the length of the simulation time and the duration of the transient. The problem has been solved by several authors, in particular by Le Boudec [16], who used Palm distributions, and Noble [28]. However, all mobility models considered so far are Short Range Dependent (SRD). This means that every mobile chooses its velocity independently by the others. In the case of VANETs, this assumption is clearly not valid anymore, especially in the case of highway traffic. We show this fact by means of basic simulations performed with CAVENET. For instance, we show that the traffic model strongly affects the statistical structure of the average velocity.

In the particular case of deterministic traffic models, the average velocity is SRD and the transient state depends on the density of the vehicles. In general, the mobility model of VANETs for the simulated variable of interest (e.g. the average velocity) can be Long Range Dependent (LRD) in some cases. This fact poses some problems on how long the simulation should be and how many samples from the starting time should be discarded.

In literature, vehicular mobility models are usually classified as either macroscopic or microscopic. The macroscopic description models gross quantities of interest, such as vehicular density or mean velocity, treating vehicular traffic according to fluid dynamics, while the microscopic description considers each vehicle as a distinct entity, modelling its behaviour in a more precise, but computationally more expensive way. Yet, a micro-macro approach may be seen more as a broad classification schema than a formal description of the models' functionalities in each class [10].

In this work, we consider the vehicular mobility model as a microscopic model. Our simulator is based on 1-dimensional CA model. The CA is a discrete time model of the vehicular traffic. The first version of CAVENET had some problems. For this reason, we improved the CAVENET, by changing the movement pattern of the vehicles from the straight line to a circle. We also implemented three routing protocols for Ad Hoc networks: Optimized Link State Routing (OLSR) [3], Ad-hoc On-demand Distance Vector (AODV) [20], Dynamic MANET On Demand (DYMO) [4,22], and investigated their performance for VANETs.

The rest of the paper is structured as follows. In Section 2, we discuss the related work for VANETs. In Section 3, we present CAVENET structure and description. In Section 4, we discuss the simulation results. Finally, the conclusions and future work are presented in Section 5.

2. Related work

In general, a simulator should have the following properties.

1. It should be open source, in order to let other users criticize the validity of the model and the implementation.
2. The code should be clear, in order to let others performing the task in 1.
3. The structure should be modular, in order to analyse single pieces of the simulation process.

In the recent years, a lot of simulators for VANETs have been emerging [10]. For example, the IMPORTANT framework has been one of the first attempt to understand the dependence between vehicular traffic and communication performance [2,23]. The authors analyzed the impact of the node mobility on the duration of communication paths. However, the author implemented the code in C, which is difficult to debug and extend without the support of a detailed documentation. Moreover, it seems that their Freeway model is not as realistic as the model we study here.

In [8], the authors present a simulator written in Java, which can generate mobility traces in several formats. The details of the implementation are not open. There are also other powerful traffic simulators, like TranSim [24], which makes use of a cellular automaton for simulating the interaction of vehicles. Unfortunately, the code is not conceived for network protocols simulation, and the software is commercially licensed. Also, SUMO is another powerful traffic simulator, intended for traffic planning and road design optimization. There is an attempt to interface SUMO with ns-2 [21]. However, in our opinion, it is very expensive to understand the SUMO language and also unnecessary, because the communications engineer needs only a parsimonious model, easy to extend and/or modify.

There are many other works which consider the possibility of using ad-hoc and MANET protocols for VANET scenarios. A car taking part in a MANET scenario could establish connections using the public hotspots while driving in the city. Also, the deployment of access points along highways in the near future seems feasible. Thus, it is important to investigate the application of MANET routing protocols for VANETs [6,11,12]. In [6], the authors present only the way of generating the vehicle movement pattern. They did not evaluate the performance of routing protocols. While, in [11], the authors used the simulation in [12] and present the performance evaluation for AODV and OLSR protocols. However, they use the uniform distribution for the generation of the node movement.

3. CAVENET structure and description

The mobility model for VANETs should take into account the following parameters.
- The number of lanes and their directions

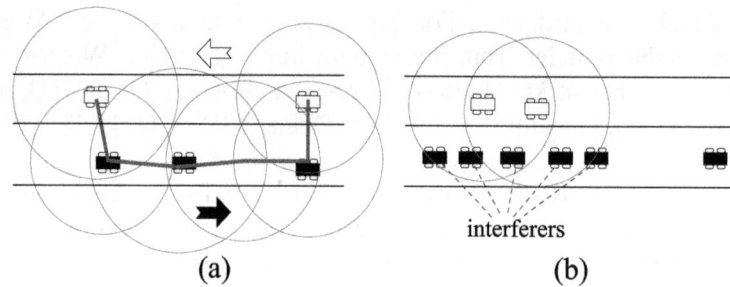

Fig. 1. Impact of multi-lanes: a) on the connectivity; b) on the interference.

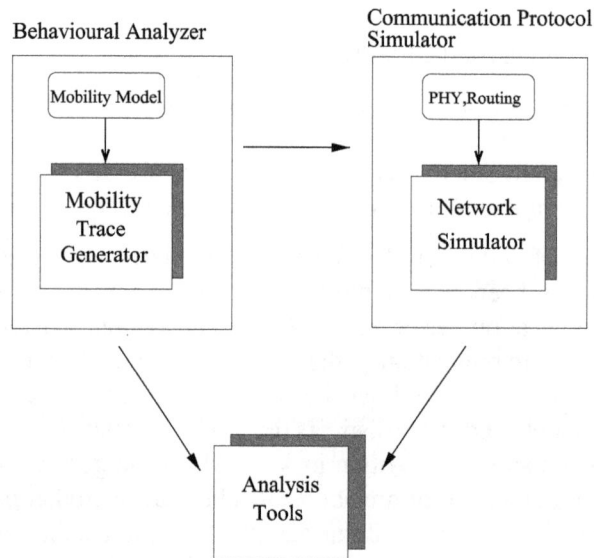

Fig. 2. Structure of CAVENET.

From the point of view of protocol operations, these parameters can affect the connectivity of the network. In particular, connectivity gaps on a lane can be filled by the presence of relay nodes on the other lanes, as shown in Fig. 1-a. On the other hand, the message penetration on a particular lane can be affected by the radio interference on the opposite lane, as in Fig. 1-b.

• The intersection of lanes

This parameter affect the traffic behaviour on the whole lane, because the crosspoint is the bottleneck for the lane.

Here, we take into account only the first parameter. With respect to the aforementioned properties, we propose to divide the simulator into two blocks, as shown in Fig. 2. The first one, which we call Behavioural Analyzer (BA) block, is concerned with the mobility model, and it should take into account the previous parameters in order to produce accurate mobility traces. The second one, which we name Communication Protocol Simulator (CPS), is the protocol simulator, and it is conceived to test the performance of communication protocols given a particular mobility trace. The BA block should be written in a high-level language, easy to understand and easy to extend. For the particular case of CAVENET, the matrix operations are needed. For this reason, we choose MATLAB. The BA block produces movement patterns which are formatted in a textual format compatible with the CPS's language.

Extending the BA block in order to export to other formats is straightforward. The CPS can be one of the many publicly available network simulators, as the well known ns-2 [15]. In principle, the two blocks could also be implemented in two separate machines, in order to speed up the simulation, as in [21].

3.1. Microscopic model

The core of our simulator is 1-dimensional CA model, which has been first studied by Nagel and Schreckenberg (NaS) [19] in a stochastic settings. The CA is a discrete time model of the vehicle traffic. It is governed by three simple rules. However, as for other CAs, these simple rules can well model and reproduce complex real systems. For this reason, the NaS model has gained a lot of attention during the last ten years.

The time is divided in discrete units Δt, so that $t_n = n\Delta t$. There are N vehicles. A lane k of the road at time $t_n, n \in \mathbb{N}$, is represented by a vector $\mathbf{L}_n^k$ of L sites. The lane is assigned an $N\mathrm{x}1$ velocity vector $\mathbf{v}_n^k = (v_{i,n}^k)_{i=1}^N$, where $v_{i,n} \in \mathbb{N}_{v_{\max}}$ is the velocity of the vehicle at time t_n and position i. If the ith site is occupied by a car, $L_{i,n} = v_{i,n}$. Otherwise, $L_{i,n} = -1$. We use the lane index only when it is explicitly required. Every cell or site of the lane has a length of s meters. By setting $v_{\max} = 135$ km/h and $\Delta t = 1$ s, we obtain $s = 7.5$ m. At every time step, the velocity v is changed according to the following rules.[1]

Deterministic, $p = 0$ **or** $p = 1, \forall i$

- 1. $v_{i,n+1} = \min(v_{i,n} + 1, v_{\max})$
- 2. $v_{i,n+1} = \min(v_{i,n}, L_{i+1,n} - L_{i,n} - 1)$
- 3. $\mathbf{L}_{n+1} = \mathbf{L}_n + \mathbf{v}_{n+1}$

Stochastic

- 2'. $v_{n+1,i} = \max(0, v_{n,i} - 1)$, with probability p.

The vehicle density is $\rho = N/L$. This simple model can recreate the footprints of real traffic scenarios, such as the $1/f$ noise of the average velocity observed in real traffic. The dynamics of the systems are regulated by three important parameters, p, ρ and L. For example, if $p = 0$ the average velocity is SRD, otherwise the system present LRD.[2]

3.2. Improvement of CAVENET

In the first version of CAVENET, the vehicles were moving in a horizontal line. When a vehicle was at the end of line, in order to continue the simulation we shifted the vehicle at the beginning of line. But,

---

[1] We assume parallel update only, i.e. the rules are applied in parallel to every vehicle on the lane.

[2] A stochastic process $\{X_n\}_{n=1}^{n=+\infty}$ is SRD if the autocorrelation is summable:

$$\sum_{k=1}^{+\infty} r(k) < +\infty,$$

where $r(k) = E[(X_n - \overline{X})(X_{n+1} - \overline{X})]/\sigma^2$. Otherwise, if $r(k)$ is not summable, the process is LRD. This means that very distant samples are not statistically independent, contrary to processes without memory, as the Poisson process which is an SRD process.

this caused a delay and the vehicles at the beginning and at the end of the line could not communicate with each other. For this reason, we improved the CAVENET, by changing the movement pattern of the vehicles from the straight line to a circle. We also implemented three routing protocols: OLSR, AODV, DYMO and investigated their performance for VANETs.

3.2.1. OLSR

The OLSR protocol is a pro-active routing protocol, which builds up a route for data transmission by maintaining a routing table inside every node of the network. The routing table is computed upon the knowledge of topology information, which is exchanged by means of Topology Control (TC) packets.

OLSR makes use of *HELLO* messages to find its one hop neighbours and its two hop neighbours through their responses. The sender can then select its Multi Point Relays (MPR) based on the one hop node which offer the best routes to the two hop nodes. By this way, the amount of control traffic can be reduced. Each node has also an MPR selector set which enumerates nodes that have selected it as an MPR node. OLSR uses TC messages along with MPR forwarding to disseminate neighbour information throughout the network. Host Network Address (HNA) messages are used by OLSR to disseminate network route advertisements in the same way TC messages advertise host routes.

OLSRv2 is currently being developed at IETF. It maintains many of the key features of the original protocol including MPR selection and dissemination. Key differences are the flexibility and modular design using shared components such as packet format packetbb and neighbourhood discovery protocol.

Recently *olsrd* has been equipped with the LQ extension, which is a shortest-path algorithm with the average of the packet error rate as metric. This metric is commonly called ETX, which is defined as $ETX(i) = 1/(NI(i) \times LQI(i))$. Given a sampling window W, $NI(i)$ is the packet arrival rate seen by a node on the i-th link during W. Similarly, $LQI(i)$ is the estimation of the packet arrival rate seen by the neighbour node which uses the i-th link. When the link has a low packet error rate, the ETX metric is higher.

3.2.2. AODV

The AODV is an improvement of DSDV to on-demand scheme. It minimize the broadcast packet by creating route only when needed. Every node in network maintains the route information table and participate in routing table exchange. When source node wants to send data to the destination node, it first initiates route discovery process. In this process, source node broadcasts Route Request (RREQ) packet to its neighbours. Neighbour nodes which receive RREQ forward the packet to its neighbour nodes. This process continues until RREQ reach to the destination or the node who know the path to destination.

When the intermediate nodes receive RREQ, they record in their tables the address of neighbours, thereby establishing a reverse path. When the node which knows the path to destination or destination node itself receive RREQ, it send back Route Reply (RREP) packet to source node. This RREP packet is transmitted by using reverse path. When the source node receives RREP packet, it can know the path to destination node and it stores the discovered path information in its route table. This is the end of route discovery process. Then, AODV performs route maintenance process. In route maintenance process, each node periodically transmits a Hello message to detect link breakage.

3.2.3. DYMO

DYMO is a new reactive (on demand) routing protocol, which is currently developed in the scope of the IETF's MANET working group. DYMO builds upon experience with previous approaches to reactive routing, especially with the routing protocol AODV. It aims at a somewhat simpler design, helping to reduce the system requirements of participating nodes, and simplifying the protocol implementation. DYMO retains proven mechanisms of previously explored routing protocols like the use of sequence numbers to enforce loop freedom. At the same time, DYMO provides enhanced features, such as covering possible MANET-Internet gateway scenarios and implementing path accumulation.

Besides route information about a requested target, a node will also receive information about all intermediate nodes of a newly discovered path. There is a major difference between DYMO and AODV. AODV only generates route table entries for the destination node and the next hop, while DYMO stores routes for each intermediate hop. To efficiently deal with highly dynamic scenarios, links on known routes may be actively monitored, e.g. by using the MANET Neighbourhood Discovery Protocol or by examining feedback obtained from the data link layer. Detected link failures are made known to the MANET by sending a route error message (RERR) to all nodes in range, informing them of all routes that now became unavailable. Should this RERR in turn invalidate any routes known to these nodes, they will again inform all their neighbours by multicasting a RERR containing the routes concerned, thus effectively flooding information about a link breakage through the MANET.

DYMO was also designed with possible future enhancements in mind. It uses a generic MANET packet and message format and offers ways of dealing with unsupported elements in a sensible way.

3.3. Vehicle model

Every vehicle is a data structure VE_i indexed by its position on the lane. The data structure for the ith vehicle stores: the gap, the velocity, and the current lane position. The relative euclidean position on the lane given by X_i is a unique identifier used for the generation of mobility trace. Moreover, for closed boundaries, i.e. if we suppose circular movement of vehicle on the lane, we check if a shift has taken place. This information will serve to properly generate the trace for ns-2. It is straightforward to arrange all these information in a vector form, what is the preferred form used in MATLAB.

3.4. Lane construction

Instead of using a particular textual language for describing the position of the lanes in the plane, we use a more general approach. Besides its length, every lane is given a lane transformation, which is used in order to set its real aspect on the plane. This information is used at the mobility trace generation stage. The transformation is a simple affine transformation of the vector $\mathbf{X}_i^k = (X_i, Y_i, 1)$, i.e. the coordinate vector of the ith vehicle on the kth road with respect to the relative reference system. For example, for the lane $\mathbf{L}^k$, we have the vehicle structure VE_i^k. This structure contains the vector $\mathbf{X}_i^k$. The real position on the plane is computed as $\widetilde{\mathbf{X}}_i^k = \mathbf{A}(k)\mathbf{X}_i^k$ where $\mathbf{A}(k)$ is the lane transformation matrix associated with the k-th lane, and $\widetilde{\mathbf{X}}_i^k$ is the vector of coordinates in the absolute reference system (i.e. that used for exporting the ns-2 traces). For example, in Fig. 3, the third lane has the following absolute coordinates:

$$\widetilde{\mathbf{X}}_i^3 = \begin{pmatrix} 0 & 1 & \frac{XS}{2} \\ 1 & 0 & \Delta \\ 0 & 0 & 1 \end{pmatrix} \begin{pmatrix} X_i \\ 0 \\ 1 \end{pmatrix},$$

where XS is the length of the simulation area.[3]

---

[3]The parameter Δ is used to avoid an apparent bug in ns-2, which fires strange errors when the absolute position is 0.

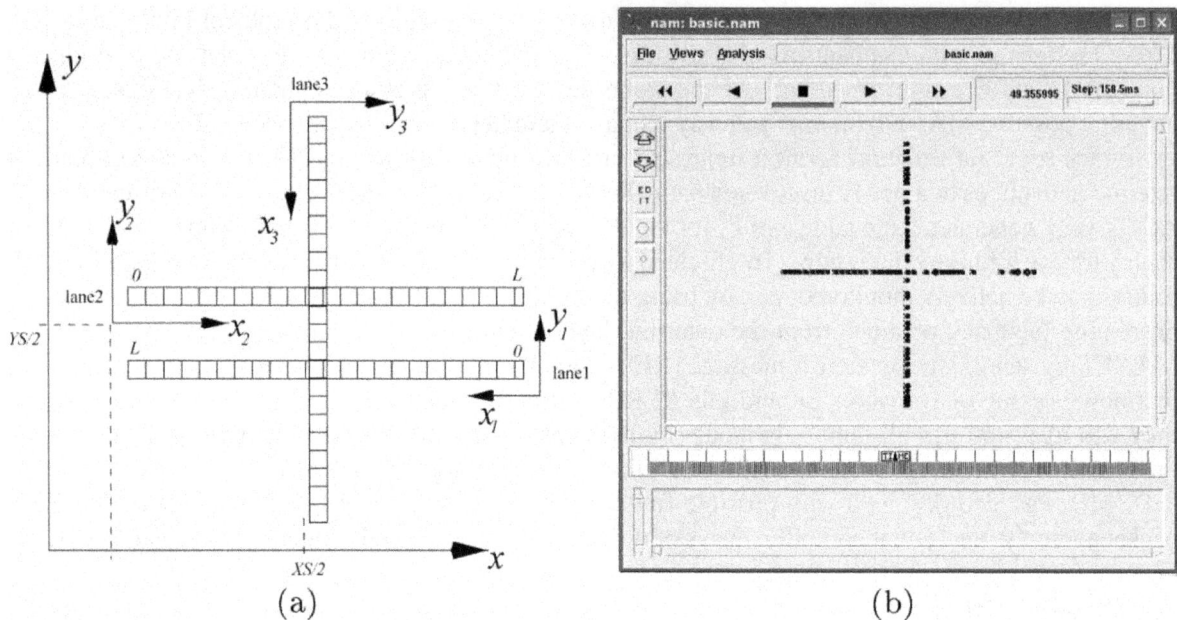

Fig. 3. Lanes construction and ns-2 trace: a) Line construction; b) Excerpt of the generated ns-2 trace for 2 lanes network.

4. Simulation results

4.1. Mobility model validation

We present here some basic simulations for the NaS model by means of CAVENET. Hereinafter, we use as simulation variable the average velocity $\bar{v}(t) = N^{-1} \sum_{i=1}^{N} v_i(t) = N^{-1} \| \mathbf{v}(\mathbf{t}) \|_1$ of all cars. CAVENET can analyze and design single and multiple lanes traces. It can also run Monte Carlo simulations. For example, in Fig. 4, we report the results for the so called fundamental diagram, i.e. the flow vs. density diagram. The flow at a particular lane section is defined as $J = \rho \bar{v}$. Each point in the figure is the ensemble average over 20 trials of a simulation trace lasting 500 iterations. Moreover, we can also visualize the space-time plot of the traffic, i.e. the evolution of the velocity for every vehicle along the road as shown in Fig. 5. We obtain the two traffic regimes, namely the laminar regime and the jammed or congested regime, as shown in Fig. 5-a and Fig. 5-b, respectively. We are interested in the the stationary distribution and transient time, which are very important to assess the next stage simulations, i.e. those related to the communication protocol analysis.

4.2. Stationary distribution

Usually, RW-like mobility models used in simulation exhibit the velocity decay problem. That means that the simulation variable slowly decays towards a steady state value as the simulation time proceeds. This is problematic, because we do not know when this transient ends. Consequently, we do not know precisely how to remove the transient values. The root of this phenomenon has to be attributed to the underlying mobility model, which has been assumed random. Every node randomly picks a velocity from a continuous uniformly distributed random variable between $[v_{\min}, v_{\max}]$. The velocity is changed at particular points called waypoints. In this way, the system has an infinite (but countable) number

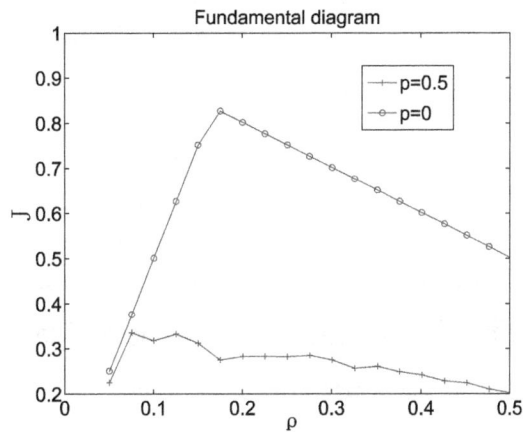

Fig. 4. Traffic flow as a function of ρ and p for $L = 400$.

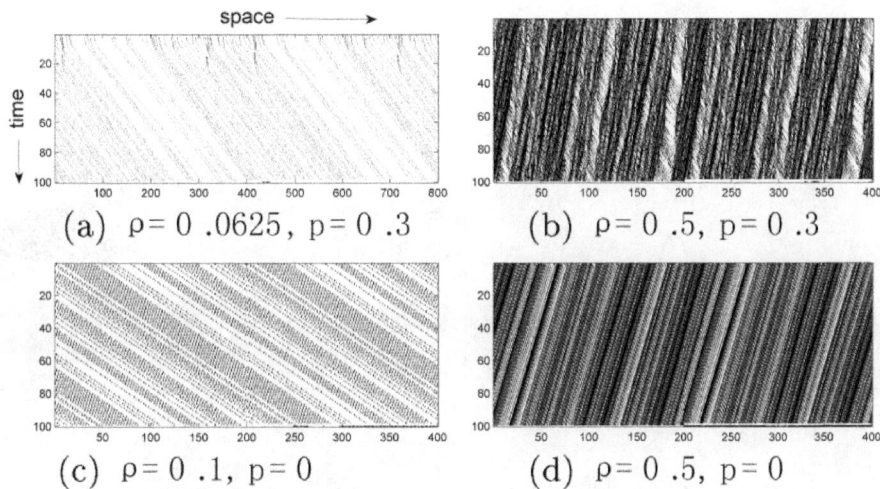

Fig. 5. Space-time plots showing the jam wave in different settings.

of states. The general solution to this problem consists in finding the steady state distribution of the simulation variable and let the system starts with that distribution. This reasoning is also equivalent to consider Palm probability distributions instead of the usual ones [16].

In our case, the system has inherently a finite state space. The automaton could be represented by a discrete-time finite-state Markov chain. We know that a Markov chain with a single class of recurrent states has always a steady state distribution. Moreover, since a Markov chain with finite state space has always at least one recurrent state, we conclude that the steady state distribution exists and is unique. The convergence rate toward this steady-state distribution depends on the eigenstructure of the transition probability matrix of the Markov chain. The problem here is that a Markov chain model is not suitable, because the process can be, in general, LRD, for $0 < p < 1$. Moreover, even in the SRD case, finding the transition probabilities is not easy.

In general, mobility models for vehicular traffic exhibits a phase transition around a particular value of ρ. As we can see in Fig. 5, for $p > 0$, the traffic is composed of jammed regions which travel on the opposite direction of movement. For low densities, these waves die out very quickly, as shown also

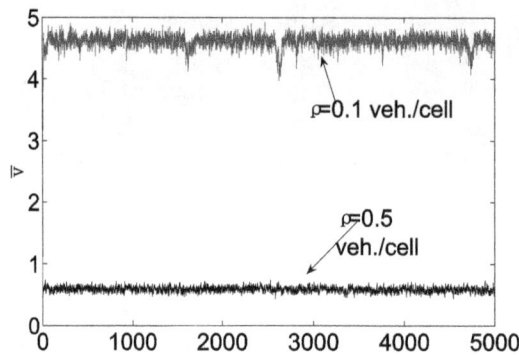

Fig. 6. Sample realizations of $\overline{v}(t)$.

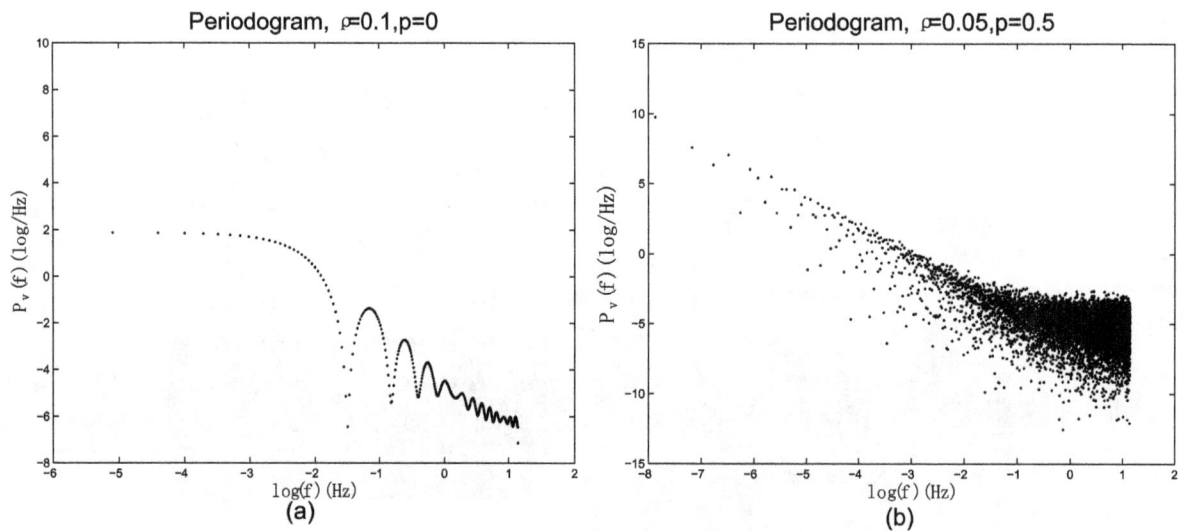

Fig. 7. Deterministic model and stochastic version: a) deterministic model; b) 1/f noise like spectrum form the the stochastic version of the NaS model.

in Fig. 6, but for higher densities there are many interconnected clusters of jammed vehicles. In this case, the steady state is reached very slowly. Therefore, it is important to investigate how many samples should be removed from the staring point in order to sample a process in its stationary regime.

In order to clarify this phenomenon, we measured the transient time τ for $p = 0$, i.e. the deterministic case. In this case, $\overline{v}(t)$ is not LRD. We can show this fact also by plotting the periodogram of $\overline{v}(t)$. In Fig. 7-a, we see that for $f \rightarrow 0$, the periodogram does not diverge. On the other hand, for $p > 0$, in Fig. 7-b, the estimated spectrum diverges at the origin, i.e. the underlying process has the LRD property.

4.3. Routing protocols evaluation

As evaluation metrics, we use the goodput and Packet Delivery Ratio (PDR). The simulation parameters are shown in Table 1. We used one line and 30 nodes for simulations. The simulation time is 100 seconds.

The receiving node is node 0 and the sending nodes are from node 1 to node 8. We prepared each scenario based on nodes ID. The mobility pattern for all scenarios is the same. In order to evaluate the

Table 1
Simulation parameters

Network Simulator	ns-2
Routing Protocol	AODV, OLSR, DYMO
Simulation Time	100 s
Simulation Area	3000 m Circuit
Number of Nodes	30
Traffic Source/Destination	Deterministic
DATA TYPE	CBR
Packets Generation Rate	5 packets/s
Packet Size	512 bytes
MAC Protocol	IEEE802.11 DCF
MAC Rate	2 Mbps
RTS/CTS	None
Transmission Range	250 m
Radio Propagation Models	Two-ray Ground
Hello$_{AODV}$ Interval	1 s
Hello$_{OLSR}$ Interval	1 s
TC$_{OLSR}$ Interval	2 s
Hello$_{DYMO}$ Interval	1 s

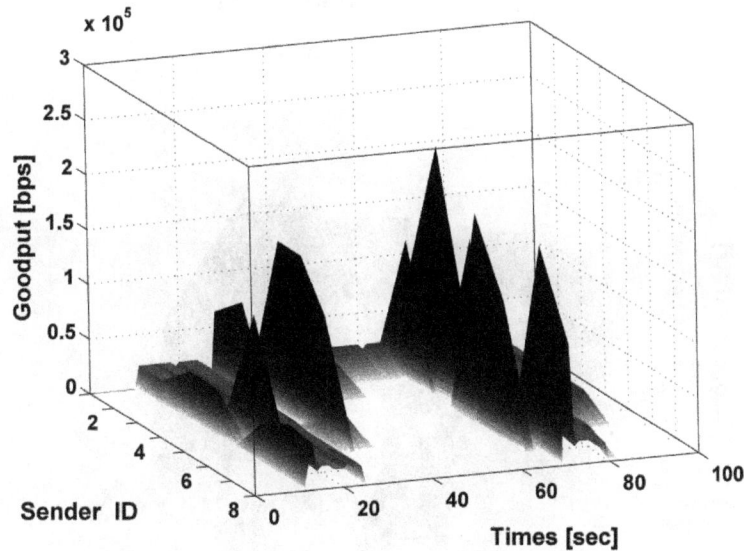

Fig. 8. AODV Goodput.

performance of each protocol, 5 packets per second as a Constant Bit Rate (CBR) traffic were transmitted between 10 seconds and 90 seconds.

The simulations results are shown from Figs 8 to 11. In Fig. 8 is shown the goodput of AODV protocol. The goodput of AODV is about ten times of CBR packet size. This is because after a back-off time all the accumulated data packets are transmitted in the discovered route. If we increase the background traffic, the number of transmitted packets will again increases and the network may be congested. Also, after 60 seconds, in AODV protocol, there is a delay caused by route finding mechanism. Comparing Fig. 8, Figs 9 and 10, we can see that reactive protocols (AODV and DYMO) have better goodput than OLSR. For AODV and DYMO, even the nodes are far from each other they can communicate between 10 seconds to 20 seconds.

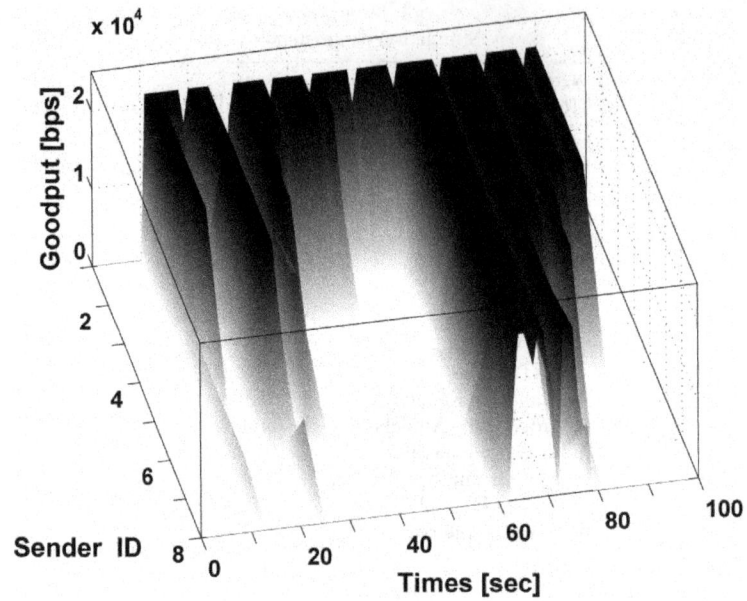

Fig. 9. OLSR Goodput.

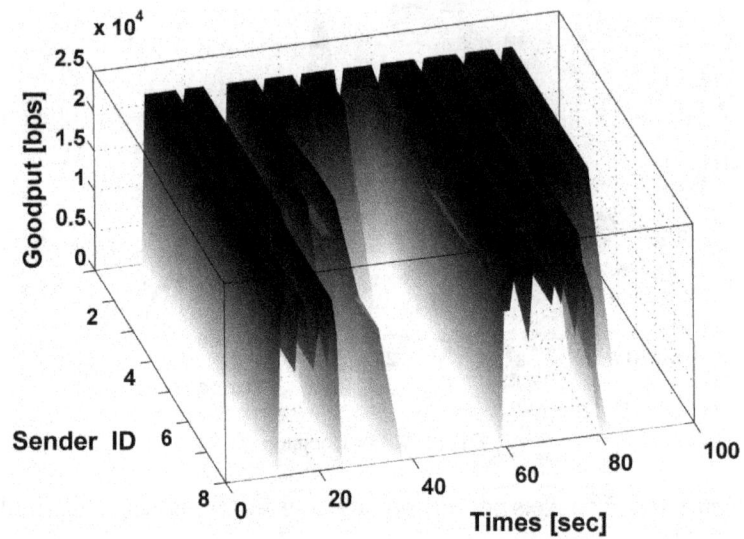

Fig. 10. DYMO Goodput.

In Fig. 11, we show the PDR for three routing protocols. We can see that among three protocols AODV has a better goodput. However, the AODV need more time for searching a new route compared with DYMO. So, the delay of AODV is higher than DYMO. The route searching time of DYMO is almost the same with OLSR protocol. However, DYMO have better goodput than OLSR. Thus, DYMO has a better performance than AODV and OLSR protocols.

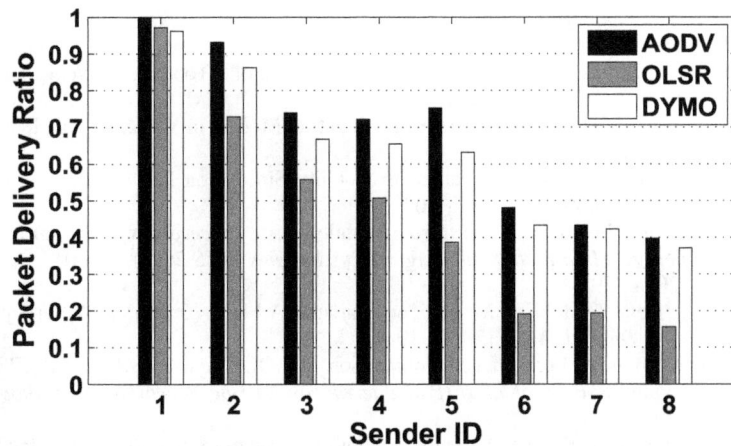

Fig. 11. PDR for AODV, OLSR and DYMO.

5. Conclusions

In this paper, we presented CAVENET, a simple simulator for VANETs. The system is modular and it separates mobility from the protocol simulation. The mobility model code is written in a language at a level as high as possible, in order to give the researcher a quick understanding of the basic properties of his/her model. For this reason, we used MATLAB. The basic structure in CAVENET is the vector representing the configuration of a linear lane. The geometry of the lanes is set by affine transformations which are stored in a text file. In such way, the user does not need to learn a particular file format, as in other traffic simulators. By means of CAVENET, we have shown some fundamental properties of vehicular traffic which should taken into account when performing network protocols simulations.

We improved the CAVENET by changing the movement pattern of the vehicles from the straight line to a circle and implemented three routing protocols: AODV, OLSR and DYMO. We evaluated the performance of these protocols in VANETs and we found that DYMO has better performance.

In this work, we evaluated AODV, OLSR and DYMO considering goodput and PDR metrics. In the future, we would like to consider other parameters such as routing overhead, traffic quantity and topology change. We also plan to extend our work for different radio propagation models and environments [1,7].

Acknowledgment

This work is support by a Grant-in-Aid for scientific research of Japanese Society for the Promotion of Science (JSPS). The authors would like to thank JSPS for the financial support.

References

[1]   L. Barolli, Gj. Mino, F. Xhafa, G. De Marco, A. Durresi, A. Koyama, Analysis of Ad-Hoc Networks Connectivity Considering Shadowing Radio Model, *Proc. of MoMM-2009*, Kuala Lumpur, Malaysia, (2009), 464–468.

[2]   F. Bai, N. Sadagopan and A. Helmy, IMPORTANT: A Framework to Systematically Analyze the Impact of Mobility on Performance of Routing Protocols for Ad-hoc Networks, *Proc of IEEE INFOCOM-2003*, (2003) 825–835.

[3]   T. Clausen and P. Jacquet, Optimized Link State Routing Protocol (OLSR), *IETF RFC 3626*, (October 2003).

[4]   I. Chakeres and C. Perkins, Dynamic MANET On-demand (DYMO) Routing, *Internet Draft (draft-ietf-manet-dymo-14)*, (June 2008).

[5]   W.P. Chen, J.C. How and L. Sha, Dynamic Clustering for Acoustic Target Tracking in Wireless Sensor Networks, *IEEE Trans. on Mobile Computing* **3**(3) (2004), 258–271.

[6]   G. De Marco, M. Tadauchi and L. Barolli, Description and Analysis of a Toolbox for Vehicular Networks Simulation, *Proc. of IEEE ICPADS/PMAC-2WN-2007* **2** (2007), 1–6.

[7]   D. Dhoutaut, A. Régis and F. Spies, Impact of Radio Propagation Models in Vehicular Ad Hoc Networks Simulations, *Proc. VANET-2006*, (2006) 40–49.

[8]   M. Fiore, J. Harri, F. Filali and C. Bonnet, Vehicular Mobility Simulation for VANETs, *Proc. of the 40-th Annual Simulation Symposium (ANSS-2007)* (2007), 301–309.

[9]   J. Goh, and D. Taniar, Mining Frequency Pattern from Mobile Users, *Proceedings of the 8th International Conference on Knowledge-Based Intelligent Information and Engineering Systems (KES-2004)*, Part III, Lecture Notes in Computer Science 3215 Springer (2004), 795–801.

[10] J. Harri, F. Filali and C. Bonnet, Mobility Models for Vehicular Ad Hoc Networks: A Survey and Taxonomy, *IEEE Communications Surveys & Tutorials* **11**(4) (2009), 19–41.

[11] J. Haerri, F. Filali and C. Bonnet, Performance Comparison of AODV and OLSR in VANETs Urban Environments under Realistic Mobility Patterns, *Proc. of 5th IFIP Mediterranean Ad-Hoc Networking Workshop (Med-Hoc-Net-2006)*, (2006) Lipari, Italy.

[12] J. Haemi, M. Fiore, F. Filali and C. Bonnet, A Realistic Mobility Simulator for Vehicular Ad Hoc Networks, *EURECOM Technical Report*, (2007) Available at: http://www.eurecom.fr/util/publidownload.en.htm?id=1811.

[13] A.M. Hanashi, I. Awan and M. Woodward, Performance Evaluation with Different Mobility Models for Dynamic Probabilistic Flooding in MANETs, *Mobile Information Systems* **5**(1) (2009), 65–80.

[14] M. Ikeda, L. Barolli, G. De Marco, T. Yang, A. Durresi and F. Xhafa, Tools for Performance Assessment of OLSR Protocol, *Mobile Information Systems* **5**(2) (2009), 165–176.

[15] Information Science Institute (ISI), *Network Simulator Version 2 (NS-2)*, http://www.isi.edu/nsnam.

[16] J.Y. Le Boudec and M. Vojnovic, The Random Trip Model: Stability, Stationary Regime, and Perfect Simulation, *IEEE/ACM Transactions on Networking* **14**(6) (2006), 1153–1166.

[17] S.S. Manvi, M.S. Kakkasageri and Jeremy Pitt, Multiagent Based Information Dissemination in Vehicular Ad Hoc Networks, *Mobile Information Systems* **5**(4) (2009), 363–389.

[18] E. Natsheh and T.C. Wan, Adaptive and Fuzzy Approaches for Nodes Affinity Management in Wireless Ad-hoc Networks, *Mobile Information Systems* **4**(4) (2009), 273–295.

[19] K. Nagel and M. Schreckenberg, A Cellular Automaton Model for Freeway Traffic, *Journal of Physics I France* **2** (December 1992), 2221–2229.

[20] C. Perkins, E. Belding-Royer and S. Das, Ad hoc On-Demand Distance Vector (AODV) Routing, *IETF RFC 3561 (Experimental)*, (July 2003).

[21] M. Piorkowski, M. Raya, A.L. Lugo, M. Grossglauser and J.P. Hubaux, Joint Traffic and Network Simulator for VANETs, *Proc. of Mobile and Information Communication Systems Conference (MICS-2006)*, (October 2006) Available on line at: http://www.mics.ch/.

[22] C. Sommer and F. Dressler, The DYMO Routing Protocol in VANET Scenarios, *Proc. of 66-th IEEE Vehicular Technology Conference (VTC2007-Fall)*, (2007), 16–20.

[23] N. Sadagopan, F. Bai, B. Krishnamachari and A. Helmy, PATHS: Analysis of Path Duration Statistics and Their Impact on Reactive MANET Routing Protocols, *MobiHoc-2003: Proc. of the 4-th ACM International Symposium on Mobile Ad Hoc Networking & Computing* (2003), 245–256.

[24] L. Smith, R. Beckan, R. Anson, K. Nagel and M. Williams, TRANSIMS: Transportation Analysis and Simulation System, *Proc. of the 5-th National Transportation Planning Methods Applications Conference*, (1995), LA-UR 95-1664.

[25] D. Taniar and J. Goh, On Mining Movement Pattern from Mobile Users *International Journal of Distributed Sensor Networks* **3**(1) (2007), 69–86.

[26] K. Xuan, G. Zhao, D. Taniar and B. Srinivasan, Continuous Range Search Query Processing in Mobile Navigation, *Proc. of IEEE ICPADS-2008 International Conference* (2008), 361–368.

[27] T. Yang, G. De Marco, M. Ikeda and L. Barolli, Impact of Radio Randomness on Performances of Lattice Wireless Sensors Networks Based on Event-reliability Concept, *International Journal of Mobile Information Systems* **2**(4) (2006), 211–227.

[28] J. Yoon, M. Liu and B. Noble, A General Framework to Construct Stationary Mobility Models for the Simulation of Mobile Networks, *IEEE Transactions on Mobile Computing* **5**(7) (2006), 860–871.

Evjola Spaho received her B.S and M.S degrees at Faculty of Information Technology, Polytechnic University of Tirana (PUT) in 2008 and 2010, respectively. Presently, she is a Ph.D Student at Graduate School of Engineering, Fukuoka Institute of Technology (FIT), Japan. Her research interests include P2P networks, vehicular networks, ad-hoc networks and robot control.

Leonard Barolli received B.S and Ph.D degrees from Tirana University and Yamagata University in 1989 and 1997, respectively. From April 1997 to March 1999, he was a JSPS Post Doctor Fellow Researcher at Department of Electrical and Information Engineering, Yamagata University. From April 1999 to March 2002, he worked as a Research Associate at the Department of Public Policy and Social Studies, Yamagata University. From April 2002 to March 2003, he was an Assistant Professor at Department of Computer Science, Saitama Institute of Technology (SIT). From April 2003 to March 2005, he was an Associate Professor and presently is a Full Professor, at Department of Information and Communication Engineering, Fukuoka Institute of Technology (FIT). Dr. Barolli has published about 300 papers in referred Journals, Books and International Conference proceedings. He was an Editor of the IPSJ Journal and has served as a Guest Editor for many International Journals. Dr. Barolli has been a PC Member of many International Conferences and was the PC Chair of IEEE AINA-2004 and IEEE ICPADS-2005. He was General Co-Chair of IEEE AINA-2006, AINA-2008, AINA-2010, CISIS-2009 and CISIS-2010, Workshops Chair of iiWAS-2006/MoMM-2006 and iiWAS-2007/MoMM-2007, Workshop Co-Chair of ARES-2007, ARES-2008, IEEE AINA-2007 and ICPP-2009. Dr. Barolli is the Steering Committee Chair of CISIS and BWCCA International Conferences and is serving as Steering Committee Co-Chair of IEEE AINA, NBiS and 3PGCIC International Conferences. He is organizers of many International Workshops. Dr. Barolli has won many Awards for his scientific work and has received many research funds. He got the "Doctor Honoris Causa" Award from Polytechnic University of Tirana in 2009. His research interests include network traffic control, fuzzy control, genetic algorithms, agent-based systems, ad-hoc networks and sensor networks. He is a member of SOFT, IPSJ, and IEEE.

Gjergji Mino was graduated in Electronics at Tirana Polytechnic University, Albania in 1995. After the graduation, he was working for different telecommunication companies in Albania. From 1997, he worked for different companies such as Digital Corporation, Compaq, Computer Technology Solutions and Peripheral Computer Support in US. Presently, he is a Ph.D Student at Graduate School of Engineering, Fukuoka Institute of Technology (FIT), Japan. His research interests include wireless networks, wireless cellular networks, vehicular networks, CAC, handover and fuzzy logic.

Fatos Xhafa holds a PhD in Computer Science (1998) from the Department of Languages and Informatics Systems, Polytechnic University of Catalonia, Barcelona, Spain. He was graduated in Mathematics from the Faculty of Natural Sciences (FNS), University of Tirana (UT), Albania, in 1988. During 1989–1993 he was an Assistant Professor at the Department of Applied Mathematics of the FNS, UT (Albania). He joined the Polytechnic University of Catalonia as a Visiting Professor at the Applied Mathematics Department I, and later he entered the Department of Languages and Informatics Systems (LSI) as a PhD Student and in 1996 as an Assistant Professor. Currently, he is an Associate Professor at the Department of Languages and Informatics Systems and member of the ALBCOM Research Group He is also a collaborator of the Open University of Catalonia (Barcelona, Spain). His current research interests include parallel and distributed algorithms, combinatorial optimization and approximation, meta-heuristics for complex problems, distributed programming, Grid and P2P computing. His research is supported by several research projects from Spain, European Union and NSF/USA. He collaborates with different international research groups. He has published in leading international journals and conferences and has served in the Organizing Committees of many conferences and workshops. He served as the Organizing Chair of ARES-2008, PC Chair of CISIS-2008, Workshops Co-chair of CISIS-2008, General Co-chair of HIS-2008, General Co-Chair of CISIS-2010 and PC Co-Chair of IEEE AIAN-2010 conferences. He is also member of editorial board of several international journals.

Vladi Kolici received his B.S and M.S degrees in Telecommunication Engineering from Polytechnic University of Tirana (PUT) in 1997 and 2005, respectively. He obtained his Ph.D from PUT in May 2009. From 1997 to 2004, he was a Research Associate and from 2005 to present he is a Lecturer at Department of Electronics and Telecommunications, Faculty of Information Technology, PUT. He is teaching several courses in the areas of wireless and mobile networking, P2P systems and quality of services. Dr. Kolici has published several papers in International and National Conference Proceedings in the areas of P2P and Ad-Hoc networks. Dr. Kolici received the Best Application Paper Award at the 6th International Conference on Advances in Mobile Computing and Multimedia (MoMM-2008) in 2008, Linz, Austria. His research interests include P2P networks, wireless and mobile networks, and high speed networks.

Rozeta Miho received her B.S and M.S degrees in Electronic and Telecommunication Engineering from Polytechnic University of Tirana (PUT), Albania, in 1985 and 1989, respectively. She obtained her Ph.D from PUT in November 1995. From 1985 to 1995, she was a lecturer, from 1996 to 2001 Assistant Professor, from 2001 to 2007 Associate Professor, and presently she is a Full Professor of PUT. From May 2009, she is the Dean of Faculty of Information Technology, PUT. She is teaching several courses in the areas of telecommunication networks, optical communications and optical fibre networks. Prof. Miho has published several papers in International and National Conference Proceedings in the areas of optical WDM networks, P2P and Ad-Hoc networks. She is the co-author of the Best Application Paper Award at the 6th International Conference on Advances in Mobile Computing and Multimedia (MoMM-2008) in 2008, Linz, Austria. Her research interests include P2P networks, optical networks, wireless networks and high speed networks.

10

RFID-based human behavior modeling and anomaly detection for elderly care

Hui-Huang Hsu* and Chien-Chen Chen

Department of Computer Science and Information Engineering, Tamkang University Taipei, Taiwan

Abstract. This research aimed at building an intelligent system that can detect abnormal behavior for the elderly at home. Active RFID tags can be deployed at home to help collect daily movement data of the elderly who carries an RFID reader. When the reader detects the signals from the tags, RSSI values that represent signal strength are obtained. The RSSI values are reversely related to the distance between the tags and the reader and they are recorded following the movement of the user. The movement patterns, not the exact locations, of the user are the major concern. With the movement data (RSSI values), the clustering technique is then used to build a personalized model of normal behavior. After the model is built, any incoming datum outside the model can be viewed as abnormal and an alarm can be raised by the system. In this paper, we present the system architecture for RFID data collection and preprocessing, clustering for anomaly detection, and experimental results. The results show that this novel approach is promising.

Keywords: RFID, behavior modeling, anomaly detection, elderly care, clustering

1. Introduction

Elderly care is an important issue in an aging society. There are many elderly people who live alone. Even though some elderly people live with their children, they are at home alone most of the time since young people need to work or go to school. Quite a few systems have been developed for health care [1–3] in general and elderly care in specific [4,5]. To monitor the movements of the elderly at home is one of the major parts of elderly care. Mining meaningful information from human movement data is also an essential research issue [6–8]. This research would like to utilize information technology to unobtrusively detect abnormal behavior of elderly people at home according to their movements. Abnormal behavior includes not only emergencies like faints and falls, but also others like "not eating," "not going to toilet regularly," "lack of movement," and "trotting around." Since everyone can have very different behaviors at home, it is not possible to predefine the so-called abnormal behaviors. Thus machine learning technology is needed to build a personalized behavior model for each individual [9, 10].

To answer the two raised issues, first, we propose to use the RFID (Radio Frequency Identification) technology for movement detection of the elderly. The reason is that the RFID system is relatively cheap and easy to deploy in the home environment. Furthermore, RFID sensors are unobtrusive comparing to the popular solution by video cameras. Secondly, to deal with the behavior modeling, we propose to

---

*Corresponding author: Department of Computer Science and Information Engineering, Tamkang University, 151 Ying-Chuan Rd. Tamsui, Taipei, 25137, Taiwan. E-mail: h_hsu@mail.tku.edu.tw.

use the clustering analysis [11,12] since presumably only positive data (normal behavior) are collected. Abnormal behavior is hard to define and a complete collection of such negative data is not possible. It is not feasible to use a classifier [13,14] which needs both positive and negative examples for training. Thus, clustering should be the way for modeling the user movements.

Sensor networks have been a popular research issue due to many useful applications [15–17]. Valuable information can be collected through various kinds of sensors. In this research, RFID sensors were chosen for monitoring human behavior at home. An RFID system includes two parts: tags and readers. The readers are used to detect the deployed tags which usually are attached to objects and store serial numbers corresponding to the objects. There are basically two kinds of RFID tags: passive and active. Passive tags simply reflect the signal sent from the reader for the transmission of the stored information. They draw power from the reader and need no battery. On the other hand, active tags have their own transmitter and power source (battery). RFID tags are designed to operate in different frequencies. General speaking, high frequency (HF) tags are used in a detecting distance shorter than 0.5 meter (usually passive tags); ultra high frequency (UHF) tags are used for a wider range (several meters to one hundred meters; usually active tags). In this research, active tags are used since they are of an ultra high frequency and can be detected in a wider range. When the reader detects the signals from the tags, received signal strength indication (RSSI) values can be obtained. The RSSI value has a negative relationship with the distance of the tag from the reader. Thus a relative distance of the tag and the reader in a measurement of signal strength can be obtained [18].

It is desired to build a behavior model (viewed as normal) for the elderly by the collected RSSI data. One or two week data are needed for such modeling. We then can use this model to detect subsequent abnormal behavior of the user. The system architecture includes deploying active tags in the living environment, e.g., the living room, the dining room, the kitchen, the rest room, and the bed rooms. The reader is to be carried by an elderly person. The detected RSSI values are recorded following the movement of the person. A clustering technique is then used to build a model of normal behavior, and the model can be used to determine if the subsequent behavior is normal or not. Since only data of normal behavior are collected, the behavior model can be build with only positive examples via clustering analysis. The key to a successful behavior modeling would be on the setting of cluster boundaries. Moreover, short-term behavior (for a few minutes) and long-term behavior (for a couple of hours) are quite different. So we actually need two models for short-term behavior and long-term behavior, respectively.

This novel approach, different from computer-vision-based approaches, not just detects predefined events like if the elderly person falls, but finds the living patterns of the person. The machine learning technique can be used to learn all the patterns without defining all of them in advance. Also, it is not necessary to install cameras at home. This can relieve the concern of personal privacy issues [1]. Researches related to using RFID and/or other sensors for elderly care and home safety can also be found in other researches [4,5,19]. But all of them do not focus on general behavior modeling as we do.

In the next section, we give an overview on related work. The third section illustrates behavior modeling, including environment settings, RFID data collection, data preprocessing, and clustering for normal behavior modeling. The fourth section shows experimental results of abnormal behavior detection. The final section draws a brief conclusion and discusses future development.

2. Related work

In this section, we first discuss a few papers related to using the RFID technology in elderly care and home care. Ho et al. gave a progress report in using RFID and sensor networks for elderly care in

2005 [4]. The goal was to build an in-home elderly care system that monitors the medication-taking behavior of the elderly. In their prototype system, a simulated HF RFID reader, a simulated UHF RFID reader, a weight scale and a base station PC were used. A HF RFID tag is attached to each medicine bottle which is placed on top of the scale within the detection range of the HF RFID reader. Each elder patient wears a UHF RFID tag. The UHF RFID reader is also placed near the medicine bottles. When a bottle is removed from the scale, the UHF RFID reader can identify which elder patient is taking the bottle and the HF RFID reader can decide which medicine bottle is removed. When the bottle is put back to the top of the scale, the scale can tell the system to record the amount of medicine the patient has taken. The system design is very interesting and the preliminary results showed that the system could record medication-taking behavior of the elder patient. Our system aims at understanding general behaviors/movements of the elderly at home, not just specifically in medication-taking behaviors.

Jih et al. intended to build a smart environment at home for the elderly in 2006 [5]. It presented a context-aware service integration system using multi-agent technologies. The system can provide context-aware healthcare services to the elderly at home and the remote caretaker can monitor and attend the elder's well-being at anytime from anywhere. It tracks the location and activities of the elderly via sensors like pressure-sensitive floors, cameras, smart furniture, and bio-sensors. The paper gave several context-aware service scenarios to demonstrate such a system can enhance the quality of care for the elder's daily life. However, the paper only showed the design of the proposed framework and discussed the usefulness of such a platform. It focused on using the agent technology in processing the sensor data from various sensors, but no real implementation and experimental results were given.

Lee and Kim proposed a context-aware home safety model which applied the RFID and sensor network in 2007 [19]. The model focused on surveillance and intrusion detection. Networking and privacy issues along with the RFID middleware were also presented. RFID as well as light and temperature sensors were used. The paper utilized the RFID technology for home safety. It is for general purpose, not just elderly care.

Quite a few researches can be found in the literature for RFID tracking/positioning [20–23]. However, we do not need such complicated systems since exact location of the elderly person is not needed. Time sequences of RSSI values collected from active tags deploying in various corners at home are a sufficient representation of the movement of the elderly person who carries the reader. Other indoor positioning systems used wireless LAN [24], floor pressure/load sensors [25], and infrared-based small motion detectors (SMDs) [26]. Furthermore, video tracking systems can also be used to monitor indoor human behavior [27] or outdoor traffic [28]. The RFID technology has become popular and the cost of the readers and tags are going down rapidly. This makes the RFID technology more appealing than the wireless LAN, SMDs, and video cameras for detecting the movements of the user. Furthermore, object tracking in video is a much more complex task. It also draws concern on personal privacy.

Another major issue of this research is the behavior modeling by only positive examples. The goal for such modeling is to detect abnormal behaviors. In the literature, we have found applications of anomaly detection in information security [29,30], credit fraud detection [31], and document selection [32]. All of them used a machine learning approach to build a classifier with both positive and negative examples. One-class SVM (support vector machines) that finds the hyperplane between the positive examples and the origin is used in [29]. The origin is used to represent the negative examples. Such a classifier is biased and not suitable in our system. On the other hand, artificial negative examples were used in [32]. This is also impractical since negative examples are difficult to define and generate. Therefore, clustering analysis is a much more feasible way for building a normal behavior model.

(a)

(b)

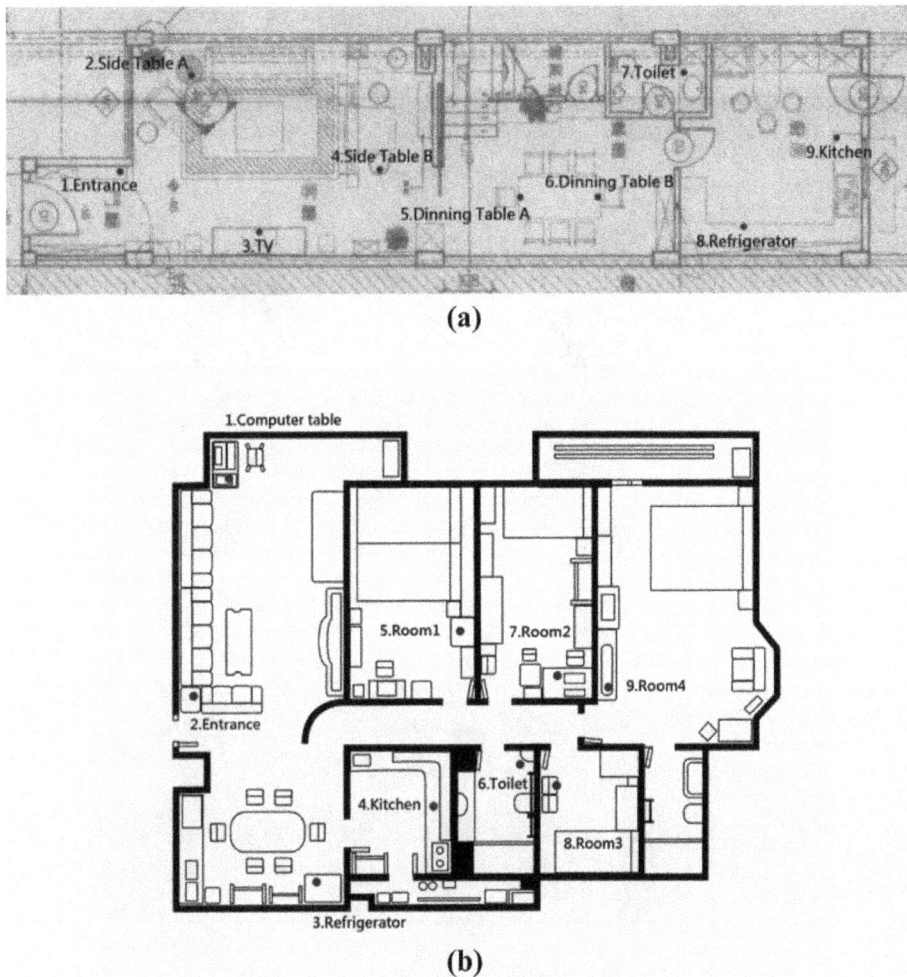

Fig. 1. Deployment of tags in testing environment: (a) Area A, (b) Area B.

3. Behavior modeling

In this section, behavior modeling by RFID is divided into three parts: environment settings and data collection, data preprocessing, and behavior modeling by clustering.

3.1. Environment settings and data collection

In our setting, a notebook PC, a wireless LAN access point (AP), a CF-card RFID reader, a PDA with a CF-card slot, and 10 active RFID tags with various sizes were used. The user carried the PDA with an RFID reader. The tags were deployed in the testing environment. Two areas were used in our tests. The first area was the first floor of a house (Area A) and the second one was an apartment (Area B). Please refer to Fig. 1. The dots in the Figure show the location of the deployed tags. Totally nine tags were deployed and the 10th tag was a bracelet tag carried by the user.

After the RFID and wireless environment were constructed, signals from 10 tags were transmitted to the reader, and the PDA transmitted the received RSSI values from the tags to the notebook PC every

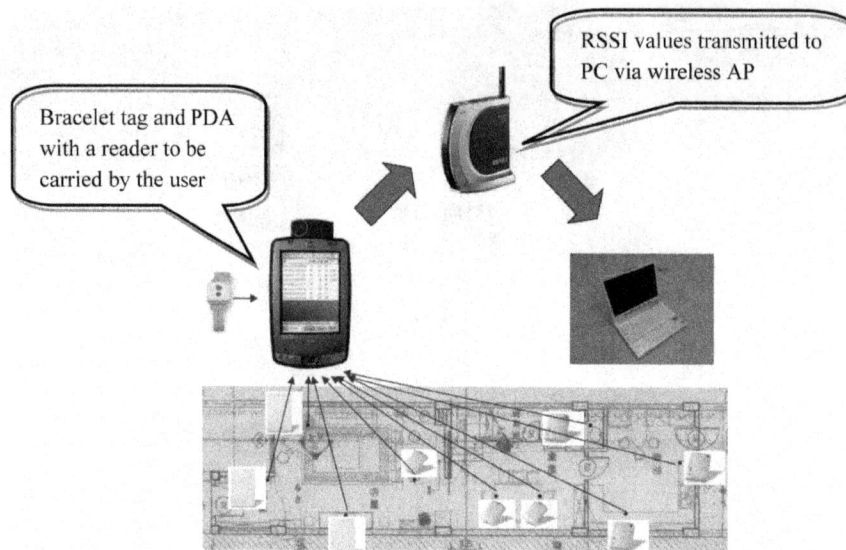

Fig. 2. System architecture for data collection.

```
ID:0001000107520133,RSSI:133,LQI:219,DI:255,T1:-,T2:-
ID:0001000107520125,RSSI:132,LQI:73,DI:255,T1:-,T2:-
ID:0001000107520124,RSSI:111,LQI:107,DI:255,T1:-,T2:-
ID:0001000107462018,RSSI:0,LQI:0,DI:0,T1:-,T2:-
ID:0001000107462016,RSSI:103,LQI:213,DI:255,T1:-,T2:-
ID:0001000107462011,RSSI:0,LQI:0,DI:0,T1:-,T2:-
ID:0001000107291249,RSSI:0,LQI:0,DI:0,T1:-,T2:-
ID:0001000107291246,RSSI:0,LQI:0,DI:0,T1:-,T2:-
ID:0001000107291241,RSSI:0,LQI:0,DI:0,T1:-,T2:-
ID:0001000107503013,RSSI:150,LQI:227,DI:255,T1:29.25,T2:29.87
```

Fig. 3. Samples of collected data.

second through the wireless AP. The collected data were then stored and preprocessed in the notebook PC for further analysis. We could have collected the movement data in a reverse way, meaning, the person carried a tag and readers were deployed in the environment. But the cost would be much higher. The system architecture for data collection can be seen in Fig. 2. In the system, active tags were deployed in the home environment. The user carried a PDA with the CF-card reader and wore a bracelet tag. (A custom-designed RFID reader which is light weighted and easy to carry should be used in real cases.) The bracelet tag can also measure and transmit body temperature of the user. This can help ensure that the reader is with the person at all times. This can also be integrated into the customized design.

The following data were transmitted: the tag ID number, the RSSI value, the connectivity quality (LQI), the battery indicator (DI) and detected temperatures (only for Tag #10 that has temperature sensors). At the current stages, only the first two items were used to build the behavior models. A set of sample data collected and transmitted in one second is shown in Fig. 3 for illustration.

To do a primary test, we collected the RSSI values for half an hour in Area A. During the test, the user stayed in the couch for 10 minutes, walked to the toilet in one minute, stayed in the toilet for 90 seconds, moved to the refrigerator in 15 seconds, stayed in front of the refrigerator for another 15 seconds, walked to the kitchen tag, stayed in front of the kitchen tag for 60 seconds, walked back to the couch in

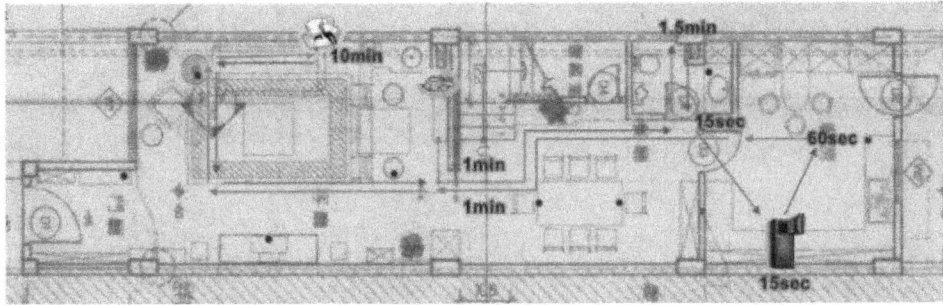

Fig. 4. Test moving route.

one minute, then repeated the whole process once (Fig. 4). The total time for this data collection was 30 minutes with 1800 data points. (The PDA transmitted detected data every second.) The user moved in the test environment alone with no interference from other people. The CF-card reader was held in an upright position at all time to ensure that signals from all tags could be more accurately detected.

The collected RSSI data sequences from five of the 10 tags are shown in Fig. 5. The five tags are *Side Table A*, *Side Table B*, *Dining Table B*, *Toilet*, and *Bracelet* (please refer to Fig. 1 (a)). In Fig. 5 (a), the tag was not detected (with an RSSI value of zero) in the two squared periods because the person was going away from the living room. Fig. 5 (b) is similar to Fig. 5 (a) except that the RSSI values are lower and noisier when the person was in the living room. The noisy signal was caused by the furniture existing between Side Table B and the couch. In Fig. 5 (c), the signal was even noisier than Fig. 5 (b) when the person was in the couch. This is because that there was a division between the living room and the dining room. On the other hand, the RSSI values of the Dining Table B tag were higher when the person moved to the toilet and the kitchen. From Fig. 5 (d), it is very obvious to know at what time the person was in the toilet. In Fig. 5 (e), it shows that the bracelet tag was with the person during the test. From these signals, we are sure that the collected RSSI values can represent the movement of the person. The RSSI values from the tags change with the distances between the tags and the reader (the person).

3.2. Data preprocessing

The RSSI values can be unstable and noisy sometimes. This is why we need to preprocess the collected data before they are sent to the clustering process. From Fig. 5, we see noises in the collected signals. Most importantly, the signals dropped to zero sometimes even when the person was not moving. The tags were not detected during that one second period. Also, furniture and divisions between rooms cause interference to the signals. Moreover, the reader is directional. Detected signals can be different when we point the reader in different directions. This is why the user had to keep the reader in an upright position to have more accurate data collection. Although this might seem to be quite inconvenient for the user, we believe a customized, miniaturized RFID reader can overcome this problem.

To provide a dataset with a better quality for clustering analysis, we need to preprocess the collected data. First, the sudden zeros are detected and smoothed out. Secondly, the RSSI values sometimes maybe unstable and noisy, so it is needed to smooth all collected data. Thirdly, the data should be sampled to reduce the total amount of data. In Fig. 6 (a), a sudden zero is shown and it can be smoothed out to 121 ($(124 + 124 + 118 + 118)/4 = 121$) as in Fig. 6 (b). Here we use a voting mechanism to decide if a zero is a sudden zero. If the majority of the previous samples and the subsequent samples are not zero, it means that the current zero should not be zero. Thus it is smoothed out by averaging the RSSI values

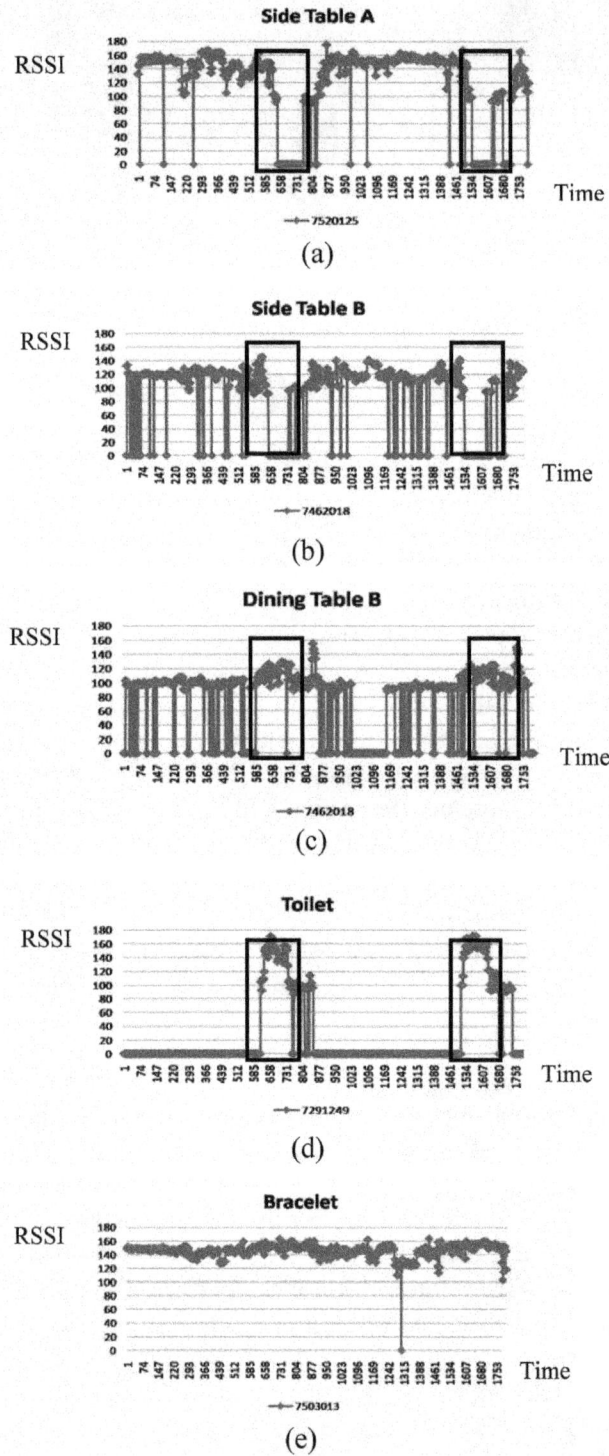

Fig. 5. Collected data from the 30-minute test movement: (a) Side Table A, (b) Side Table B, (c) Dining Table B, (d) Toilet, (e) Bracelet.

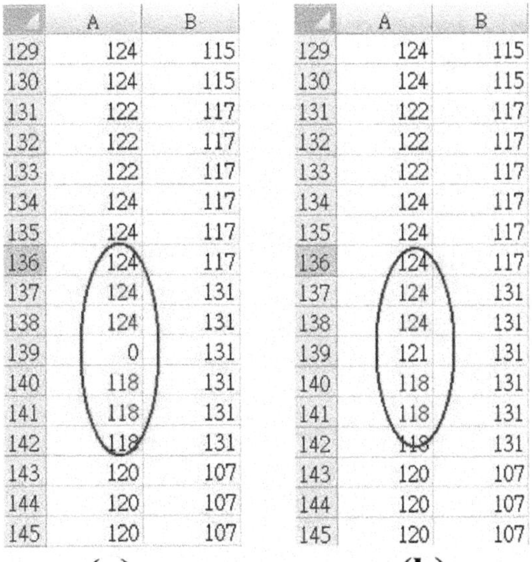

	A	B
129	124	115
130	124	115
131	122	117
132	122	117
133	122	117
134	124	117
135	124	117
136	124	117
137	124	131
138	124	131
139	0	131
140	118	131
141	118	131
142	118	131
143	120	107
144	120	107
145	120	107

(a)

	A	B
129	124	115
130	124	115
131	122	117
132	122	117
133	122	117
134	124	117
135	124	117
136	124	117
137	124	131
138	124	131
139	121	131
140	118	131
141	118	131
142	118	131
143	120	107
144	120	107
145	120	107

(b)

Fig. 6. RSSI values with a sudden zero (a) before smoothing (b) after smoothing.

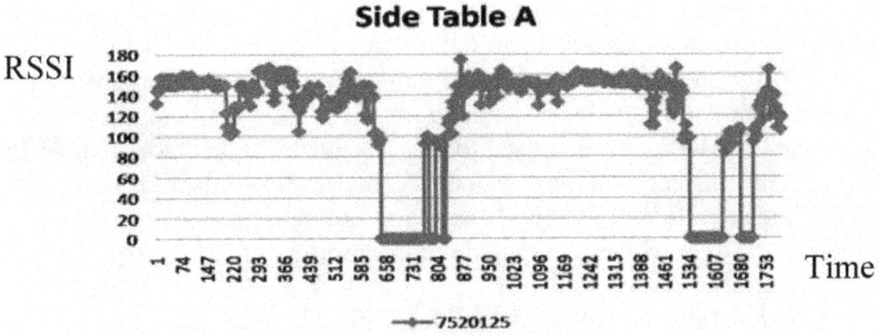

Fig. 7. Result with sudden zeros smoothed out for Side Table A.

of the neighbors. In our experiments, 10 neighbors were used to make the decision. In this example, only four neighbors were used for illustration. The sudden zeros in Fig. 5 (a) were smoothed out by this technique (Fig. 7). Next, we applied a general moving average technique to the rest of the data sequences to filter out the noises. Meaning, each data point is replaced by the average of its neighbors and itself.

For reducing the space and time complexity, it is needed to sample the collected data. This is to reduce the number of data points in a windowed data from the time sequences. The sampling interval cannot be large because the user may move from one point to another then back in a few seconds. The information of this movement may be lost if the data are sampled for a long period of time. From our experiments, it is suggested that the sampling interval is set to five seconds. Furthermore, we believe that it is better to retain the maximal RSSI value in the sampled interval so that movement information can be best preserved. Figure 8 shows an example for such a sampling process. 7520125 and 7520124 were the tag ID numbers. The data were sampled every five seconds and the maximal RSSI value during that five-second period was saved. For example, the maximal value of the 7520125 tag in the first five seconds was 144. It was 117 for the 7520124 tag in the same period.

7520125	7520124
132	111
132	111
144	111
144	117
144	117
144	117
144	111
144	111
132	111
148	98

7520125	7520124
144	117
148	117

Fig. 8. Data sampling with the maximum retained.

3.3. Behavior modeling by clustering

After the RFID data sequences are properly smoothed and sampled, the preprocessed data sequences are windowed by a sliding window. Since we want to analyze the movements of the user, a short segment of data (windowed data) from the data sequences should be assembled into a movement datum for behavior modeling, not just the data at a particular time. Two behavior (movement) models are then built. They are models for short-term behavior and long-term behavior. The short-term model is used to detect anomalies like falls (unusually staying in a place for too long) and anxiety (trotting around). On the other hand, the long-term model is used to detect anomalies like "not going to the toilet for all day" and "not opening the refrigerator for a few hours." The windowed datasets for short-term modeling and long-term modeling are different in window size and sampling interval. The dataset for long-term behavior modeling has a larger window to include more information and a larger sampling interval to reduce processing time.

The datasets for short-term modeling and long-term modeling are clustered respectively through standard clustering techniques like K-means [33]. The K-means algorithm randomly selects K centers. Each data point then chooses to join the cluster of a specific center if the distance between the data point and the center is the shortest (comparing to the distances to other cluster centers). It iteratively updates the cluster centers and members of each cluster until it is converged. Manhattan distance is used here to measure the similarity between data points.

$$d_M(x_i, x_j) = \sum_{d=1}^{N} |x_{id} - x_{jd}| \tag{1}$$

The number of clusters would be the number of different spaces in the environment. In Fig. 9, we have four clusters in the two-dimensional space. In real cases we should have more clusters in the 10-dimensional space (due to 10 RFID tags).

Since only data of normal behavior are collected, the model (a classifier) can be built with only positive examples. The key for this behavior modeling is to determine the boundaries of all clusters. Once the boundaries are properly determined, we can have a model to detect anomalies. In Fig. 9, point A is outside the boundary of either one of the four clusters. So it is viewed as an anomaly. In our experiments, the boundary of a cluster was set to the maximal distance from all member points to the cluster center.

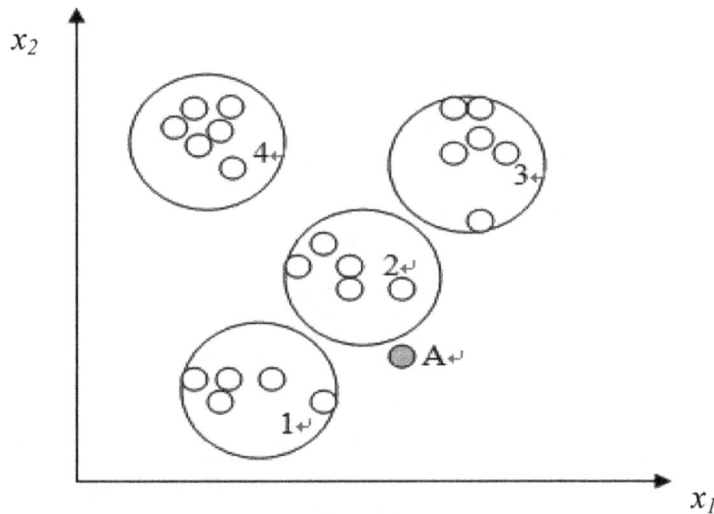

Fig. 9. Clustering for behavior modeling.

Fig. 10. Eight-hour training data of normal behavior.

4. Experimental results

In the following experiments, RFID RSSI data were collected in Area B (Fig. 1 (b)). An eight-hour dataset was collected as the training data for building the normal behavior models (Fig. 10). Two other test datasets were designed to simulate possible abnormal short-term behaviors, which will be shown later.

Before doing the tests, we need to first set the criteria for triggering the alarm. We cannot trigger the alarm simply by the appearance of one single abnormal data point because there would be too many false alarms. The number of continuous abnormal data points should be greater than a preset threshold, say five points, to be considered as an abnormal event. When an abnormal event occurs, it would generally last for a period of time. For example, if the user falls down at the entrance of the toilet, he/she would stay there for a period of time longer than he/she usually does. The alarm should be triggered in a short period of time after continuous abnormal data points are observed.

Table 1
Detected abnormal data points of Route 1

Abnormal points	Time interval	Number of points
160~164	13m50s~14m10s	5
445~451	37m35s~38m05s	7
460~486	38m50s~41m	28
491~493	41m25s~41m35s	3
501~504	42m15s~42m30s	4
518~529	43m40s~44m35s	12
545~547	45m55s~46m05s	3
566~572	47m40s~48m10s	7
577~581	48m35s~48m55s	5
589~592	49m35s~49m50s	4

Table 2
Detected abnormal data points of Route 2

Abnormal points	Time interval	Number of points
119~120	10m25s~10m30s	2
255~257	21m45s~21m55s	3
279~285	23m45s~24m15s	7
293~298	24m55s~25m20s	6
302~320	25m40s~27m10s	19

4.1. Short-term anomaly detection

The following two routes are the designed testing data for anomaly detection of short-term behaviors.

1. Living room (15 minutes) → 30 seconds → Room1 (15 minutes)→ 30 seconds → Refrigerator (20 minutes)
2. Room1 (10 minutes) → 30 seconds → Toilet (10 minutes) → 20 seconds → Toilet entrance (20 minutes)

The experimental results of Route 1 are shown in Table 1. In Route 1, the user stopped in front of the refrigerator for 20 minutes at the end. Since in the training data (Fig. 10) the user has stayed in front of the refrigerator for at most one minute, we would expect to receive an alarm from the system after 32 minutes 30 seconds (in Route 1, the sum of 15 minutes, 30 seconds, 15 minutes, 30 seconds, 1 minute, and 30 seconds). 30 seconds were added because the abnormal event criterion was set at six consecutive abnormal points and the time interval between two points is 5 seconds. From Table 1, abnormal events are detected at 38 minutes (37 minutes and 35 seconds plus 25 seconds). It is a few minutes later than expected, but still acceptable. A false alarm could have been raised at 14 minutes 10 seconds, but owing to the preset criterion (over five consecutive abnormal points), the alarm was not issued by the system.

In Route 2, the user stopped at the toilet entrance for 20 minutes at the end. From the training data (Fig. 10), the user has stayed in the toilet three times, for 15, 3, and 10 minutes respectively, but he/she actually never stopped at the toilet entrance. We would expect the system to issue an alarm a short time after 21 minutes and 20 seconds (in Route 2, the sum of 10 minutes, 30 seconds, 10 minutes, 20 seconds, and 30 seconds). In Table 2, an abnormal event is detected at 24 minutes and 10 seconds (23 minutes and 45 seconds plus 25 seconds). It is a bit later than expected. Abnormal points 255~257 (3 points, ends at 21 minutes and 55 seconds) were viewed as normal because of the suppression from the alarm-triggering criterion. Otherwise, we could have got an alarm at a time even closer to the expected time (22 minutes and 10 seconds).

Table 3
Clustering analysis of long-term testing data

Clusters	Number of data
1.Computer	1251
2.Living room	3688
3.Restaurant	2006
4.Refrigerator	4206
5.Kitchen	1306
6.Room1	1874
7.Toilet	0
Abnormal points	5

4.2. Long-term anomaly detection

In the experiment for long-term behavior modeling, we intentionally removed the data, collected when the user was in the toilet, from the original training data (Fig. 10) for the purpose of testing. This is to simulate the situation that the elder person did not go to toilet for the whole day. We classified the testing data (without toilet data) by the long-term behavior model which was built by the training data with toilet data. The results in Table 3 show that no data point was classified into the toilet cluster. We can conclude from the results that the user did not go to the toilet. We cannot simply make this judgment from the RSSI values of the toilet tag because the threshold would be hard to set. With this long-term anomaly detection, long-term behaviors for not going to a certain place, e.g., the kitchen and the toilet, for hours or even a whole day can be detected. Not eating much and not going to toilet are both serious problems for the elderly.

5. Conclusion

In this paper, we present a novel RFID-based approach for modeling human behavior, which we believe has never been tried before. The aim is to detect abnormal behavior by the personalized behavior model. This can be very useful for elderly care at home since it is unobtrusive and easy to deploy. Only one custom-designed RFID reader and a dozen of active RFID tags are needed for detecting the movements of the user. The custom-designed device should certainly be light weighted and easy to carry. Moreover, the primary test results show that this approach is very promising. When the user stays at a location longer than he/she usually does (simulating falls and faints), the system can detect the situation and issue an alarm. If the user does not go to the toilet for hours, a warning message can also be issued.

References

[1] Y. Xiao, X. Shen, B. Sun and L. Cai, Security and privacy in RFID and applications in telemedicine, *IEEE Communications Magazine* **44** (2006), 64–72.
[2] A. Koyama, J. Arai, S. Sasaki and L. Barolli, Design, field experiments and evaluation of a web-based remote medical care support system, *International Journal of Web and Grid Services* **4** (2008), 80–99.
[3] A. Yamazaki, A. Koyama, J. Arai and L. Barolli, Design and implementation of a ubiquitous health monitoring system, *International Journal of Web and Grid Services* **5** (2009), 339–355.
[4] L. Ho, M. Moh, Z. Walker, T. Hamada and C.-F. Su, A prototype on RFID and sensor networks for elder healthcare: progress report, Proceedings 2005 ACM SIGCOMM Workshop on Experimental Approaches to Wireless Network Design and Analysis, 2005, pp. 70–75.
[5] W.-R. Jih, J.Y. Hsu and T.-M. Tsai, Context-aware service integration for elderly care in a smart environment, Proceedings 2006 AAAI Workshop on Modeling and Retrieval of Context Retrieval of Context, 2006, pp. 44–48.

[6] J.Y. Goh and D. Taniar, Mobile data mining by location dependencies, Proceedings of the 5th International Conference on Intelligent Data Engineering and Automated Learning, *Lecture Notes in Computer Science* **3177** (2004), 225–231.

[7] J. Goh and D. Taniar, Mining frequency pattern from mobile users, Proceedings of the 8th International Conference on Knowledge-Based Intelligent Information and Engineering Systems, *Lecture Notes in Computer Science* **3215** (2004), 795–801.

[8] D. Taniar and J. Goh, On mining movement pattern from mobile users, *International Journal of Distributed Sensor Networks* **3** (2007), 69–86.

[9] O. Daly and D. Taniar, Exception rules mining based on negative association rules, Proceedings of the International Conference on Computational Science and Its Applications, *Lecture Notes in Computer Science* **3046** (2004), 543–552.

[10] D. Taniar, J.W. Rahayu, V.C.S. Lee and O. Daly, Exception rules in association rule mining, *Applied Mathematics and Computation* **205** (2008), 735–750.

[11] T. Kwok, K.A. Smith, S. Lozano and D. Taniar, Parallel fuzzy c-means clustering for large data sets, Proceedings of the 8th International Euro-Par Conference on Parallel Processing, *Lecture Notes in Computer Science* **2400** (2002), 365–374.

[12] L. Tan, D. Taniar and K.A. Smith, A clustering algorithm based on an estimated distribution model, *International Journal of Business Intelligence and Data Mining* **1** (2005), 229–245.

[13] L. Tan, D. Taniar and K.A. Smith, Maximum-entropy estimated distribution model for classification problems, *International Journal of Hybrid Intelligent Systems* **3** (2006), 1–10.

[14] L. Tan and D. Taniar, Adaptive estimated maximum-entropy distribution model, *Information Sciences* **177** (2007), 3110–3128.

[15] A. Aikebaier, T. Enokido and M. Takizawa, Design and evaluation of reliable data transmission protocol in wireless sensor networks, *Mobile Information Systems* **4** (2008), 237–252.

[16] S. Kumar and S.-J. Park, Probability model for data redundancy detection in sensor networks, *Mobile Information Systems* **5** (2009), 195–204.

[17] W. Wu, X. Li, S. Xiang, H.-B. Lim and K.-L. Tan, Sensor relocation for emergent data acquisition in sparse mobile sensor networks, *Mobile Information Systems* **6** (2010), 155–176.

[18] R. Crepaldi, A.F. Harris, R. Kooper, R. Kravets, G. Maselli, C. Petrioli and M. Zorzi, Managing heterogeneous sensors and actuators in ubiquitous computing environments, *Proceedings International Conference on Mobile Computing and Networking*, 2007, pp. 35–42.

[19] B. Lee and H. Kim, A design of context aware smart home safety management using networked RFID and sensor, *Proceedings First IFIP WG6.2 Home Networking Conference*, 2007, pp. 215–224.

[20] R.S. Sangwan and R.G. Qiu, Using RFID tags for tracking patients, charts and medical equipment within an integrated health delivery network, Proceedings 2005 IEEE Networking, *Sensing and Control* (2005), 1070–1074.

[21] W. Jiang, D. Yu and Y. Ma, A tracking algorithm in RFID reader network, *Proceedings Japan-China Joint Workshop on Frontier of Computer Science and Technology*, 2006, pp. 164–171.

[22] E. Aydin, R. Oktem, Z. Dincer and I.K. Akbulut, study of an RFID based moving object tracking system, *Proceedings RFID Eurasia* (2007), 1–5.

[23] S. Jia, J. Sheng, D. Chugo and K. Takase, Obstacle recognition for a mobile robot indoor environment using RFID and stereo vision, *Proceedings 2007 International Conference on Mechatronics and Automation*, 2007, pp. 2789–2794.

[24] P. Prasithsangaree, P. Krishnamurthy and P.K. Chrysanthis, On indoor position location with wireless LANs, *Proceedings 13th IEEE International Symposium On Personal, Indoor, & Mobile Radio Communications* **2** (2002), 720–724.

[25] W.H. Liau, Inhabitant tracking and service provision in an intelligent e-home via floor load sensors, Master's Thesis, Department of Computer Science and Information Engineering, Nation Taiwan University, Taipei, Taiwan, 2005.

[26] K. Hara, T. Omori and R. Ueno, Detection of unusual human behavior in intelligent house, *Proceedings 12th IEEE Workshop on Neural Networks for Signal Processing*, 2002, pp. 697–706.

[27] P.-C. Chung and C.-D. Liu, A daily behavior enabled hidden Markov model for human behavior understanding, *Pattern Recognition* **40** (2008), 1572–1580.

[28] C.P. Lin, J.C. Tai and K.T. Song, Traffic monitoring based on real-time image tracking, *Proceedings 2003 IEEE International Conference on Robotics and Automation* **2** (2003), 2091–2096.

[29] Y. Wang, J. Wong and A. Miner, Anomaly intrusion detection using one class SVM, *Proceedings Fifth Annual IEEE SMC Information Assurance Workshop*, 2004, pp. 358–364.

[30] C.-H. Lin, J.-C. Liu and C.-H. Ho, Anomaly detection using LibSVM training tools, *Proceedings 2008 International Conference on Information Security and Assurance*, 2008, pp. 166–171.

[31] A.K. Ghosh and A. Schwartzbard, A study in using neural network for anomaly and misuse detection, *Proceedings 8th Conference on USENIX Security Symposium* **8** (1999), 12–12.

[32] J. Zizka, J. Hroza, B. Pouliquen, C. Ignat and R. Steinberger, The selection of electronic text documents supported by only positive examples, *Proceedings 8th International Conference on Statistical Analysis of Textual Data*, 2006, pp. 993–1002.

[33] P. Mitra, C.A. Murthy and S.K. Pal, Unsupervised feature selection using feature similarity, *IEEE Transactions on Pattern Analysis and Machine Intelligence* **24** (2002), 301–312.

Hui-Huang Hsu is an Associate Professor in the Department of Computer Science and Information Engineering at Tamkang University, Taipei, Taiwan. He received both his PhD and MS Degrees from the Department of Electrical and Computer Engineering at the University of Florida, USA, in 1994 and 1991, respectively. He has published over 80 referred papers and book chapters, as well as participated in many international academic activities. His current research interests are in the areas of machine learning, data mining, bio-medical informatics, ambient intelligence, and multimedia processing. Dr. Hsu is a senior member of the IEEE.

Chien-Chen Chen received his master's degree in Computer Science and Information Engineering from Tamkang University, Taipei, Taiwan in 2009. He worked on intelligent systems and RFID applications.

11

Genetic algorithms for satellite scheduling problems

Fatos Xhafa[a,*], Junzi Sun[b], Admir Barolli[c], Alexander Biberaj[d] and Leonard Barolli[e]

[a]*Department of Languages and Informatics Systems, Technical University of Catalonia, Barcelona, Spain*
[b]*Centre de Tecnologia Aeroespacial, Barcelona, Spain*
[c]*Seikei University, Tokyo, Japan*
[d]*Polytechnic University of Tirana, Tirana, Albania*
[e]*Fukuoka Institute of Technology, Fukuoka, Japan*

Abstract. Recently there has been a growing interest in mission operations scheduling problem. The problem, in a variety of formulations, arises in management of satellite/space missions requiring efficient allocation of user requests to make possible the communication between operations teams and spacecraft systems. Not only large space agencies, such as ESA (European Space Agency) and NASA, but also smaller research institutions and universities can establish nowadays their satellite mission, and thus need intelligent systems to automate the allocation of ground station services to space missions. In this paper, we present some relevant formulations of the satellite scheduling viewed as a family of problems and identify various forms of optimization objectives. The main complexities, due highly constrained nature, windows accessibility and visibility, multi-objectives and conflicting objectives are examined. Then, we discuss the resolution of the problem through different heuristic methods. In particular, we focus on the version of ground station scheduling, for which we present computational results obtained with Genetic Algorithms using the STK simulation toolkit.

Keywords: Genetic algorithms, satellite scheduling, ground station, mission operations, multi-objective optimization, STK toolkit

1. Introduction

Mission operations arise by the need to coordinate communications of spacecrafts (extra-planetary crafts including satellites, space stations, etc.) with ground stations. Operation teams request an antenna at a ground station for a specific time window. The number of requests could be large but most importantly different requests maybe conflicting making it very complex to manually compute the time windows for communication of spacecrafts with ground stations.

Clear examples of the real need for automation of the allocation process are ESA (European Space Agency) and NASA. ESA manages several ground stations to support its own missions as well as other missions upon industry costumers request. The number of ground stations [2] is rather limited, actually less than ten, supporting the ESA's missions (Kourou (French Guiana), Maspalomas, Villafranca and Cebreros (Spain), Redu (Belgium), Santa Maria (Portugal), Kiruna (Sweden), Perth and New Norcia

---

*Corresponding author: Fatos Xhafa, Department of Languages and Informatics Systems, Technical University of Catalonia, Jordi Girona 1–3, 08034 Barcelona, Spain. E-mail: fatos@lsi.upc.edu.

(Australia), Malargue (Argentina)). ESA missions are mainly scientific missions. As per the study of the mission planning, it is not needed to know specific characteristics of the missions but their information can be used in simulation phase of scheduling systems. Examples of ESA missions include CLUSTER II,, GIOVE-A/GIOVE-B, INTEGRAL, METEOSAT-6/METEOSAT-7, and XMM-NEWTON [10]. ESA uses the ESA tracking station network (ESTRACK) [8], which is a worldwide system of ground stations providing links between satellites in orbit and Operations Control Center. The scheduling of operations for spacecraft-ground stations communications is currently used but has shown several limitations, the most significant one being the need for human coordination and a labour intensive activity for manual organization of the mission schedules. Certainly, with the increasing number of satellites and the growing number of requests for more observations, manual organization cannot tackle the complexity of the problem.

In the USA, besides NASA whose fleet of Earth Observing Satellites (EOSs) keeps growing to meet demands of assisting scientists in their research activities, other examples where satellite scheduling arises include the Air Force Satellite Control Network (AFSCN). The AFSCN accounts for more than 100 satellites and 16 antennae located at nine ground stations. Interested users request a ground station for use during a preferred time window, while specifying alternative time slots due to the large number of requests received and possible time-window overlapping conflicts among them (typically about a few hundreds requests/a hundred of conflicts per day) [6].

Finally, there have been reported a growing number of other satellite missions from projects in research institutions and universities (e.g. Berkeley mission supported by NASA) that need intelligent systems to automate the allocation of ground station services to space missions. Indeed, with fast advancements in satellite networks [9] the number of applications based on satellite networks is increasing in many domains [18–20].

Despite their particularities, common to all these mission operations projects is the requirement to make maximum usage of ground stations (reducing their idle time) in order to support the largest possible number of space missions. Scheduling problem then arises in a variety of forms, including ground station scheduling (with its variants of one ground station, multiple ground station), satellite range scheduling, AFSCN scheduling, LEO satellite scheduling [3,14,21]. All scheduling variants, in their general formulations are highly constrained problems and have been shown computationally hard [16]. Other scheduling models related to on-air information which uses satellite technology has also been reported in the literature [22].

In this paper, we present and analyse some relevant formulations of the satellite scheduling viewed as a family of problems and identify various forms of optimization objectives. The main complexities, due highly constrained nature, windows accessibility and visibility, multi-objectives and conflicting objectives are examined. Then, we discuss the resolution of the problem through different heuristic methods. In particular, we focus on the version of ground station scheduling, for which we present computational results obtained with Genetic Algorithms using the STK simulation toolkit.

The rest of the paper is organized as follows. In Section 2 we describe some basic terminology and concepts from satellite scheduling domain. Different versions of the problem are presented in Section 3. For each version, we identify the main characteristics and its computational complexity. We consider in more details the case of ground station scheduling as a multi-objective optimization problem. The heuristic resolution methods that can be used for near-optimally solving the problem are given in Section 5. We present a GA for the problem and computational results for its evaluation in Subsection 5.1. We end the paper in Section 6 with some conclusions and remarks for future work.

2. Terminology and preliminary concepts

2.1. Tasks

A task is an object which is scheduled. In satellite scheduling tasks refer to observations, communications, manoeuvres, imaging, taking measurements, uplinks, downlinks, etc. Tasks have specified a starting time and a duration time for their completion. Tasks could have specific requirements on the type of the resource they need to be completed (a resource may not execute all types of tasks). Tasks could be given priorities and, in some cases (e.g. imaging tasks) could be specified location information.

2.2. Resources

In satellite scheduling resources comprise satellites, earth observing satellites, ground stations, instruments, recording devices, transmitters, etc.

The ground station and spacecraft are the main resources in satellite scheduling.

2.3. Ground Station (GS) and Spacecraft (SC)

GSs are terrestrial terminals designed for extra-planetary communications with SCs and in data processing [12]. SCs are extra-planetary crafts, such as satellites, probes, space stations, orbiters, etc. Ground stations communicate with a spacecraft by transmitting and receiving radio waves in high frequency bands (e.g. microwaves). A ground station usually contains more than one satellite dish. Each dish is usually assigned to a specific space mission. With the scheduling from control center, dishes are able to handle and switch among mission spacecrafts.

2.4. Mission operation

These are activities that include payload operations (e.g., using a sensor on the satellite to collect data), bus operations (e.g., maintaining the health and status of the vehicle) and communications operations (e.g., transmitting data between satellites or to the ground and receiving information or commands from a ground station) [14].

2.5. Mission requirements

Time duration is a main requirement of mission operations. Depending on the mission, the time required for links can possibly range from less than an hour to 8 hours per day. Besides the time required, there are also other kinds of mission operation requirements, which should be taken into account in the scheduling process. Following is an example of partial requirements for two missions M1 and M2, for which we assume that they share the same ground station G1. Due to the usage of other SCs, it also involves the use of another ground station G2:

1. Lunar occultation periods have to be avoided
2. The minimum pass duration is 3-hours for M1, selected from within the physical station visibility
3. The minimum pass duration is 3-hours for M2, selected from within the physical station visibility
4. The separation of pass should be 24 hours $\pm$ 30 minute
5. The maximum separation of passes should be 27 hours
6. The minimum pass elevation should be 5°

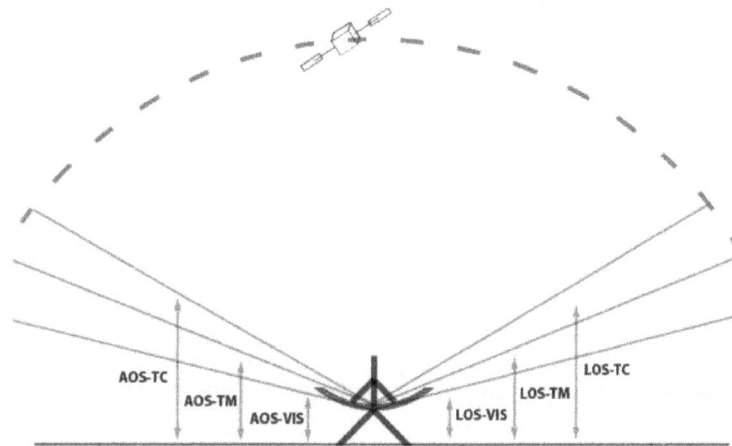

Fig. 1. Illustration of a GS AOS and LOS angles.

7. M1 and M2 pass should be scheduled within a period of 8 hours. It is expected a ground station reconfigure time between spacecrafts of less than 30 minutes, including the pre-pass test
8. The order M1-M2 or M2-M1 should be retained until a change is requested
9. During some period, due to the load of G1, it may required to support of M1 and M2 spacecraft from G2. The selection of SC support by G2 should be maintained for the full duration of the further analysis.

2.6. Visibility

A ground station can communicate with a SC only when the satellite is within the transmitting angle of the ground station. A spacecraft has three types of *visibility* to a ground station, namely:

– AOS-VIS: Acquisition of Signal, Visible. This indicates the time when the SC appears in the line of sight of the GS.
– AOS-TM: Acquisition of Signal, Telemetry. This is time when GS can start receiving telemetry signals from the SC.
– AOS-TC: Acquisition of Signal, Tele-command. This is time when GS is allowed to send signal to SC.

The visibility angle of each case is defined with the regulations of GS, so that the high amount of transmitting energy will not harm the organisms on the ground. Similarly defined, we also have Loss of Signals, i.e. LOS-TC, LOS-TM, LOS-VIS. In this work we are concerned with scheduling the events ranging from TAOS-TC to TLOS-TC. The scheduling is concerned with planning of the events ranging from TAOS-TC to TLOS-TC. In the simulation process, to get this data, we will set the elevation angle to 10 degrees for all the spacecraft to simplify the process. Figure 1 shows the relations among all those angles.

2.7. Spacecraft visibility clash

The most common constraint in ground station scheduling is the clash of visibility windows caused by multiple spacecrafts to a single ground station. A visibility clash of two spacecraft happens when the AOS time of second spacecraft starts before the LOS time of first one. Figure 2 shows a two-day visibility for 5 spacecraft (the horizontal axis is the time, and vertical are the SCs).

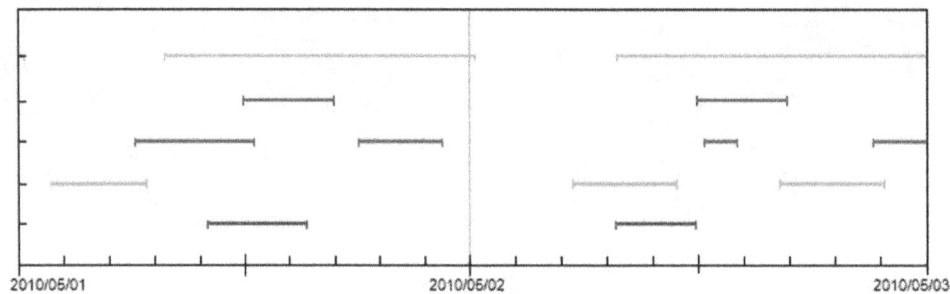

Fig. 2. Visibility clashes of 2 days for 5 spacecrafts.

3. The family of mission operations scheduling problems

Despite the large body of research on scheduling, mission operations scheduling can be seen as a family of problems in its own. We describe next some representative formulations of the problem, identify the complexities and resolution methods proposed in the literature.

3.1. Satellite scheduling

3.1.1. Description and characteristics
Satellite scheduling is the problem of planning tasks to resources. The particularities of this scheduling problem with regard to traditional scheduling problems from manufacturing or distributed systems consists in the fact that tasks cannot be executed at any time, but at specific times, usually specified as time-windows. Additionally, it is an over-constrained problem because not all requests can be scheduled within the specified slot times and thus conflicts should be minimized. One particular case of satellite scheduling arises when satellites are Low Earth Orbit satellites. In this case the time window may vary from a few minutes to only a small fraction of a second.

3.1.2. Constraints
Constraints arise due when and where tasks can be scheduled as well as due to limited resource capacity and availability. In [14], there have been identified three types of satellite scheduling constraints: (a) task constraints, (b) resource constraints and (c) event constraints. Task constraints refer to possible dependencies among tasks (e.g. a task should be scheduled after another task – see Section 2). Resource constraints refer to their processing capabilities (the type of the tasks they can execute). Event constraints refer to time windows in which a task can be executed. This kind of constraint is due spacecrafts have windows visibility to a ground station, and thus can communicate only during concrete window times.

3.1.3. Modelling
Given that the problem is highly constraints, the CSP – Constraint Satisfaction Problem is among first used for modelling and resolution purposes. The resulting instances, nevertheless, cannot be solved to optimality due to the computational hardness of CSP.

3.2. Satellite range scheduling

3.2.1. Description and characteristics
Satellite range scheduling [6] is another flavour of the satellite scheduling problem for a rather concrete scenario: coordinate communications between civilian and military organizations using more

than 100 satellites, 16 antennas located at nine ground stations around the globe. Customers request to reserve an antenna at a ground station for a specified time window to complete two types of task(s): low altitude (usually requiring access to low attitude satellites and having short duration) and high altitude (requiring access to high altitude satellites and having longer window times for completion).

3.2.2. Constraints

As in the case of general satellite scheduling, the constraints are task constraints, resource constraints and time-window (events) constraints. Additionally, some works in literature have considered requests with priorities, fixed start times and durations, that is a kind of strict scheduling requirement. In some other cases, requests are more flexible by specifying earliest start time, latest final time and the minimum and maximum duration.

3.2.3. Objectives

The costumers' requests may cause conflicts due more request can arrive than possible to accommodate. The main objective is then minimize the number of request conflicts when scheduling costumers requests. This objective could equivalently be formulated as maximizing the number of non-conflict requests that can be scheduled. In fact, by maximizing the number of scheduled requests, the usage of resources (satellites) could be maximized as well.

3.3. Task scheduling for satellite based imagery

3.3.1. Description and characteristics

This kind of scheduling (also known as photo-reconnaissance satellite scheduling) arises for a particular type of tasks, namely, when it is needed to schedule tasks to acquire images of the earth surface and scheduling the transmission of the image files to a set of ground stations [11,23]. Two modes of this scheduling can be addressed: static scheduling and dynamic scheduling. In the former, the imaging satellites are to execute tasks according to a schedule *a priori* computed. In the later case, the objective is to increase the resource (satellite) usage dynamically according to the tasks requirements and resource availability.

3.3.2. Constraints

The time windows for task completion in this kind of scheduling are of about one-day horizon. Again resource constraints should be taken into account as well as possible preferences of visibility time windows. Additionally, there are constraints about imaging tasks such as transition time between consecutive downloads, memory capacity, etc.

3.3.3. Objectives

The main objective is to maximize the amount of images acquired and transmitted, maximizing resource usage and minimizing conflicts. Indeed, the problem is over-constrained, in that there are more imaging tasks than can be possibly completed in one pass of the satellite, thus in general only a subset of tasks will be completed. Imaging tasks could be given priorities, in which case the objective would be to to maximize the total priority of the tasks.

3.4. Ground station scheduling

3.4.1. Description and characteristics

Ground station scheduling problem arises in spacecraft operations and aims to allocate ground stations to spacecraft to make possible the communication between operations teams and spacecraft systems. The

Table 1
Time eequirements of spacecrafts

SC	From (min)	To (min)	Require (min)	Meaning
1	1	2880	60	1 hour/2 days
1	2881	5760	60	1 hour/2 days
1	5761	8640	60	1 hour/2 days
1	8641	12960	60	1 hour/2 days
2	1	2880	80	80 mins/2 days
2	2881	5760	80	80 mins/2 days
2	5761	8640	80	80 mins/2 days
2	8641	12960	80	80 mins/2 days
3	1	1440	120	2 hours/day
3	1441	2880	120	2 hours/day
3	2881	4320	120	2 hours/day
3	4321	5760	120	2 hours/day
3	5761	7200	120	2 hours/day
3	7201	8640	120	2 hours/day

problem belongs to the family of satellite scheduling for the specific case of mapping communications to ground stations.

3.4.2. Objectives

Different types of objectives can be formulated, namely, maximizing matching of visibility windows of spacecrafts to communicate with ground stations, minimizing the clashes of different spacecrafts to one ground station, maximizing the communication time of the spacecraft with the ground station, and maximizing the usage of ground stations.

3.4.3. Constraints

The most common constraint in ground station scheduling is the clash of visibility windows caused by multiple spacecrafts to a ground station (see Table 1 for an example of time requirements of three spacecrafts).

3.5. System assumptions

In order to make a practical system environment for a Ground Station (GS) – SpaceCraft (SC) schedule system [8], we make the following system assumptions:

– A maximum amount of communication time is defined for each spacecraft.
– 7 real mission spacecrafts orbit information are used (which are Cluster, Integral, XMM, etc. – see Section 1). The SCs are defined as $\bigcup_{i=1}^{7} SC(i)$.
– 4 real ground station locations are selected. (Cebreros, Kourou, Malargue, and New Norica). These GS are defined as $\bigcup_{g=1}^{4} GS(g)$.
– The relation of GSs and SCs are hypothetical.
– Each SC has one communication with GS each day.
– During the 8-day period, one SC will only use one GS.
– The goal of is to allocate the maximum possible amount of communication time for each SC.

3.6. One vs. multiple ground stations

In general, we may assume that there are multiple ground stations available to communicate with spacecrafts. It might be the case that some spacecraft might have no visibility with some ground stations (see Fig. 3 for an example of four ground stations and their window time to spacecrafts).

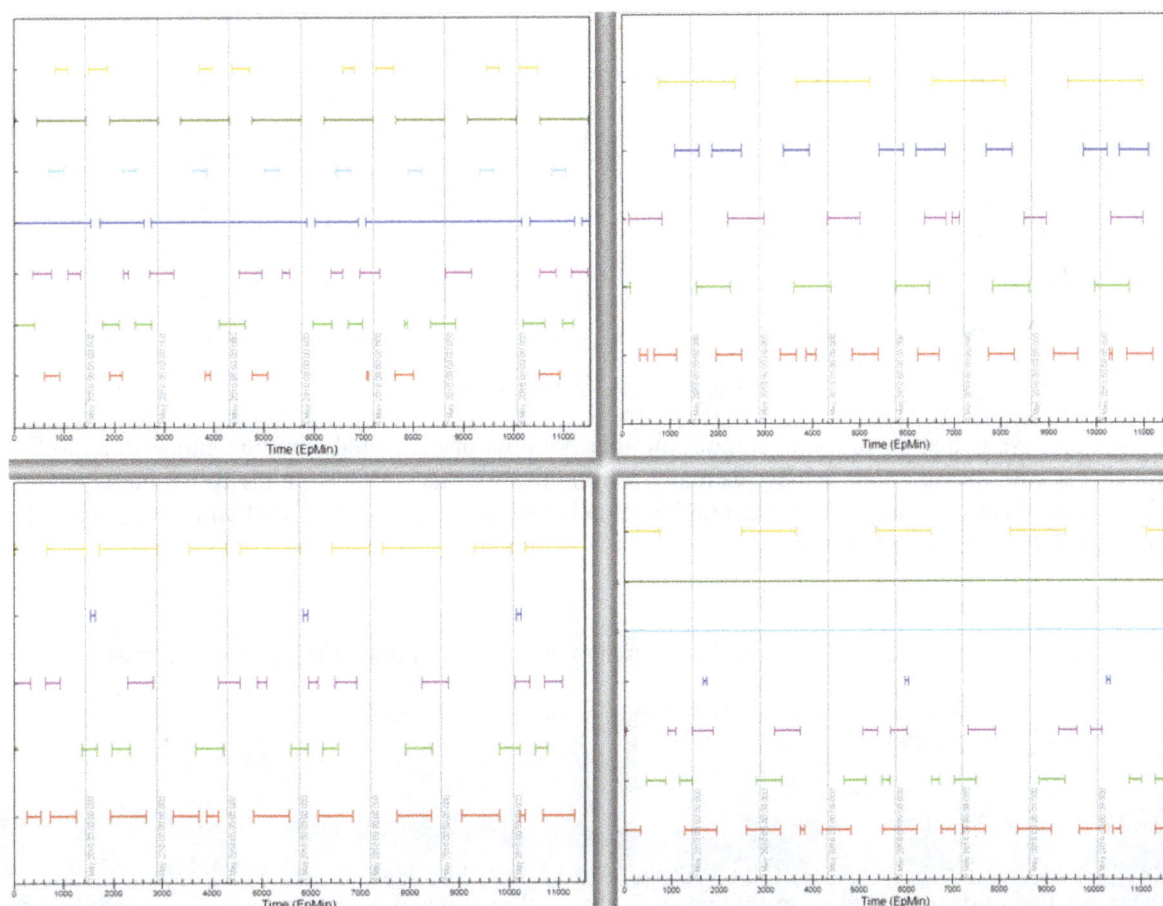

Fig. 3. Four ground stations and their window time to spacecrafts.

4. Scheduling fitness types

4.1. Scheduling fitness

One of the major complexities of the mission operations scheduling comes from the many objectives that can be sought for the problem. These objectives are related to visibility window, communication clashes, communication time, resource usage, among others. The total fitness function, besides being composed of multiple objectives, poses the challenge of how to combine them and in which order to evaluate them. For the combination, one can adopt a hierarchical optimization approach based on the priority of the objectives or a simultaneous optimization approach.

We define next the four objectives that would compose the fitness function and then show possible combinations of them into one fitness function.

4.1.1. Access window fitness

Visibility windows are the time periods when a GS has the possibility to setup a communication link with a SC.

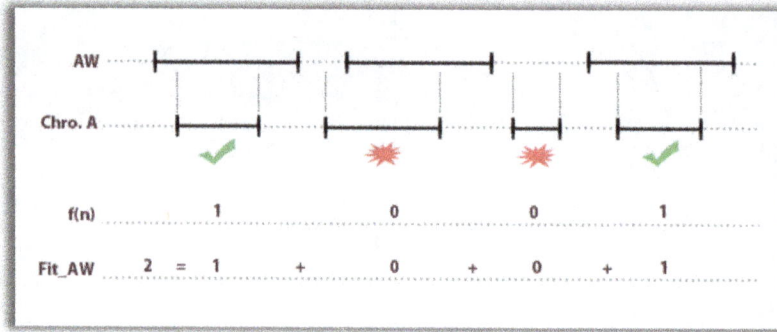

Fig. 4. Access windows fitness.

The objective is that all or the maximum number possible of generated communication links to fall into access windows and thus achieve as many communications as possible. In the following equation, $W_{(g,i)}$ is the Access Window set for Ground Station g and Spacecraft i, $T_{Start}(s)$ and $T_{End}(s)$ are the start and end of each access window.

$$AW(g, i) = \cup_{s=1}^{S}[T_{AOS(g,i)}(s), T_{LOS(g,i)}(s)]$$

Then we define the final Access Window fitness of the scheduling solution (Fit_{AW}) calculated as follows:

$$f(n) = \begin{cases} 1, \text{ if } [T_{Start}(n), T_{Start}(n) + T_{Dur}(n)] \subseteq AW(ng, ni), \\ 0 \text{ otherwise.} \end{cases}$$

$$Fit_{AW} = \sum_{n=1}^{N} f(n)$$

where n value corresponds to an event and N is the total number of events of an entire schedule (see Fig. 4).

4.1.2. Communication clashes fitness

Communications clash represents the event when the start of one communication task happens before the end of another one on the same ground station. The objective is to minimize the clashes of different spacecrafts to one ground station. To compute the number of clashes, SCs are sorted by their start time. If as a result of the sorting:

$$T_{Start}(n + 1) < T_{Start}(n) + T_{Dur}(n), \quad 1 \leqslant n \leqslant N - 1$$

where n value corresponds to an event and N is the total number of events of an entire schedule, then there is a clash. The fitness will be reduced, and one of the clashed entries has to be removed from the solution. The total fitness of communication clashes is then:

$$f(n) = \begin{cases} -1, \text{ if } T_{Start}(n + 1) < T_{Start}(n) + T_{Dur}(n), \\ 0 \quad \text{otherwise.} \end{cases}$$

$$Fit_{CS} = N + \sum_{n=1}^{N} f(n)$$

Figure 5 shows an example.

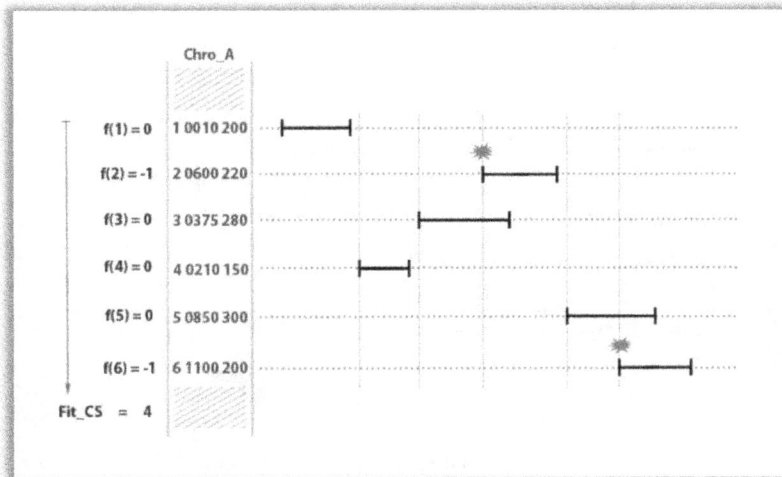

Fig. 5. Fitness communication clashes.

$$\begin{array}{cccc} SC & From & To & T_{REQ}[Min] \end{array}$$

$$\begin{bmatrix} SC[1] & T_{From}[1]_1 & T_{TO}[1]_1 & T_{REQ}[1]_1 \\ SC[1] & T_{From}[1]_2 & T_{TO}[1]_2 & T_{REQ}[1]_2 \\ \vdots & & & \\ SC[1] & T_{From}[1]_N & T_{TO}[1]_N & T_{REQ}[1]_N \\ SC[2] & T_{From}[2]_1 & T_{TO}[2]_1 & T_{REQ}[2]_1 \\ SC[2] & T_{From}[2]_2 & T_{TO}[2]_2 & T_{REQ}[2]_2 \\ \vdots & & & \\ SC[i] & T_{From}[i]_n & T_{TO}[i]_n & T_{REQ}[i]_n \\ \vdots & & & \\ SC[I] & T_{From}[I]_N & T_{TO}[I]_N & T_{REQ}[I]_N \end{bmatrix}$$

Fig. 6. Matrix representation of periodic tasks.

4.1.3. Fitness – Communication time requirement

The objective is to maximize the communication time of the spacecraft with the ground station so that every spacecraft $SC(i)$ will communicate at least $T_{req}(k)$ time.

A sufficient amount of time should be granted for TTC (Telemetry, Tracking and Command). Depending on the mission type, there are different amount of data needed to be downloaded from the spacecraft. For example, satellites that need to download huge amount of image data require more time for linking with ground station. These communications, especial for data download tasks are usually periodical tasks (e.g. 2 hours communication for SC1 each day, 5 hours data downlink for SC2 every 2 days, etc.) A matrix is used to define those requirements, which is used as the input for the scheduling system (see Fig. 6 for an example.)

The fitness is calculated by summing up all the communication link durations of each spacecraft, and dividing them in the required period to compare if the scheduled time matches requirements, using the

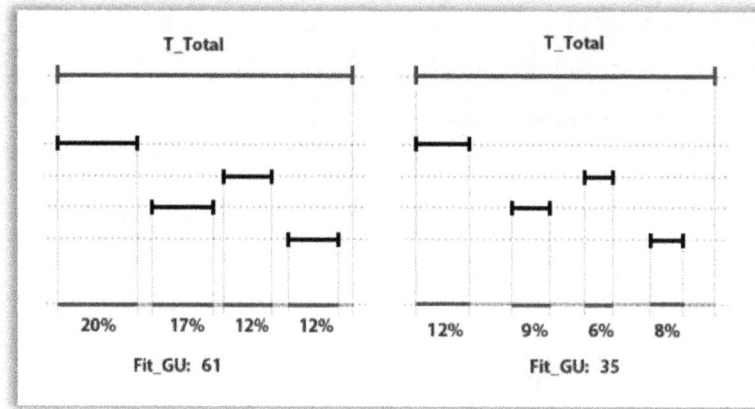

Fig. 7. Ground station usage.

following equations.

$$T_{Start}(m) > T_{From(k)}$$

$$T_{Start}(n) + T_{Dur}(n) < T_{TO}(k)$$

$$T_{Comm}(k) = T_{Dur}(j)$$

$$f(k) = \begin{cases} 1, \text{ if } T_{Comm}(k) \geqslant T_{REQ}(k), \\ 0 \text{ otherwise.} \end{cases}$$

$$FIT_{TR} = \sum_{k=1}^{K} f(k)$$

4.1.4. Ground station usage fitness

Given that the number of ground stations is usually much smaller than the number of spacecrafts missions, the objective is to maximize the usage of ground stations, that is, try to reduce the idle time of a ground station. A maximized usage would contribute to provide additional time for SC communications (Fig. 7).

This fitness value is calculated as the percentage of ground stations occupied time by the total amount of the possible communication time. The more a GS is used, the better is the corresponding schedule.

$$Fit_{GU} = \frac{\sum_{n=1}^{N} T_{Dur}(n)}{\sum_{g=1}^{G} T_{Total(g)}} \times 100$$

where N is the number of events of an entire schedule, G is the number of ground stations and $T_{Total(g)}$ is the total available time of a ground station.

4.1.5. Hierarchic optimization model

In this model, the objectives are classified (sorted) according to their priority. Thus, for a problem having k objectives sorted as follows:

$$f_1 \succ f_2 \succ \cdots \succ f_k$$

means that f_1 is the most important objective and f_k is the least important objective. The optimization procedure would first optimize according to f_1 until no further improvements are possible. Then, the algorithm optimizes according to f_2 subject to not worsening the value achieved for f_1, and so on.

This model is useful when for the design or deployment needs some parameters (objectives) are considered of more priority than others. The disadvantage of this model is that the final solution computed by the optimization procedure could be far from optimal for the less priority objectives.

4.1.6. Simultaneous optimization approach

In the simultaneous optimization approach, all objectives are simultaneously optimized. Thus, for a problem having k objectives $f_1, f_2, \ldots, f_k$, the optimization procedure tries to optimize at the same time all the objectives, which actually leads to computing the so called *Pareto front* which contains the optimal solutions. In some cases, it could be possible to apply the *sum model* in which the k objectives are reduced to two objectives:

$$f = \lambda_1 f_1 + \lambda_2 f_2 + \cdots + \lambda_k f_k, \quad \sum_{i=1}^{i=k} \lambda_i = 1, \ \lambda_i > 0.$$

4.1.7. Combination of fitness objectives

The fitness objectives defined above (FIT_{AW}, FIT_{CS}, FIT_{TR}, FIT_{GU}) are conceived as fitness modules so as to facilitate the design phase of the scheduler to easily plug-in other fitness objectives. From the definition of the fitness objectives, we can observe that some of them can be applied in serial fashion (due dependencies, denoted serial-FM), while some others can be applied in parallel (denoted parallel-FM). Thus, in a hierarchical mode, one possible way to arrange fitness checking is that of Fig. 8.

With regard to the simultaneous combination of objectives, one can either consider a proper Pareto-front approach, or combine all the fitness modules into one total fitness function using weights for different fitness module:

$$Fit = \sum_{i=1}^{n} w_i \cdot Fit_S(i) + \sum_{j=1}^{m} w_j Fit_P(j)$$

where w_i, w_j are the weights of fitness modules, $Fit_S(i)$ and $Fit_P(j)$ are the fitness values from Serial-FMs and Parallel-FMs, and n, m are the number of fitness modules, resp.

5. Resolution methods

Most formulations of the satellite scheduling problem are NP-hard, and thus it is unlikely to find polynomial time algorithms for finding optimal solutions. Several heuristics algorithms can be used to search for near-optimal scheduling solutions. Some approaches proposed in the literature include Genitor [4], Branch-and-Bound Algorithm [4], Graph Colouring [21], Tabu Search [15], Hill Climbing [4], Fuzzy techniques [1].

Genetic Algorithms are among most successful algorithms for efficiently tackling with the complexity of computationally hard problems from networking domain [7,13]. We present in the next section Genetic Algorithms for the ground station scheduling and report some computational results for the version with multiple ground stations.

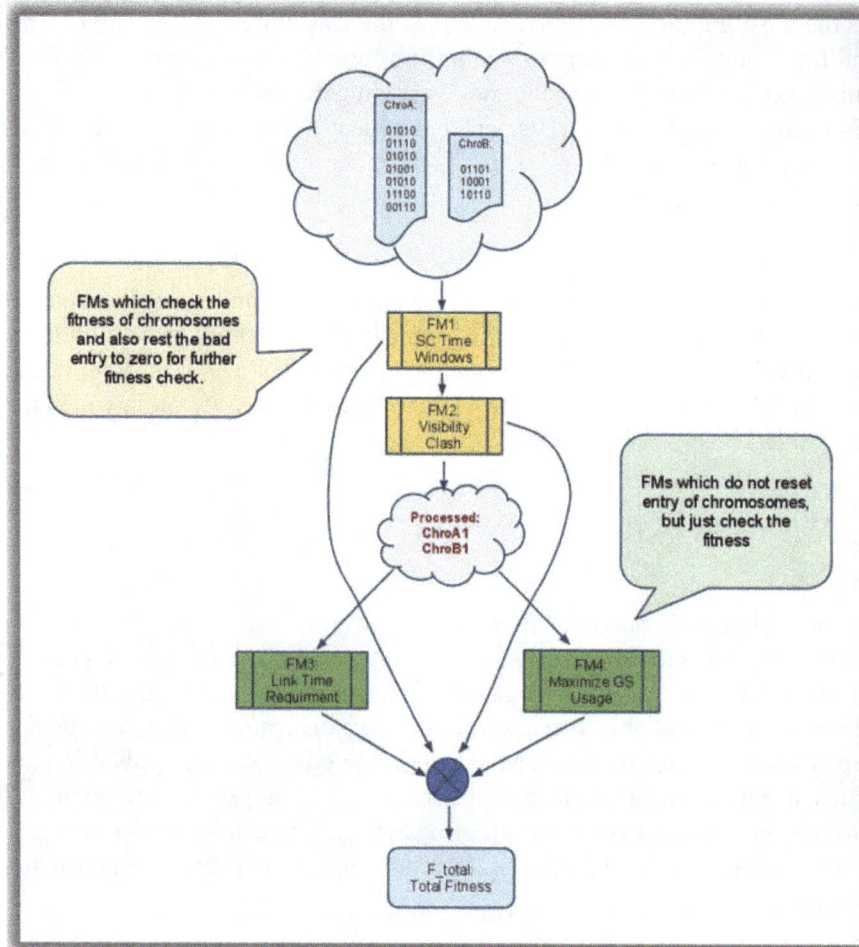

Fig. 8. Combination of fitness modules.

5.1. *Genetic algorithm for ground station scheduling*

We have used the *template* given in algoritm 1.
We briefly present next the main features of GAs.

Algorithm 1 Genetic algorithm template

Generate the initial population P^0 of size μ;
Evaluate P^0;
while not termination-condition **do**
   Select the parental pool T^t of size λ; $T^t := Select(P^t)$;
   Perform crossover procedure on pairs of individuals in T^t with probability p_c; $P_c^t := Cross(T^t)$;
   Perform mutation procedure on individuals in P_c^t with probability p_m; $P_m^t := Mutate(P_c^t)$;
   Evaluate P_m^t ;
   Create a new population P^{t+1} of size μ from individuals in P^t and/or P_m^t;
   $P^{t+1} := Replace(P^t; P_m^t)$
   $t := t + 1$;
end while
return Best found individual as solution;

$$\text{Chromosome } A: \quad \text{Chromosome } B:$$

$$\begin{bmatrix} SC[1], & T_{Start}, & T_{Dur} \\ SC[2], & T_{Start}, & T_{Dur} \\ \vdots & & \\ SC[i], & T_{Start}, & T_{Dur} \end{bmatrix} \quad \begin{bmatrix} SC[1], & GS[g_1] \\ SC[2], & GS[g_2] \\ \vdots & \\ SC[i], & GS[g_i] \end{bmatrix}$$

Fig. 9. Chromosome representations.

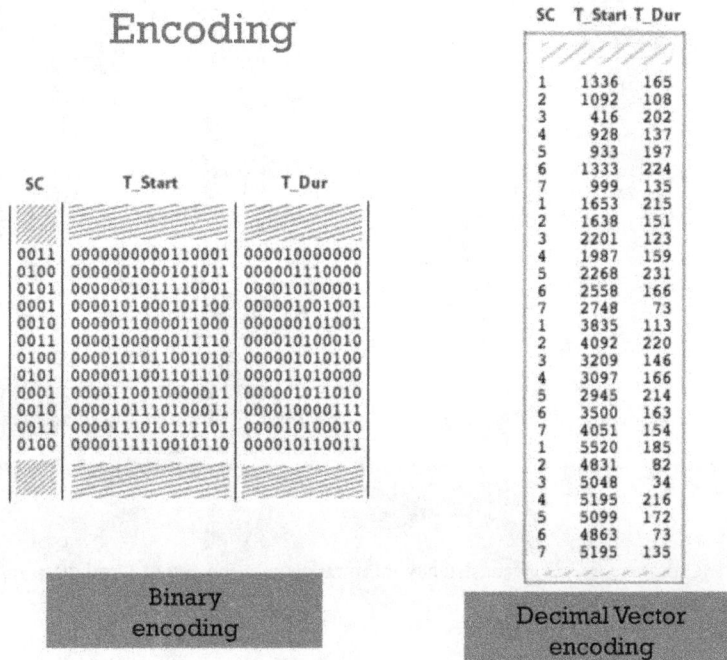

Fig. 10. Chromosome encoding.

5.1.1. Population of individuals

Unlike local search techniques that construct a path in the solution space jumping from one solution to another one through local perturbations, GAs use a population of individuals giving thus the search a larger scope and chances to find better solutions. This feature is also known as "exploration" process in difference to "exploitation" process of local search methods.

5.1.2. Fitness

The determination of an appropriate fitness function, together with the chromosome encoding are crucial to the performance of GAs. Ideally we would construct objective functions with "certain regularities", i.e. objective functions that verify that for any two individuals which are close in the search space, their respective values in the objective functions are similar.

5.1.3. Selection

The selection of individuals to be crossed is another important aspect in GAs as it impacts on the convergence of the algorithm. Several selection schemes have been proposed in the literature for selection operators trying to cope with premature convergence of GAs.

GA operator: Crossover

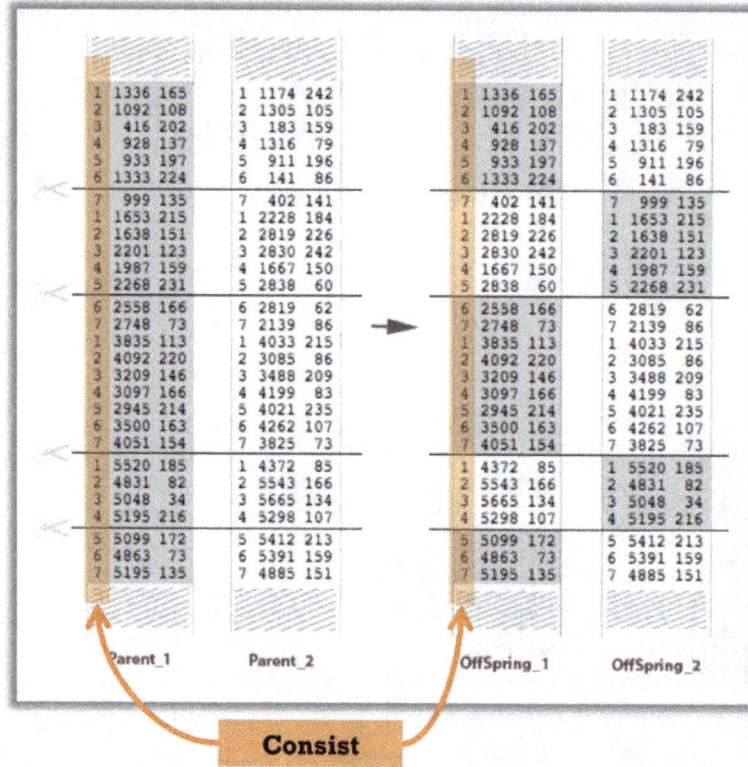

Fig. 11. Crossover operator (consistency information between parent 1 and offspring 1).

5.1.4. Crossover operators

Use of crossover operators is one of the most important characteristics. Crossover operator is the means of GAs to transmit best genetic features of parents to offsprings during generations of the evolution process.

5.1.5. Mutation operators

These operators intend to improve the individuals of a population by small local perturbations. They aim to provide a component of randomness in the neighbourhood of the individuals of the population.

5.1.6. Escaping from local optima

GAs have the ability to avoid falling prematurely into local optima and can eventually escape from them during the search process.

5.1.7. Convergence

The convergence of the algorithm is the mechanism of GAs to reach to good solutions. A premature convergence of the algorithm would cause that all individuals of the population be similar in their genetic features and thus the search would result ineffective and the algorithm getting stuck into local optima. Maintaining the diversity of the population is therefore very important to this family of evolutionary algorithms.

GA operator: Crossover

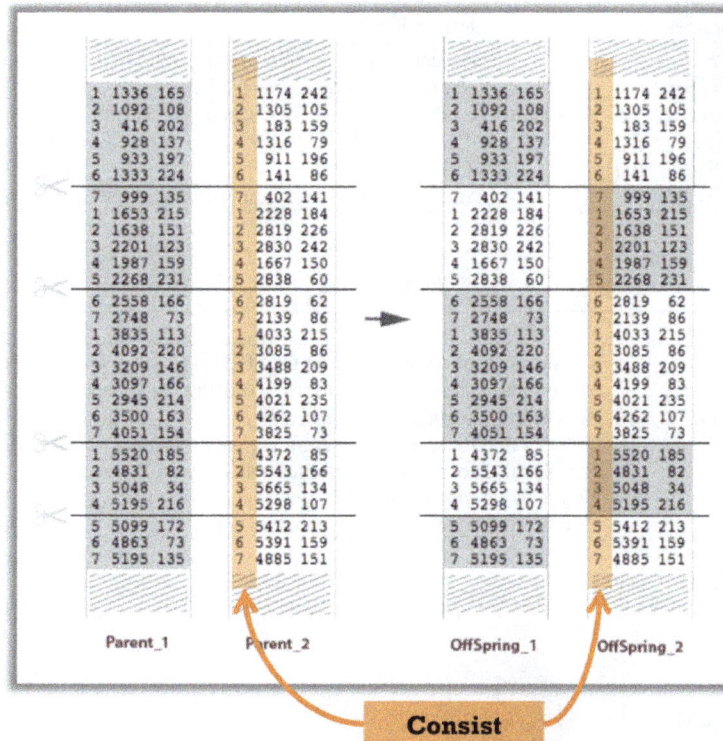

Parent_1	Parent_2	OffSpring_1	OffSpring_2

Consist

Fig. 12. Crossover operator (consistency information between parent 2 and offspring 2).

5.2. GA for ground station scheduling

5.2.1. Chromosome encoding

A chromosome (an individual of the population) encodes the scheduling solution, which consists of (a) encoding the spacecrafts timetable and (b) encoding the pairs of GS-SC. One possibility of encoding would be to encode separately into Chromosome A and Chromosome B this information using binary or decimal encoding schemes (see Figs 9 and 10).

5.2.2. Initial population

Initial population is generated randomly.

5.2.3. Crossover operator

Multi-points crossover has been selected as appropriate because of the length of chromosome we are dealing is usually large (e.g. an 8-Day, 7-SC, scheduling can have a vector chromosome size of 56.) We show in Figs 11 and 12 an example of crossover. As can be seen it is important to ensure consistency when passing the information from parents to offsprings.

5.2.4. Mutation operator

Mutation makes some small local perturbations of an individual. Specifically, we only mutate the time of communication (T_{Start} and T_{Dur}); it's worth noting that the bits corresponding to $SC[i]$ cannot be changed. We show in Fig. 13 an example.

Table 2
System input

Parameters	Descriptions
$\bigcup SC(i)$	Spacecrafts that need to be scheduled in the system
$\bigcup GS(g)$	Ground station resources for spacecrafts
$\bigcup N_{days}$	Number of days for one scheduling period
$\bigcup T_{AOS}(g)(i))$	AOS of a SC to GS
$\bigcup T_{LOS}(g)(i))$	LOS of a SC to GS
$\bigcup T_{req}(i))$	Communication time required for each SC

Table 3
System output

Parameters	Descriptions
$\bigcup_{All} T_{start}(i)$	All start time of communication for SC(i) during the schedule period
$\bigcup_{All} T_{dur}(i)$	All communication duration for SC(i) during the schedule period
$\bigcup SC/GS(i,g)$	Pairs of ground station and spacecraft
Fit_{CC}	Fitness value for minimizing communication clashes
Fit_{VW}	Fitness value for meeting visibility windows of all spacecrafts
Fit_{TR}	Fitness value for fulfilling mission time requirements
Fit_{GU}	Fitness value for maximizing ground station usage

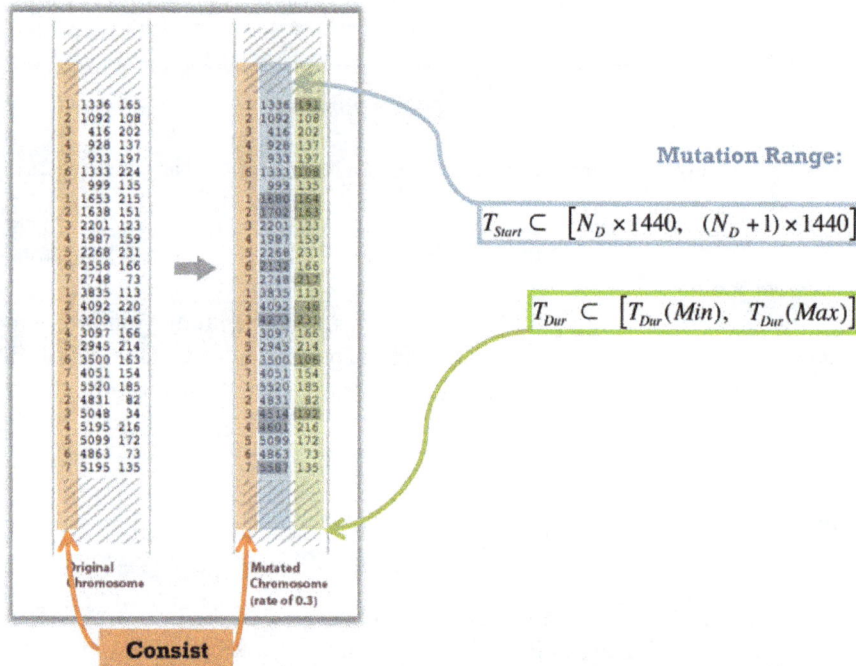

Fig. 13. Mutation operator (feasibility for visibility windows).

5.2.5. Selection operators

The selection method determines which individuals will survive and produce offspring according to their fitness score. The selection methods considered in this work are Fitness Proportionate Selection, Stochastic Universal Sampling, and Tournament Selection.

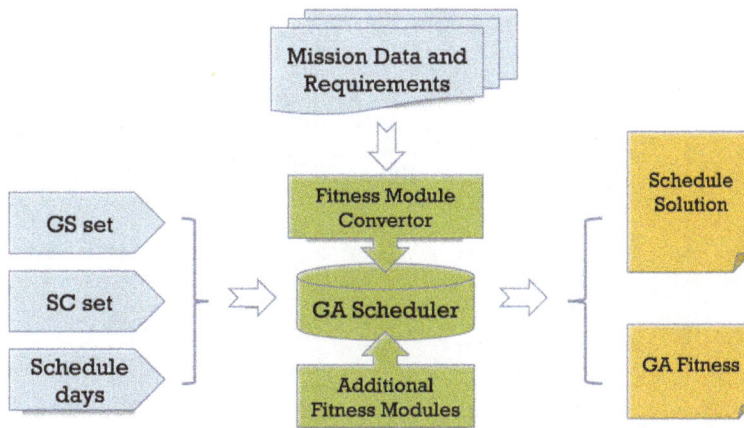

Fig. 14. Scheduling system model.

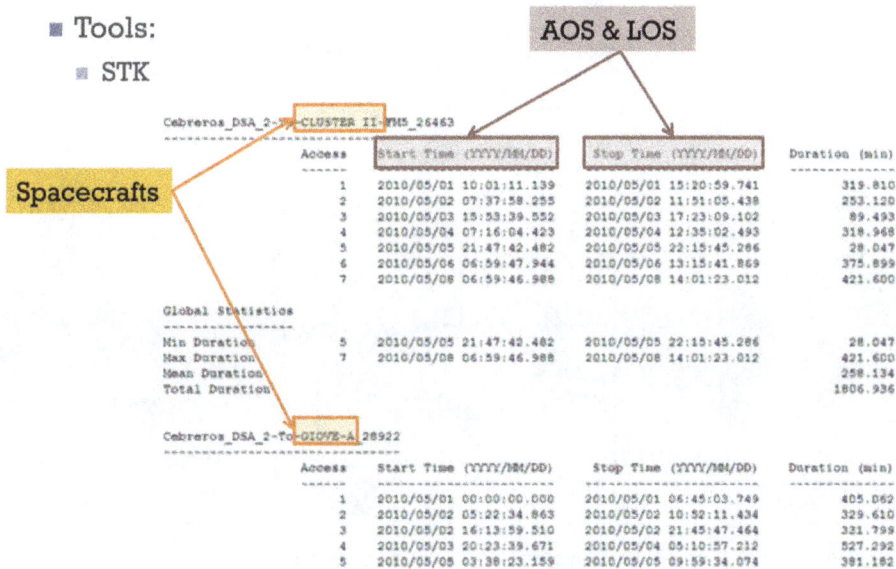

Fig. 15. Visibility windows.

5.3. Scheduling system model

The scheduling system, using Genetic Algorithms as solver, can be modelled as shown in Fig. 14.

As can be seen from the model, the scheduling system receives in input a set of parameters, which consists of ground stations information, spacecrafts information and information on the schedule days (see Table 2). Additionally, the scheduling system receives other constraints (spacecraft constraints, ground station constraints, and mission data requirements).

Then, the scheduling system, using GA as solver, outputs the pairs of ground stations – spacecrafts, the starting time of communication among ground stations and spacecrafts, the duration time of the communication scheduled (see Table 3). The scheduling system also outputs different fitness values associated with the scheduling such as fitness value that minimizes the communication clashes, fitness value of matching the visibility windows of all spacecrafts and fitness value of maximizing the ground

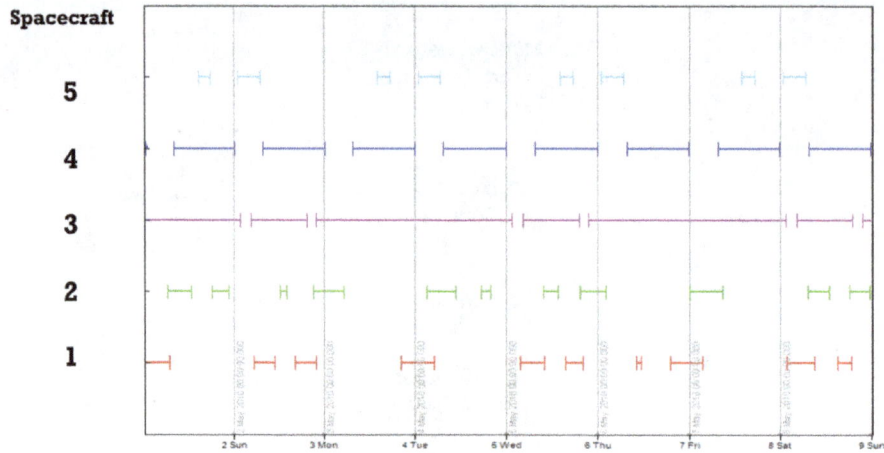

Fig. 16. Ground station and its satellites visibility windows.

$$
\begin{array}{lll}
SC & AOS & TOS \\
\hline
SC[1], & T_{AOS}[1]_1, & T_{LOS}[1]_1 \\
SC[1], & T_{AOS}[1]_2, & T_{LOS}[1]_2 \\
\vdots \\
SC[1], & T_{AOS}[1]_N, & T_{LOS}[1]_N \\
SC[2], & T_{AOS}[2]_1, & T_{LOS}[2]_1 \\
SC[2], & T_{AOS}[2]_2, & T_{LOS}[2]_2 \\
\vdots \\
SC[i], & T_{AOS}[i]_n, & T_{LOS}[i]_n \\
\vdots \\
SC[I], & T_{AOS}[I]_N, & T_{LOS}[I]_N
\end{array}
$$

Fig. 17. The extracted information for spacecrafts.

station usage. This last fitness value is important given that the number of ground stations is a limited resource.

5.4. Data simulation model for ground station scheduling

We have used the Satellite Tool Kit (STK) [17] that allows engineers and scientists to design and develop complex dynamic simulations of real-world problems. It is a powerful software tool to solve problems of Earth-orbiting satellites. We used it to simulate a ground station and its satellites. In order to achieve realistic simulation, we are based on ESA ground stations, and create them with exact position on Earth. Then, based on real ESA space mission, orbits of spacecrafts are added to the scenario.

STK generates a report of the time stamp of visibilities of the ground to its entire associated spacecraft (see Fig. 15.)

At the same time, we can also generate the graphic view of the visibility windows in the entire scheduling period. In Fig. 16, from the bottom to the top, the lines represent the visibility windows of

Table 4

GA parameter values

GA parameter	Value
Size of population	20
Num. of crossover points	15
Mutation rate (Start time)	10%
Mutation rate (Comm. Duration)	10%

Table 5

Best fitness results (numeric values)

Objective	Fitness
Less link clashed	34
Fit to access windows	33
Link time required	24 [4 4 8 4 4]
Ground station usage	65

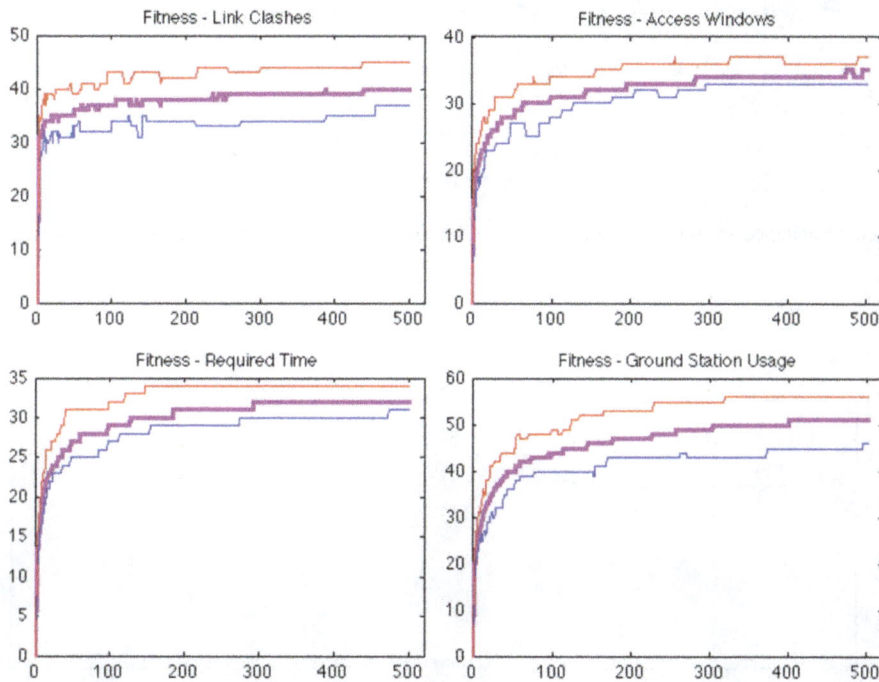

Fig. 18. Fitness functions values during 500 generations for a single ground stations and 5 spacecraft.

spacecrafts. From that information we can extract the necessary information for a chromosome (see the corresponding Fig. 17).

5.5. Computational results

We present some computational results obtained with GA for the ground stations scheduling for the case of single ground station scenario. Within this scenario, we present results for 5 spacecrafts and 7 spacecrafts. In both cases, a total of 20 independent runs of the GA were performed (under the same parameter configuration) and average results are reported.

5.6. The case of one ground station and five spacecrafts

In the case of one ground station and five spacecrafts, for the evaluation of the GA the parameter values used are given in Table 4.

We show in Fig. 18 the graphical representation of different fitness functions. The corresponding schedule is graphically shown in Fig. 19. In the figure, the timeslots of squares are the Access Window

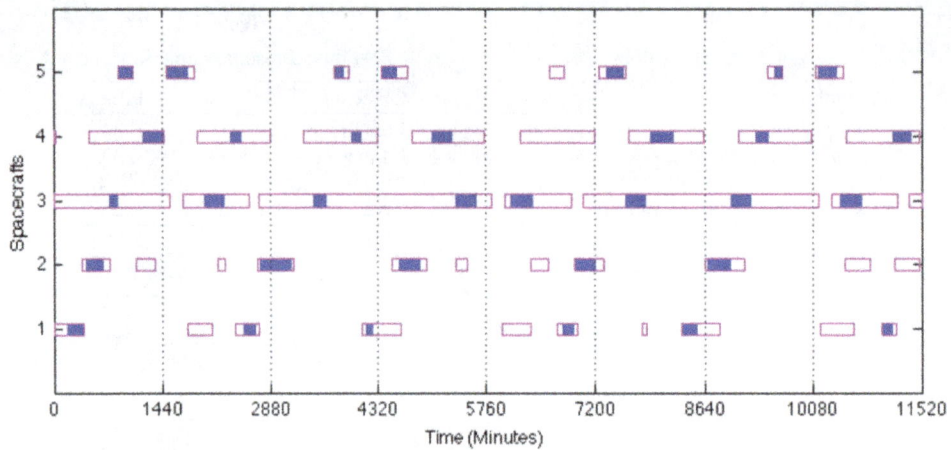

Fig. 19. The schedule obtained from best solution computed by GA for a single ground stations and 5 spacecraft for an eight days period.

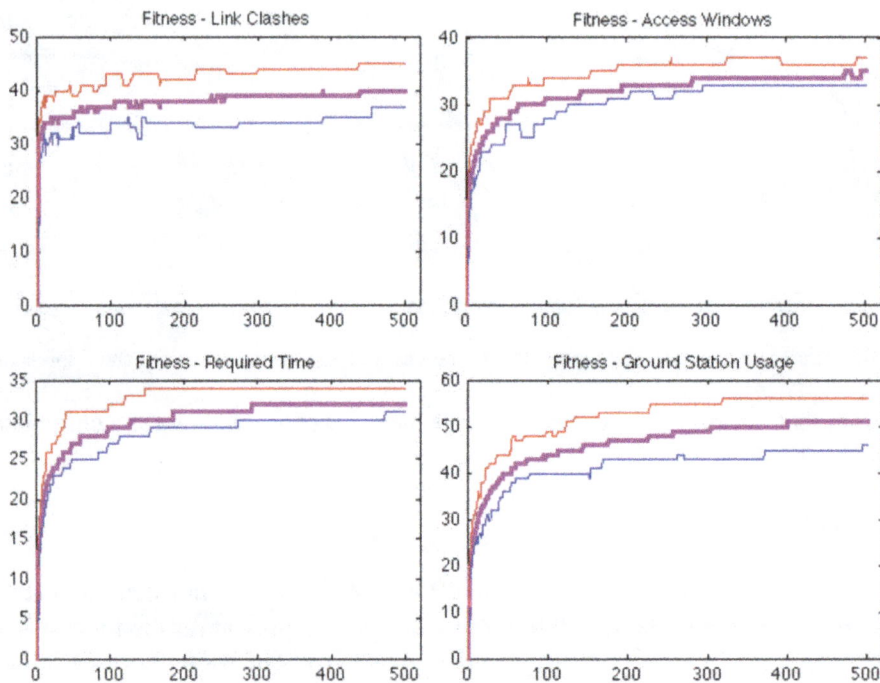

Fig. 20. Fitness functions values during 500 generations for a single ground stations and 7 spacecrafts.

of the ground station to the spacecrafts. While, the solid bars are actually scheduled communication times for each spacecrafts.

The numeric values of the fitness functions are given in Table 5 and the corresponding schedule in Table 6.

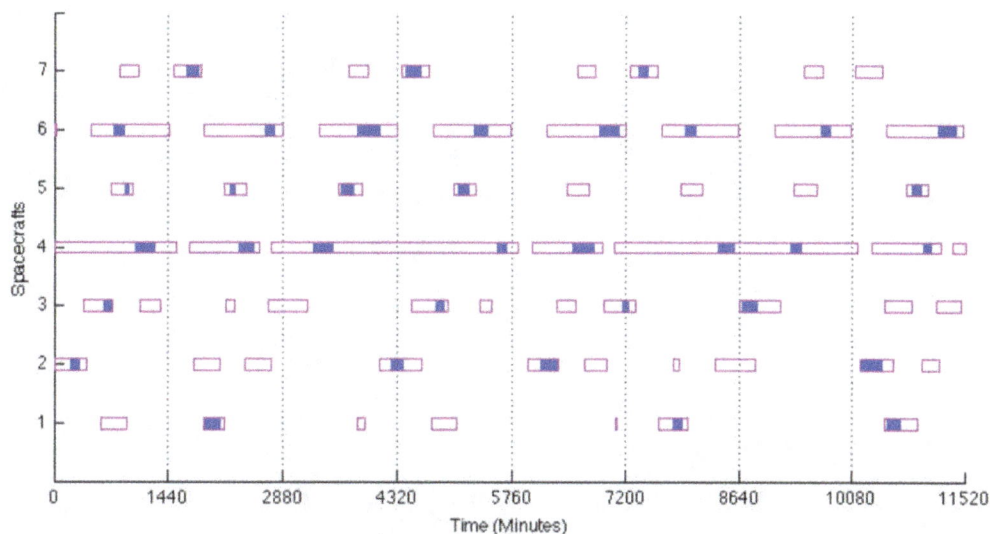

Fig. 21. The schedule obtained from best solution computed by GA for a single ground stations and 7 spacecraft for an eight days period.

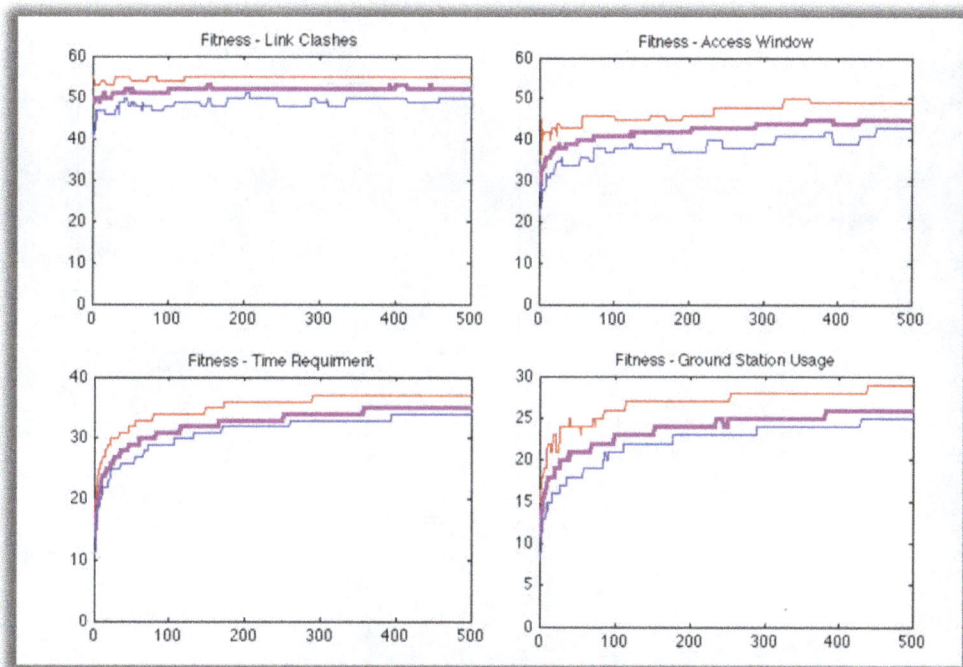

Fig. 22. Fitness functions values during 500 generations for multiple ground stations.

5.7. The one ground station case

In this case, for the evaluation of the GA, the parameter values used are given in Table 7.

We show in Fig. 20 the graphical representation of different fitness functions. The corresponding

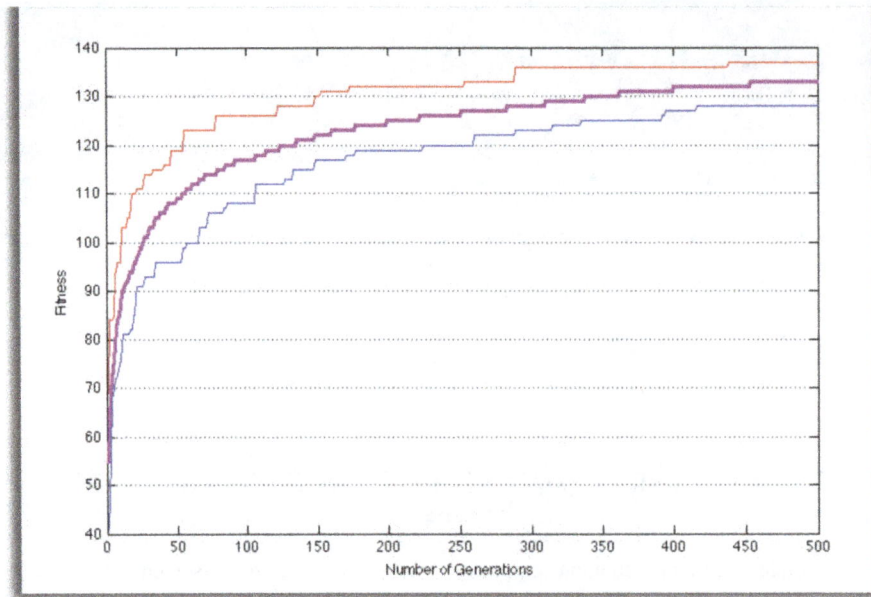

Fig. 23. Selection fitness functions values during 500 generations for multiple ground stations.

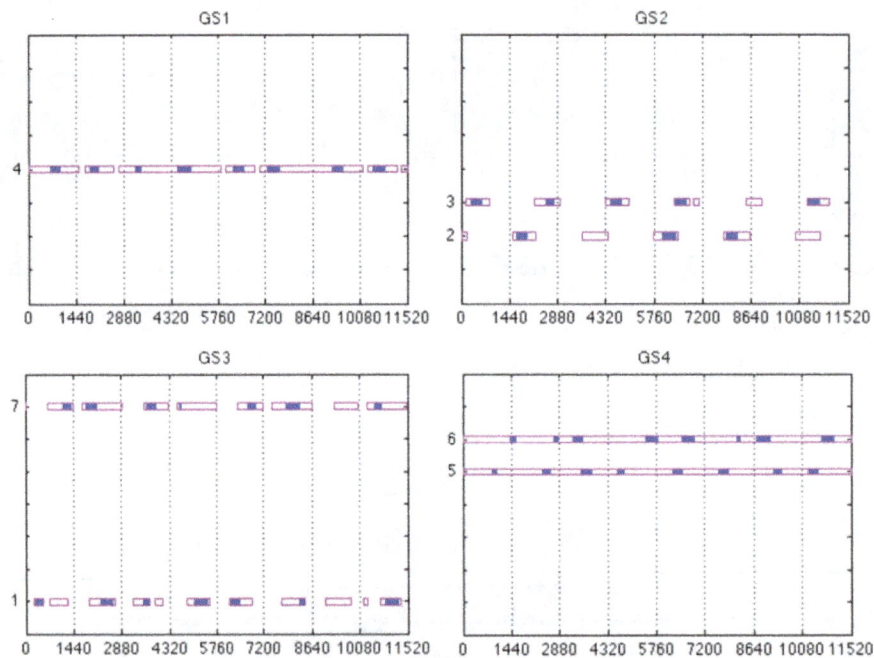

Fig. 24. The schedule obtained from best solution computed by GA for multiple ground stations.

schedule is graphically shown in Fig. 21.

The numeric values of the fitness functions are given in Table 8 and the corresponding schedule in Table 9.

Table 6
The generated schedule for a single GS and 5 SCs

Spacecraft	Link start	Link end
1	183	214
	2525	170
	4165	106
	6769	168
	8350	205
	11000	141
2	437	233
	2760	143
	2910	270
	4612	292
	6943	276
	8694	307
3	735	123
	1996	268
	3472	180
	5361	298
	6089	297
	7601	271
	9001	266
	10447	295
4	1191	261
	2345	152
	3973	141
	5046	287
	7926	309
	9323	184
	11142	255
5	871	185
	1513	283
	3739	141
	4396	186
	7346	243
	9574	95
	10154	259

Table 7
GA parameter values

GA parameter	Value
Size of Population	20
Num. of crossover points	15
Mutation rate	10%

Table 8
Best fitness results (numeric values)

Objective	Fitness
Less link clashed	38
Fit to access windows	35
Link time required	33 [3 4 4 8 3 8 3]
Ground Station Usage	65

Table 9
The generated schedule for a single GS and 5 SCs

Spacecraft	Link start	Link end
1	1915	209
	7806	133
	10540	187
2	205	128
	4255	157
	6127	232
	10192	300
3	637	108
	4821	97
	7159	105
	8692	193
4	1033	256
	2338	196
	3283	256
	5592	125
	6546	274
	8371	227
	9289	143
	11006	97
5	908	62
	2231	73
	3619	178
	5082	154
	10842	137
6	753	152
	2671	118
	3831	286
	5285	187
	6870	254
	7948	153
	9683	124
	11183	231
7	1672	171
	4438	198
	7359	141

Table 10
GA parameter values

GA parameter	Value
Size of population	20
Num. of crossover points (Chromosome A)	10
Num. of crossover points (Chromosome B)	3
Mutation rate (Chromosome A)	10%
Mutation rate (Chromosome B)	40%

5.8. The multiple ground station case

We present some computational results obtained with GA for the ground stations scheduling for the case of multiple ground station scenario, namely, 7 spacecraft with 4 ground stations. The Satellite Tool Kit (STK) [17] is used to generate some simulation scenarios. A total of 20 independent runs of the GA were performed (under the same parameter configuration –see Table 10) and average results are reported.

We show in Fig. 22 the graphical representation of different fitness functions and in Fig. 23 for the particular case of selection fitness. The corresponding schedule is graphically shown in Fig. 24. In the figure, the timeslots of squares are the Access Window of the ground station to the spacecrafts while, the solid bars are actually scheduled communication times for each spacecraft.

6. Conclusions

Mission operations scheduling problem arises in management of satellite/space missions. In most real case projects from ESA and NASA, mission planning is still done under human intervention. But, because the problem is very complex, its automation is very important to handle increasing demands for mission planning. We have presented some variants of the problem and have analysed its complexities due the large number of constraints and multi-objective nature. The problem is known to be intractable and therefore meta-heuristics are the *de facto* choice to solve the problem for practical interest. In this paper we evaluate the effectiveness of Genetic Algorithms (GAs) for near-optimally solving the problem. A data simulation model based on STK (Satellite Tool Kit) is used for the experimental study and computational results are presented for the cases of one and multiple ground stations scheduling obtained with Genetic Algorithms using the STK simulation toolkit. Genetic Algorithms showed to efficiently solve the problem as an important resolution technique for the family of satellite scheduling problems.

In our future work we plan to implement other versions of GAs, such as Steady State GA and Struggle GA, and parallel versions of GAs to speed up the GA computations.

References

[1]   S. Badaloni, M. Falda and M. Giacomin, Solving temporal over-constrained problems using fuzzy techniques, *Journal of Intelligent and Fuzzy Systems* **18**(3) (2007), 255–265.

[2]   L. Barbulescu, A. Howe, J. Watson and L. Whitley, Satellite Range Scheduling: A Comparison of Genetic, Heuristic and Local Search, *Parallel Problem Solving from Nature –PPSN*, VII: 611–620, 2002.

[3]   L. Barbulescu, J.-P. Watson, L.D. Whitley and A.E. Scheduling, space-ground communications for the air force satellite control network, *Journal of Scheduling* **7**(1) (2004), 7–34.

[4]   L. Barbulescu, A.E. Howe, L.D. Whitley and M. Roberts, Trading places: How to schedule more in a multi-resource oversubscribed scheduling problem. In *Proceedings of the International Conference on Planning and Scheduling*, 2004.

[5]   L. Barbulescu, J.P. Watson, L.D. Whitley and A.E. Howe, Scheduling Space-Ground Communications for the Air Force Satellite Control Network, *J of Scheduling* **7**(1) (2004), 7–34.

[6]   L. Barbulescu, A.E. Howe and D. Whitley, AFSCN scheduling: How the problem and solution have evolved, *Mathematical and Computer Modelling* **43**(9–10) (2006), 1023–1037.

[7]   A. Barolli, E. Spaho, L. Barolli, F. Xhafa and M. Takizawa, QoS routing in ad-hoc networks using GA and multi-objective optimization, *Mobile Information Systems* **7**(3) (2011), 169–188, IOS Press.

[8]   S. Damiani, H. Dreihahn, J. Noll, M. Nizette and G.P. Calzolari, A Planning and Scheduling System to Allocate ESA Ground Station Network Services, *The Int'l Conference on Automated Planning and Scheduling*, USA, 2007.

[9]   A. Durresi, M. Durresi, L. Barolli and F. Xhafa, MPLS Traffic Engineering for Multimedia on Satellite Networks, *Journal of Mobile Multimedia* **5**(1) (2009), 3–11.

[10] ESA Science and Technology, http://www.esa.int/.

[11] S.A. Harrison, M.E. Price and M.S. Philpott, Task Scheduling for Satellite Based Imagery. In The *Eighteenth Workshop of the UK Planning and Scheduling, Special Interest Group*, University of Salford, UK, 1999, 64–78.

[12] J. Jayaputera and D. Taniar, Data retrieval for location-dependent queries in a multi-cell wireless environment, *Mobile Information Systems* **1**(2) (2005), 91–108.

[13] T. Oda, A. Barolli, F. Xhafa, L. Barolli, M. Ikeda and M. Takizawa, Performance evaluation of WMN-GA for different mutation and crossover rates considering number of covered users parameter, *Mobile Information Systems* **8**(1) (2012), 1–16, IOS Press.

[14] J.C. Pemberton and F. Galiber, A constraint-based approach to satellite scheduling. In *DIMACS workshop on on Constraint programming and large scale discrete optimization*, E.C. Freuder and R.J. Wallace, eds, American Mathematical Society, Boston, MA, USA, 2000, 101–114.

[15] A. Sarkheyli, B.G. Vaghei and A. Bagheri, New tabu search heuristic in scheduling earth observation satellites. In *Proceedings of 2nd International Conference on Software Technology and Engineering (ICSTE)*, V2-199–V2-203, 2010.

[16] W.T. Scherer and F. Rotman, Combinatorial optimization techniques for spacecraft scheduling automation, *Annals of Operations Research* **50**(1) (1994), 525–556.

[17] Satellite Tool Kit: http://www.agi.com/products/by-product-type/applications/stk/.

[18] K. Xuan, G. Zhao, D. Taniar and B. Srinivasan, Continuous Range Search Query Processing in Mobile Navigation, *Proceedings of the 14th International Conference on Parallel and Distributed Systems* (ICPADS 2008), IEEE, 2008, pp. 361–368.

[19] K. Xuan, G. Zhao, D. Taniar, M. Safar and B. Srinivasan, Voronoi-based multi-level range search in mobile navigation, *Multimedia Tools Appl* **53**(2) (2011), 459–479.

[20] G. Zhao, K. Xuan, W. Rahayu, D. Taniar, M. Safar, M. Gavrilova and B. Srinivasan, Voronoi-Based Continuous k Nearest Neighbor Search in Mobile Navigation, *IEEE Transactions on Industrial Electronics* **58**(6) (2011), 2247–2257.

[21] N. Zufferey, P. Amstutz and P. Giaccari, Graph Colouring Approaches for a Satellite Range Scheduling Problem, *Journal of Scheduling* **11**(4) (2008), 263–277.

[22] A.B. Waluyo, W. Rahayu, D. Taniar and B. Srinivasan, A Novel Structure and Access Mechanism for Mobile Broadcast Data in Digital Ecosystems, *IEEE Transactions on Industrial Electronics* **58**(6) (2011), 2173–2182.

[23] P. Wang, G. Reinelt, P. Gao and Y. Tan, A model, a heuristic and a decision support system to solve the scheduling problem of an earth observing satellite constellation, *Comput Ind Eng* **61**(2) (2011), 322–335.

Fatos Xhafa holds a PhD in Computer Science from the Department of Languages and Informatics Systems (LSI) of the Technical University of Catalonia (UPC), Barcelona, Spain. He was a Visiting Professor at the Department of Computer Science and Information Systems, Birkbeck, University of London, UK (2009/2010) and a Research Associate at College of Information Science and Technology, Drexel University, Philadelphia, USA (2004/2005). He holds a permanent position of Professor Titular at the Department of LSI, UPC (Spain). His research interests include parallel and distributed algorithms, combinatorial optimization, approximation and meta-heuristics, networking and distributed computing, Grid and P2P computing. He has widely published in peer reviewed international journals, conferences/workshops, book chapters and edited books and proceedings in the field. He is Editor in Chief of the International Journal of Space-based and Situated Computing, and of International Journal of Grid and Utility Computing, Inderscience Publishers. He is an associate/member of Editorial Board of several international peer-reviewed scientific journals. He has also guest co-edited several special issues of international journals. He is actively participating in the organization of several international conferences.

Junzi Sun holds a MSc degree from the Technical University of Catalonia (Barcelona), Spain. He is currently with CTAE-The Aerospace Research and Technology Center, Barcelona, Spain for Information Technology and Aerospace R+D. He is interested in space and computer science. During his time in university, he participated in the Space Study Program (SSP) of the International Space University (ISU) three times as an IT staff member, in Vancouver, Strasbourg and Beijing. And finally he participated as a student in SSP 09 in NASA Ames Research Center. His research field is Information Technology, Netwoking, and Software development.

Admir Barolli holds a PhD from Fukuoka Institute of Technology, Fukuoka, Japan. He has participated in several research projects related to advanced networking systems. He has published in many international conferences and in high quality international journals. From October 2009 to June 2010, he was a Visiting Researcher at Curtin University of Technology, Australia. In 2011, he was a Visiting Researcher at the Department of Computer and Information Science, Seikei University, Japan. His research interests include Optimization, Genetic Algorithms, intelligent systems, wireless mesh networks, ad hoc networks, networking simulation software and P2P systems.

Alexander Biberaj, PhD is an Assistant Professor at the Faculty of Information Technology of the Polytechnic University of Tirana, Tirana, Albania. He has participated in several ICT programs for advances information society development. His research interests include Networking Systems, Resource Management, Management Information Systems, Information Technology and its application to e-Government and e-Society.

Leonard Barolli received BE and PhD degrees from Tirana University and Yamagata University in 1989 and 1997, respectively. From April 1997 to March 1999, he was a JSPS Post Doctor Fellow Researcher at Department of Electrical and Information Engineering, Yamagata University. From April 1999 to March 2002, he worked as a Research Associate at the Department of Public Policy and Social Studies, Yamagata University. From April 2002 to March 2003, he was an Assistant Professor at Department of Computer Science, Saitama Institute of Technology (SIT). From April 2003 to March 2005, he was an Associate Professor and presently is a Full Professor, at Department of Information and Communication Engineering, Fukuoka Institute of Technology (FIT). Dr. Barolli has published more than 250 papers in referred Journals and International Conference proceedings. He has served as a Guest Editor for many International Journals. Dr. Barolli has been a PC Member of many International Conferences and was the PC Chair of IEEE AINA-2004 and IEEE ICPADS-2005. He was General Co-Chair of IEEE AINA-2006 and IEEE AINA-2008, Workshops Chair of iiWAS-2006/MoMM-2006 and iiWAS-2007/MoMM-2007, Workshop Co-Chair of ARES-2007, ARES-2008 and IEEE AINA-2007 and AINA-2009. Presently, he is General Co-Chair of CISIS-2009. Dr. Barolli is the Steering Committee Chair of CISIS International Conference and is serving as Steering Committee Member in many International Conferences. He is organizers of many International Workshops. His research interests include network traffic control, fuzzy control, genetic algorithms, agent-based systems, ad-hoc networks and sensor networks, Web-based applications, distance learning systems and P2P systems. He is a member of SOFT, IPSJ, and IEEE.

Permissions

The contributors of this book come from diverse backgrounds, making this book a truly international effort. This book will bring forth new frontiers with its revolutionizing research information and detailed analysis of the nascent developments around the world.

We would like to thank all the contributing authors for lending their expertise to make the book truly unique. They have played a crucial role in the development of this book. Without their invaluable contributions this book wouldn't have been possible. They have made vital efforts to compile up to date information on the varied aspects of this subject to make this book a valuable addition to the collection of many professionals and students.

This book was conceptualized with the vision of imparting up-to-date information and advanced data in this field. To ensure the same, a matchless editorial board was set up. Every individual on the board went through rigorous rounds of assessment to prove their worth. After which they invested a large part of their time researching and compiling the most relevant data for our readers.

The editorial board has been involved in producing this book since its inception. They have spent rigorous hours researching and exploring the diverse topics which have resulted in the successful publishing of this book. They have passed on their knowledge of decades through this book. To expedite this challenging task, the publisher supported the team at every step. A small team of assistant editors was also appointed to further simplify the editing procedure and attain best results for the readers.

Apart from the editorial board, the designing team has also invested a significant amount of their time in understanding the subject and creating the most relevant covers. They scrutinized every image to scout for the most suitable representation of the subject and create an appropriate cover for the book.

The publishing team has been an ardent support to the editorial, designing and production team. Their endless efforts to recruit the best for this project, has resulted in the accomplishment of this book. They are a veteran in the field of academics and their pool of knowledge is as vast as their experience in printing. Their expertise and guidance has proved useful at every step. Their uncompromising quality standards have made this book an exceptional effort. Their encouragement from time to time has been an inspiration for everyone.

The publisher and the editorial board hope that this book will prove to be a valuable piece of knowledge for researchers, students, practitioners and scholars across the globe.

List of Contributors

Kazunori Uchida
Department of Information and Communication Engineering, Fukuoka Institute of Technology, Higashi-Ku, Fukuoka, Japan

Masafumi Takematsu
Graduate School of Engineering, Fukuoka Institute of Technology, Higashi-Ku, Fukuoka, Japan

Jun-Hyuck Lee
Graduate School of Engineering, Fukuoka Institute of Technology, Higashi-Ku, Fukuoka, Japan

Junichi Honda
Surveillance and Communications Department, Electronic Navigation Research Institute, Tokyo, Japan

Najmeh Neysani Samany
Department of Surveying and Geomatics Engineering, College of Engineering, University of Tehran, Tehran, Iran

Mahmoud Reza Delavar
Center of Exellence in Geomatic Engineering in Disaster Management, Department of Serveying and Geomatic Engineering, College of Engineering, University of Tehran, Tehran, Iran

Nicholas Chrisman
Department of Geomatic Science, Laval University, Québec, QC, Canada

Mohammad Reza Malek
Department of GIS, Faculty of Geodesy and Geomatic Engineering, K.N. Toosi University of Technology, Tehran, Iran

Pedro M.P. Rosa
Instituto de Telecomunicações, University of Beira Interior, Covilhã, Portugal

Joel J.P.C. Rodrigues
Instituto de Telecomunicações, University of Beira Interior, Covilhã, Portugal

Filippo Basso
Zirak s.r.l., Italy

Jianwei Niu
State Key Laboratory of Software Development Environment, Beihang University, Beijing, China

Mingzhu Liu
State Key Laboratory of Software Development Environment, Beihang University, Beijing, China

Han-Chieh Chao
Institute of Computer Science and Information, National Ilan University, I-Lan, Taiwan

Hyung-Ju Cho
Department of Information & Computer Engineering, Ajou University Woncheon-dong, Suwon Si Yeongtong-gu, Gyeonggi-Do, South Korea

Se Jin Kwon
Department of Information & Computer Engineering, Ajou University Woncheon-dong, Suwon Si Yeongtong-gu, Gyeonggi-Do, South Korea

Tae-Sun Chung
Department of Information & Computer Engineering, Ajou University Woncheon-dong, Suwon Si Yeongtong-gu, Gyeonggi-Do, South Korea

Anitha Manikandan
Department of Information Science and Technology, College of Engineering, Anna University, Guindy, Chennai, Tamilnadu 600 025, India

Yogesh Palanichamy
Department of Information Science and Technology, College of Engineering, Anna University, Guindy, Chennai, Tamilnadu 600 025, India

Tatjana Kapus
Faculty of Electrical Engineering and Computer Science, University of Maribor, Smetanova ul. 17, SI-2000 Maribor, Slovenia

Yusuke Gotoh
Okayama University, Okayama, Japan
La Trobe University, Melbourne, VIC, Australia

Tomoki Yoshihisa
Osaka University, Osaka, Japan

Hideo Taniguchi
Okayama University, Okayama, Japan

Masanori Kanazawa
The Kyoto College of Graduate, Studies for Informatics, Kyoto, Japan

Wenny Rahayu
La Trobe University, Melbourne, VIC, Australia

Yi-Ping Phoebe Chen
La Trobe University, Melbourne, VIC, Australia

Dinh-Thuan Do
Duy Tan University (DTU), Da Nang 550000, Vietnam

Lian-Fen Huang
Department of Communication Engineering, Xiamen University, Xiamen, Fujian 361005, China

Sha-Li Zhou
Department of Communication Engineering, Xiamen University, Xiamen, Fujian 361005, China

Yi-Feng Zhao
Department of Communication Engineering, Xiamen University, Xiamen, Fujian 361005, China

Han-Chieh Chao
Institute of Computer Science & Information Engineering and Department of Electronic Engineering, National Ilan University, I-Lan, Taiwan
Department of Electrical Engineering, National Dong Hwa University, Hualien, Taiwan

Ashutosh Bhatia
Department of Computer Science and Automation, Indian Institute of Science, Bangalore 560012, India

R. C. Hansdah
Department of Computer Science and Automation, Indian Institute of Science, Bangalore 560012, India

Fatos Xhafa
Department of Languages and Informatics Systems, Technical University of Catalonia, Barcelona, Spain

Junzi Sun
Centre de Tecnologia Aeroespacial, Barcelona, Spain

Admir Barolli
Seikei University, Tokyo, Japan

Alexander Biberaj
Polytechnic University of Tirana, Tirana, Albania

Leonard Barolli
Fukuoka Institute of Technology, Fukuoka, Japan